P. R. HALMOS

# SELECTA
# EXPOSITORY WRITING

P. R. Halmos

P. R. HALMOS

# SELECTA
# EXPOSITORY WRITING

Edited by
Donald E. Sarason
Leonard Gillman

Springer-Verlag
New York Heidelberg Berlin

P. R. Halmos
Department of Mathematics
Indiana University
Swain Hall East
Bloomington, IN 47405
U.S.A.

*Editors*

Donald E. Sarason
Department of Mathematics
University of California
Berkeley, CA 94720
U.S.A.

Leonard Gillman
Department of Mathematics
University of Texas
Austin, TX 78712
U.S.A.

AMS Classification (1980): 00A10

With One Illustration

Library of Congress Cataloging in Publication Data
Halmos, Paul R. (Paul Richard), 1916–
Selecta: expository writing.
Bibliography: p.
1. Mathematics—Collected works. I. Sarason, Donald. II. Gillman, Leonard. III. Title.
QA3.H24 1982 510 82-19580


Printed in the United States of America

9 8 7 6 5 4 3 2 1

ISBN 0-387-**90756**-4 Springer-Verlag New York Heidelberg Berlin
ISBN 3-540-**90756**-4 Springer-Verlag Berlin Heidelberg New York

# TABLE OF CONTENTS



## CHAPTER I



## CHAPTER II



## CHAPTER III

## CHAPTER IV

# EXPOSITORY ARTICLES

Paul Halmos has been famous for several decades as a passed master in all forms of mathematical exposition: articles, books, lectures. This volume comprises 28 articles: 27 of his expository papers and one article about him.

Halmos's success as an expositor is not just a matter of technical writing skill but results from his sustained interest in truly helping the reader. He loves his subject, and is consumed with an urge to communicate, and teach and explain. He particularly enjoys "giving away the trade secrets" (see the preface to "Finite Dimensional Vector Spaces," Princeton University Press, Princeton, [1942f]; not published here) by pointing out how a general or abstract concept, expressed in its own terminology, is related to the special or concrete one. He also abhors a communications gap and enjoys stepping in to close it. He seems to write effortlessly.

Halmos has won the Chauvenet Prize and two Lester R. Ford awards; these are prizes for exposition awarded by the Mathematical Association of America. Halmos always knows whom he is writing for, and he adapts to his audience. I have classified this collection into four chapters, in roughly decreasing order of mathematical stringency, indicated approximately as follows:

Chapter I, somewhat technical;
Chapter II, for the *American Mathematical Monthly*;
Chapter III, advice and comments;
Chapter IV, popular.

The boundaries are pliant.

## CHAPTER I

"Measurable Transformations" [1949d] and "Recent Progress in Ergodic Theory" [1961a] are invited addresses before the American Mathematical Society. They provide a review of the progress made in the study of measurable transformations during the thirty years 1931–1960. These papers are models of relaxed prose; it is remarkable how much mathematics a skilled writer can get across without plunging into a lot of specialized notation. "Entropy in Ergodic Theory" [1959d] is a set of lecture notes for a graduate course at the University of Chicago. In contrast to most lecture notes I have seen, these are carefully put together and the tone is friendly. Their purpose was to close a communications gap (to provide a treatise on information theory written in English and addressed to the mathematician).

"A Glimpse into Hilbert Space" [1963b] is based on lectures delivered at George Washington University, as part of a series on various topics by various speakers. It is a gem of exposition. "What Does the Spectral Theorem Say?" [1963a] is Halmos the teacher busy giving away the trade secrets. "Finite-Dimensional Hilbert Spaces" [1970b] is devoted to some problems in finite-dimensional linear algebra whose original impetus came from operator theory in infinite-dimensional Hilbert spaces; this paper won a Ford award.

## CHAPTER II

The *American Mathematical Monthly* has long been the world's foremost journal for expository articles addressed to the college mathematics teacher. Many famous research mathematicians have contributed to it. Halmos himself has published 15 articles in the *Monthly*. Appropriately, he has just begun a five-year term (1982–1986) as its editor.

Eight of the nine papers in Chapter II appeared in the *Monthly*. The first one finds Halmos closing a communications gap: "The Foundations of Probability" [1944c] was written to convince mathematicians that probability is a branch of mathematics. The need may seem strange today, but it wasn't then. This article won the Chauvenet Prize, MAA's highest award for expository writing (at the time, given only every three years).

"American Mathematics from 1940 to the Day before Yesterday" [1976b] won a Ford award for Halmos and his five co-authors. It is based on the talk Halmos gave at the MAA annual meeting in January 1976, in San Antonio, whose theme was the U.S. Bicentennial. This paper describes ten of the high points in mathematical research that were achieved during the period, and it is thrilling. Each topic gets about one printed page. The bulk of each article is devoted to explaining the problem and recounting its history. Definitions are stated when practical and hinted at when not; proofs are given when practical and outlined or simply omitted when not. Here is a good example (pp. 119–120; compressed):

> Computable functions are the ones obtained from certain easy functions (constant, successor, coordinate) by three procedures (composition, minimalization, primitive recursion). The details do not matter here; it might be comforting to know, however, that they are not at all difficult. A set will be called computable in case its characteristic function is computable. Consequence: a set is computable if and only if its complement is computable.

> The central concept of the proof is that of a Diophantine set [defined earlier in the article], and the major step proves that every computable set is Diophantine.
>
> If $S$ [the index set of the solvable equations] were computable, then it would follow (by a slight bit of additional argument) that the complement of every Diophantine set is Diophantine. The contradiction is derived by exhibiting a Diophantine set whose complement is not Diophantine. The last step uses a version of the familiar Cantor diagonal argument.

These devices are familiar to everyone who has thought seriously about what to say in an expository article or lecture, but they are rarely used so effectively. It is easy to see why the paper won a prize and why those in the San Antonio audience raved so about the talk.

I think I get points for some of this: I was the program chairman, the choice of Halmos as speaker was my suggestion, and I personally extended the invitation to him. (Halmosian response: he asked for a month to think it over, then phoned three days later to accept.) Halmos's audience was without doubt the largest ever to hear a mathematical talk in this country; according to the official AMS tally, it numbered one thousand seven hundred seventy five. By an ironic twist, I did not hear the talk. (I was in my room resting for a musical performance to follow.) But surely I count as an honorary member of the audience, and AMS should add 1 to their total.

In 1977 the *Monthly* inaugurated a column called "Progress Reports," under Halmos's editorship. I feel certain that Halmos was spurred to this at least in part by the outstanding success of the format I have just described [1976b]. Here is how "Progress Reports" is introduced:

> It is easy to be too busy to pay attention to what anyone else is doing, but not good. All of us should know, and want to know, what has been discovered since our formal education ended, but new words, and relations between them, are growing too fast to keep up. ...
>
> "Progress Reports" is to be an almost periodic column intended to increase everyone's mathematical information about what others have been up to. Each column will report one step forward: ...what is the name of the subject, what are some of the words it uses, what is a typical question, what is the answer, who found it....

To lead the way, Halmos contributed the first six reports (two with co-authors): "Bernoulli Shifts" [1977b], "Fourier Series" [1978a], "Arithmetic Progressions" [1978b], "Invariant Subspaces" [1978c], "Schauder Bases" [1978d], and "The Serre Conjecture" [1978e]. Like their precursors in [1976b], they are miniature masterpieces. Only three or four other authors have since come through with "Progress Reports"; apparently very few people possess the right combination of interest, willingness, mathematical expertise, and expository skill.

"The Work of F. Riesz" [1983], which closes the chapter, is a warmly written, informative work. It tells a little about Riesz himself and discusses his chief contributions to mathematics: in topology, analysis, and functional analysis. The spirit is the same as the "Progress Reports" but a little more is expected of the reader.

## CHAPTER III

This chapter and the next contain a lot of advice for mathematicians and comments about mathematics. What immediately stands out is Halmos's willingness to go out on a limb; he even seems to relish the prospect that someone will try to saw it off. I don't agree with everything he says, and I doubt that he still does (if he ever did). But vigorous writing is more valuable than pussyfooting.

"How to Write Mathematics" [1970a] is superb. It is presumably addressed to the beginner, but it contains all sorts of sound advice that is too often overlooked even by the experienced mathematician. Almost all mathematical writers should look it over from time to time. (Halmos belongs to the set of measure zero.) "How to Talk Mathematics" [1974b], likewise, is tactfully addressed to the young mathematician, but should be considered seriously by "experts" as well. Unfortunately, as is well known, the people who need the advice are often the ones who do not learn from it. "What to Publish" [1975a] is based on a talk at the MAA annual meeting in January 1973, in Dallas, as part of a panel discussion. It offers some helpful rules on what *not* to publish and philosophizes about the hopelessly difficult problem of explaining what one *should* publish. It therefore lacks the bite of the two preceding articles. Nevertheless, the talk itself was a great success: I heard it, as the next following panelist. To follow Halmos on a program in front of several hundred mathematicians is a mistake. I have made it twice. On this occasion, I squirmed out (I think) by giving my speech in one minute flat and sitting down. (The other time I sought refuge at the piano.)

The next two articles are about problem solving and are addressed to the mathematics teacher. "The Teaching of Problem Solving" [1975b] asserts that solving problems is what mathematics is all about, and extols the Moore method for fostering the right problem-solving attitude. "The worst way to teach is to talk." "Don't preach facts—stimulate acts." "The Heart of Mathematics" [1980c] maintains that that heart is problem solving, and goes on to review several interesting problem books.

"Logic from *A* to *G*" [1977a] is an excellent little piece telling students about the development of mathematical logic.

"Does Mathematics Have Elements?" [1981c] is an unusually interesting article: a serious attempt to pick out those basic principles ("elements") upon which mathematics is built. The suggested list ranges "from dim analogy (everything is continuous), through conventional wisdom (structural constituents must be compatible) and universal algebra (look for invariants, form quotient structures), to computational trickery (sum the geometric series)". "The Thrills of Abstraction" [1982b], a light-spirited affair, also gets to considering basic principles, but at a simpler level. (The first paper talks about quotient structures, this one about equivalence relations.)

## CHAPTER IV

"Nicolas Bourbaki" [1957b] is a popular article about a fictitious person; it contains the appropriate anecdotes. "The Legend of John von Neumann" [1973a] also contains appropriate anecdotes, but it is a warm account of a real person by one who had been close to him.

"Mathematics as a Creative Art" [1968d] is based on a public lecture; the stated goal was to persuade the audience that the subject of mathematics exists. By mathematics, Halmos means pure mathematics ("mathology"), as opposed to applied mathematics ("mathophysics"). If you had to address an audience of laymen on mathematics as a creative art, would you know what to say? I suggest you simply crib from this lecture. In "Applied Mathematics is Bad Mathematics" [1981b], Halmos climbs out on a limb (the title) and then hands the reader his choice of saw (for instance: pure mathematics is to applied mathematics as Mozart is to military marches).

"Paul Halmos: A Maverick Mathologist" [1982a] is an interview conducted by Donald J. Albers, editor of the *Two-Year College Mathematics Journal*. It turns out that Halmos works very hard at his writing. That is surely a lesson for all the rest of us.

Austin, Texas
January, 1983

LEONARD GILLMAN

# BIBLIOGRAPHY OF THE PUBLICATIONS OF P. R. HALMOS[†]

* in front of a title denotes a book

[1938] Note on almost-universal forms, *Bull. Am. Math. Soc.* **44** (1938) 141–144.

[1939 a] Invariants of certain stochastic transformations; the mathematical theory of gambling systems, *Duke Math. J.* **5** (1939) 461–478.

[1939 b] On a necessary condition for the strong law of numbers, *Ann. Math.* **40** (1939) 800–804.

[1941 a] Statistics, set functions, and spectra, *Mat. Sb.* **9** (1941) 241–248.

[1941 b] The decomposition of measures, *Duke Math. J.* **8** (1941) 386–392.

[1942 a] The decomposition of measures, II, (With W. Ambrose and S. Kakutani), *Duke Math. J.* **9** (1942) 43–47.

[1942 b] Square roots of measure preserving transformations, *Am. J. Math.* **64** (1942) 153–166.

[1942 c] (Review) An introduction to linear transformations in Hilbert space, By F. J. Murray, *Bull. Am. Math. Soc.* **48** (1942) 204–205.

[1942 d] On monothetic groups, (With H. Samelson), *Proc. Natl. Acad. Sci. U.S.A.* **28** (1942) 254–257.

[1942 e] Operator methods in classical mechanics, II, (With J. von Neumann), *Ann. Math.* **43** (1942) 332–350.

[1942 f] **Finite dimensional vector spaces*, Princeton Univ. Press, Princeton (1942).

[1943] On automorphisms of compact groups, *Bull. Am. Math. Soc.* **49** (1943) 619–624.

[1944 a] Approximation theories for measure preserving transformations, *Trans. Am. Math. Soc.* **55** (1944) 1–18.

[1944 b] Random alms, *Ann. Math. Stat.* **15** (1944) 182–189.

[†]As of the publication of this volume.

[1944 c] The foundations of probability, *Am. Math. Monthly* **51** (1944) 493–510.

[1944 d] In general a measure preserving transformation is mixing, *Ann. Math.* **45** (1944) 786–792.

[1944 e] Comment on the real line, *Bull. Am. Math. Soc.* **50** (1944) 877–878.

[1946 a] The theory of unbiased estimation, *Ann. Math. Stat.* **17** (1946) 34–43.

[1946 b] An ergodic theorem, *Proc. Natl. Acad. Sci. U.S.A.* **32** (1946) 156–161.

[1947 a] Functions of integrable functions, *J. Indian Math. Soc.* **11** (1947) 81–84.

[1947 b] On the set of values of a finite measure, *Bull. Am. Math. Soc.* **53** (1947) 138–141.

[1947 c] Invariant measures, *Ann. Math.* **48** (1947) 735–754.

[1948] The range of a vector measure, *Bull. Am. Math. Soc.* **54** (1948) 416–421.

[1949 a] On a theorem of Dieudonné, *Proc. Natl. Acad. Sci. U.S.A.* **35** (1949) 38–42.

[1949 b] Application of the Radon–Nikodym theorem to the theory of sufficient statistics, (With L. J. Savage) *Ann. Math. Stat.* **20** (1949) 225–241.

[1949 c] A non-homogeneous ergodic theorem, *Trans. Am. Math. Soc.* **66** (1949) 284–288.

[1949 d] Measurable transformations, *Bull. Am. Math. Soc.* **55** (1949) 1015–1034.

[1950 a] Normal dilations and extensions of operators, *Summa Brasil. Math.* **2** (1950) 125–134.

[1950 b] Commutativity and spectral properties of normal operators, *Acta Sci. Math.* (1950) 153–156.

[1950 c] The marriage problem, (With H. E. Vaughan) *Am. J. Math.* **72** (1950) 214–215.

[1950 d] Measure theory, *Proc. Int. Cong. Math.*, 1950, Volume II, p. 114.

[1950 e] **Measure theory*, Van Nostrand, New York (1950).

[1951 a] Algunos problemas actuales sobre operadores en espacios de Hilbert, *UNESCO Symp.*, Punta del Este, Montevideo (1951) 9–14.

[1951 b] (Review) Gelöste und ungelöste mathematische Probleme aus alter und neuer Zeit, By H. Tietze, *Bull. Am. Math. Soc.* **57** (1951) 502–503.

[1951 c] **Introduction to Hilbert space and the theory of spectral multiplicity*, Chelsea, New York (1951).

[1952 a] Spectra and spectral manifolds, *Ann. Soc. Math. Pol.* **25** (1952) 43–49.

[1952 b] Commutators of operators, *Am. J. Math.* **74** (1952) 237–240.

[1953 a] Square roots of operators, (With G. Lumer and J. J. Schäffer) *Proc. Am. Math. Soc.* **4** (1953) 142–149.

[1953 b] (Review) Introducción a los métodos de la estadística, By S. Ríos, *J. Am. Stat. Assoc.* **48** (1953) 154–155.

[1953 c] (Review) Intégration, By N. Bourbaki, *Bull. Am. Math. Soc.* **59** (1953) 249–255.

[1953 d] (Review) Abstract set theory, By A. A. Fraenkel, *Bull. Am. Math. Soc.* **59** (1953) 584–585.

[1953 e] (Review) Lezioni sulla teoria moderna dell'integrazione, By M. Picone and T. Viola, *Bull. Am. Math. Soc.* **59** (1953) 94.

[1954 a] Commutators of operators, II, *Am. J. Math.* **76** (1954) 191–198.

[1954 b] (Review) Les systèmes axiomatiques de la théorie des ensembles, By H. Wang and R. McNaughton, *Bull. Am. Math. Soc.* **60** (1954) 93–94.

[1954 c] Polyadic Boolean algebras, *Proc. Natl. Acad. Sci. U.S.A.* **40** (1954) 296–301.

[1954 d] (Review) Linear Analysis, By A. C. Zaanen, *Bull. Am. Math. Soc.* **60** (1954) 487–488.

[1954 e] Polyadic Boolean algebras, *Proc. Int. Cong. Math.* 1954 Volume II, pp. 402–403.

[1954 f] Square roots of operators, II, (With G. Lumer) *Proc. Am. Math. Soc.* **5** (1954) 589–595.

[1955 a] Algebraic logic, I, Monadic Boolean algebras, *Compos. Math.* **12** (1955) 217–249.

[1955 b] (Review) Mathematics and plausible reasoning, By G. Pólya, *Bull. Am. Math. Soc.* **61** (1955) 243–245.

[1955 c] (Review) Elements of algebra, By H. Levi, *Bull. Am. Math. Soc.* **61** (1955) 245–247.

[1955 d] (Review) Topological dynamics, By W. H. Gottschalk and G. A. Hedlund, *Bull. Am. Math. Soc.* **61** (1955) 584–588.

[1955 e] (Review) Theorie der linearen Operatoren im Hilbert-Raum, By N. I. Achieser and I. M. Glasmann, *Bull. Am. Math. Soc.* **61** (1955) 588–589.

[1955 f] (Review) Introducción a los métodos de la estadística (Segunda parte) By S. Ríos, *J. Am. Stat. Assoc.* **50** (1955) 1002.

[1956 a] Predicates, terms, operations, and equality in polyadic Boolean algebras, *Proc. Natl. Acad. Sci. U.S.A.* **42** (1956) 130–136

[1956 b] (Review) Einführung in die Verbandstheorie, By H. Hermes, *Bull. Am. Math. Soc.* **62** (1956) 189–190.

[1956 c] Algebraic logic (II). Homogeneous locally finite polyadic Boolean algebras of infinite degree, *Fund. Math.* **43** (1956) 255–325.

[1956 d] The basic concepts of algebraic logic, *Am. Math. Monthly* **63** (1956) 363–387.

[1956 e] Algebraic logic, III. Predicates, terms, and operations in polyadic algebras, *Trans. Am. Math. Soc.* **83** (1956) 430–470.

[1956 f] **Lectures on ergodic theory*, Math. Soc. Japan, Tokyo (1956).

[1957 a] Algebraic logic, IV. Equality in polyadic algebras, *Trans. Am. Math. Soc.* **86** (1957) 1–27.

[1957 b] Nicolas Bourbaki, *Scientific American* **196** (1957) 88–99.

[1957 c] (Review) Logic, semantics, metamathematics, By A. Tarski, *Bull. Am. Math. Soc.* **63** (1957) 155–156.

[1958 a] Innovation in mathematics, *Scientific American* **199** (1958) 66–73.

[1958 b] Products of symmetries, (With S. Kakutani) *Bull. Am. Math. Soc.* **64** (1958) 77–78.

[1958 c] Von Neumann on measure and ergodic theory, *Bull. Am. Math. Soc.* **64** (1958) 86–94.

[1958 d] **Finite-dimensional vector spaces*, Second ed., Van Nostrand, Princeton (1958).

[1959 a] Free monadic algebras, *Proc. Am. Math. Soc.* **10** (1959) 219–227.

[1959 b] (Review) Linear operators. Part I: General theory, By N. Dunford and J. T. Schwartz, *Bull. Am. Math. Soc.* **65** (1959) 154–156.

[1959 c] The representation of monadic Boolean algebras, *Duke Math. J.* **26** (1959) 447–454.

[1959 d] Entropy in ergodic theory, (Mimeographed notes) University of Chicago (1959).

[1960] **Naive set theory*, Van Nostrand, Princeton (1960).

[1961 a] Recent progress in ergodic theory, *Bull. Am. Math. Soc.* **67** (1961) 70–80.

[1961 b] Injective and projective Boolean algebras, *Proc. Symp. Pure Math., Lattice theory* Volume II (1961) 114–122.

[1961 c] Shifts on Hilbert spaces, *J. reine angew. Math.* **208** (1961) 102–112.

[1962 a] (Review) Neurere Methoden und Ergebnisse der Ergodentheorie, By K. Jacobs, *Bull. Am. Math. Soc.* **68** (1962) 59–60.

[1962 b] **Algebraic logic*, Chelsea, New York (1962).

[1963 a] What does the spectral theorem say?, *Am. Math. Monthly* **70** (1963) 241–247.

[1963 b] A glimpse into Hilbert space, *Lectures on Modern Mathematics*, Wiley, New York, Volume I (1963) 1–22.

[1963 c] Partial isometries, (With J. E. McLaughlin) *Pacific J. Math.* **13** (1963) 585–596.

[1963 d] Algebraic properties of Toeplitz operators, (With A. Brown) *J. reine angew. Math.* **213** (1963) 89–102.

[1963 e] **Lectures on Boolean algebras*, Van Nostrand, Princeton (1963).

[1964 a] Numerical ranges and normal dilations, *Acta Sci. Math.* **25** (1964) 1–5.

[1964 b] On Foguel's answer to Nagy's question, *Proc. Am. Math. Soc.* **15** (1964) 791–793.

[1964 c] (Review) Lectures on invariant subspaces, By H. Helson, *Bull. Am. Math. Soc.* **71** (1965) 490–494.

[1965 a] Cesàro operators, (With A. Brown and A. L. Shields) *Acta Sci. Math.* **26** (1965) 125–137.

[1965 b] Commutators of operators on Hilbert space, (With A. Brown and C. Pearcy) *Canad. J. Math.* **17** (1965) 695–708.

[1966] Invariant subspaces of polynomially compact operators, *Pacific J. Math.* **16** (1966) 433–437.

[1967] **A Hilbert space problem book*, Van Nostrand, Princeton (1967).

[1968 a] Invariant subspaces, *Abstract spaces and approximation*, Birkhäuser, Basel (1968) 26–30.

[1968 b] Irreducible operators, *Mich. Math. J.* **15** (1968) 215–223.

[1968 c] Quasitriangular operators, *Acta Sci. Math.* **29** (1968) 283–293.

[1968 d] Mathematics as a creative art, *Am. Sci.* **56** (1968) 375–389.

[1968 e] Permutations of sequences and the Schröder–Bernstein theorem, *Proc. Am. Math. Soc.* **19** (1968) 509–510.

[1969 a] Invariant subspaces 1969, *Seventh Brazil. Math. Colloq.*, Poços de Caldas (1969) 1–54.

[1969 b] Two subspaces, *Trans. Am. Math. Soc.* **144** (1969) 381–389.

[1970 a] How to write mathematics, *Enseign. Math.*, **16** (1970) 123–152.

[1970 b] Finite dimensional Hilbert spaces, *Am. Math. Monthly* **77** (1970) 457–464.

[1970 c] Powers of partial isometries, (With L. J. Wallen) *J. Math. Mech.* **19** (1970) 657–663.

[1970 d] Ten problems in Hilbert space, *Bull. Am. Math. Soc.* **76** (1970) 887–933.

[1971 a] Capacity in Banach algebras, *Indiana Univ. Math. J.* **20** (1971) 855–863.

[1971 b] Reflexive lattices of subspaces, *J. London Math. Soc.* **4** (1971) 257–263.

[1971 c] Eigenvectors and adjoints, *Linear algebra appl.* **4** (1971) 11–15.

[1972 a] Continuous functions of Hermitian operators, *Proc. Am. Math. Soc.* **31** (1972) 130–132.

[1972 b] Positive approximants of operators, *Indiana Univ. Math. J.* **21** (1972) 951–960.

[1972 c] Products of shifts, *Duke Math. J.* **39** (1972) 779–787.

[1973 a] The legend of John von Neumann, *Am. Math. Monthly* **80** (1973) 382–394.

[1973 b] Limits of shifts, *Acta Sci. Math.* **34** (1973) 131–139.

[1974 a] Spectral approximants of normal operators, *Proc. Edinburgh Math. Soc.* **19** (1974) 51–58.

[1974 b] How to talk mathematics, *Notices Am. Math. Soc.* **21** (1974) 155–158.

[1974 c] (Review) Creative teaching: heritage of R. L. Moore, By R. D. Traylor, *Hist. Math.* **1** (1974) 188–192.

[1975 a] What to publish, *Am. Math. Monthly* **82** (1975) 14–17.

[1975 b] The teaching of problem solving, *Am. Math. Monthly* **82** (1975) 466–470.

[1976 a] Products of involutions, (With W. H. Gustafson and H. Radjavi) *Linear algebra appl.* **13** (1976) 157–162.

[1976 b] American mathematics from 1940 to the day before yesterday, (With J. H. Ewing, W. H. Gustafson, S. H. Moolgavkar, W. H. Wheeler, and W. P. Ziemer) *Am. Math. Monthly* **83** (1976) 503–516.

[1976 c] Some unsolved problems of unknown depth about operators on Hilbert space, *Proc. Royal Soc. Edinburgh* **76** A (1976) 67–76.

[1977 a] Logic from A to G, *Math. Mag.* **50** (1977) 5–11.

[1977 b] Bernoulli shifts, *Am. Math. Monthly* **84** (1977) 715–716.

[1978 a] Fourier series, *Am. Math. Monthly* **85** (1978) 33–34.

[1978 b] Arithmetic progressions, (With C. Ryavec) *Am. Math. Monthly* **85** (1978) 95–96.

[1978 c] Invariant subspaces, *Am. Math. Monthly* **85** (1978) 182–183.

[1978 d] Schauder bases, *Am. Math. Monthly* **85** (1978) 256–257.

[1978 e] The Serre conjecture, (With W. H. Gustafson and J. M. Zelmanowitz) *Am. Math. Monthly* **85** (1978) 357–359.

[1978 f] **Bounded integral operators on* $L^2$ *spaces*, (With V. S. Sunder) Springer-Verlag Berlin (1978).

[1978 g] Integral operators, Hilbert space operators, Proceedings, Long Beach, California (1977) 1–15; *Lecture Notes in Math.* 693, Springer-Verlag Berlin (1978).

[1979 a] (Review) Panorama des mathématiques pures. Le choix Bourbachique. By. J. Dieudonné, *Bull. Am. Math. Soc.* **1** (1979) 678–681.

[1979 b] Ten years in Hilbert space, *Integral Equations Oper. Theory* **2** (1979) 529–564.

[1980 a] Limsups of Lats, *Indiana Univ. Math. J.* **29** (1980) 293–311.

[1980 b] Finite-dimensional points of continuity of Lat, (With J. B. Conway) *Linear algebra appl.* **31** (1980) 93–102.

[1980 c] The heart of mathematics, *Am. Math. Monthly* **87** (1980) 519–524.

[1981 a] (Review) The William Lowell Putnam Mathematical Competition, Problems and Solutions: 1938–1964, By A. M. Gleason, R. E. Greenwood, and L. M. Kelly, *Am. Math. Monthly* **88** (1981) 450–451.

[1981 b] Applied mathematics is bad mathematics, *Mathematics Tomorrow*, Springer-Verlag New York (1981) 9–20.

[1981 c] Does mathematics have elements? *Math. Intell.* **3** (1981) 147–153, *Bull. Austral. Math. Soc.* **25** (1982) 161–175.

[1981 d] (Review) Encyclopedic Dictionary of Mathematics, Edited by Shôkichi Iyanaga and Yukiyosi Kawada; translation reviewed by K. O. May, *Math. Intell.* **3** (1981) 138–140.

[1982 a] Think it gooder, *Math. Intell.* **4** (1982) 20–21.

[1982 b] (Review) Recurrence in ergodic theory and combinatorial number theory, by H. Furstenberg, *Math. Intell.* **4** (1982) 52–54.

[1982 c] Quadratic Interpolation, *J. Oper. Theory* **7** (1982) 303–305.

[1982 d] Asymptotic Toeplitz Operators, *Trans. Am. Math. Soc.* **273** (1982) 621–630.

[1982 e] The thrills of abstraction, *Two-Year College Math. J.* **13** (1982) 243–251.

[1982 f] **A Hilbert space problem book*, second ed., Springer-Verlag New York (1982).

[1983] The work of F. Riesz, *Selecta — Expository Writing*, Springer-Verlag New York (1983).

[1984] Weakly transitive matrices (with José Barría), *Ill. J. Math.* **28** (1984).

# PERMISSIONS

Springer-Verlag would like to thank the original publishers of P. R. Halmos's scientific papers for granting permissions to reprint a selection of his papers in this volume. The following credit lines were specifically requested:

[1944 c] Reprinted from *Am. Math. Monthly* **51**, © 1944 by Math. Assoc. Am.
[1949 d] Reprinted from *Bull. Am. Math. Soc.* **55**, © 1949 by Am. Math. Soc.
[1957 b] Reprinted from *Sci. Am.* **196**, © 1975 by W. H. Freeman & Co.
[1961 a] Reprinted from *Bull. Am. Math. Soc.* **67**, © 1961 by Am. Math. Soc.
[1963 a] Reprinted from *Am. Math. Monthly* **70**, © 1963 by Math. Assoc. Am.
[1963 b] Reprinted from *Lect. Mod. Math.*, © 1963 by J. Wiley & Sons Inc.
[1968 d] Reprinted from *Am. Scientist* **56**, © 1968 by Sigma Xi.
[1970 a] Reprinted from *Enseign. Math.* **16**, © 1970 by Enseign. Math.
[1970 b] Reprinted from *Am. Math. Monthly* **77**, © 1970 by Math. Assoc. Am.
[1973 a] Reprinted from *Am. Math. Monthly* **80**, © 1973 by Math. Assoc. Am.
[1974 b] Reprinted from *Notices Am. Math. Soc.* **21**, © 1974 by Am. Math. Soc.
[1975 a] Reprinted from *Am. Math. Monthly* **82**, © 1975 by Math. Assoc. Am.
[1975 b] Reprinted from *Am. Math. Monthly* **82**, © 1975 by Math. Assoc. Am.
[1976 b] Reprinted from *Am. Math. Monthly* **83**, © 1976 by Math. Assoc. Am.
[1977 a] Reprinted from *Math. Mag.* **50**, © 1977 by Math. Assoc. Am.
[1977 b] Reprinted from *Am. Math. Monthly* **84**, © 1977 by Math. Assoc. Am.
[1978 a] Reprinted from *Am. Math. Monthly* **85**, © 1978 by Math. Assoc. Am.
[1978 b] Reprinted from *Am. Math. Monthly* **85**, © 1978 by Math. Assoc. Am.
[1978 c] Reprinted from *Am. Math. Monthly* **85**, © 1978 by Math. Assoc. Am.
[1978 d] Reprinted from *Am. Math. Monthly* **85**, © 1978 by Math. Assoc. Am.
[1978 e] Reprinted from *Am. Math. Monthly* **85**, © 1978 by Math. Assoc. Am.
[1980 c] Reprinted from *Am. Math. Monthly* **87**, © 1980 by Math. Assoc. Am.
[1982 e] Reprinted from *Two-Year Coll. Math. J.* **13**, © 1982 by Math. Assoc. Am.

"Paul Halmos: A Maverick Mathologist" reprinted from *Two-Year Coll. Math. J.* **13**, © 1982 by the Math. Assoc. Am.

# CHAPTER I

Reprinted from the
BULLETIN OF THE AMERICAN MATHEMATICAL SOCIETY
Vol. 55, No. 11, pp. 1015-1034, Nov. 1949

# MEASURABLE TRANSFORMATIONS

PAUL R. HALMOS

1. **Introduction.** The purpose of this paper is to review the progress made in the study of measurable and measure preserving transformations during the last 17 years. The interest of mathematicians in this subject was aroused at the end of 1931 by von Neumann's and Birkhoff's proofs of their respective versions of the ergodic theorem [**8, 9, 101**].[1] It was very quickly recognized that the proper general framework for von Neumann's mean ergodic theorem lay in the direction of Hilbert spaces and Banach spaces, whereas the extent of generality suitable to Birkhoff's theorem was to be found in the concept of a measure space. A measure space is a set possessing no intrinsic algebraic, analytic, or topological structure—all that is necessary is that a concept of measurability and a numerical measure be defined in it. Perhaps the best known nontrivial example of a measure space is one which, to be sure, has many essential non measure theoretic properties, but which may, nevertheless, be considered typical of measure spaces in general—namely the closed unit interval $X=[0, 1]$. For the sake of definiteness I shall begin the discussion by considering a one-to-one transformation $T$ of this space $X$ onto itself, such that, for every measurable subset $E$ of $X$, both $TE$ and $T^{-1}E$ are measurable and $\mu(E)=\mu(TE)=\mu(T^{-1}E)$ (where $\mu$ denotes Lebesgue measure in $X$). In much of what follows the space $X$ and the transformation $T$ can be replaced by more general spaces and transformations respectively. I shall indicate some of these generalizations in what might be called the geometric direction (that is, generalizations that retain something like an underlying measure space and a transformation acting on it), but I shall not enter at all into the analytic generalizations which constitute the current theory of the mean ergodic theorem.

2. **Asymptotic properties.** The problems that were first treated, and that are still of interest and importance, are connected with the behavior of the sequence $\{T^n\}$ of powers of $T$. One of the first results in this direction is the Poincaré recurrence theorem, which asserts that, for every measurable set $E$ and for almost every point $x$ in $E$,

An address delivered before the Chicago meeting of the Society on November 26, 1948, by invitation of the Committee to Select Hour Speakers for Western Sectional Meetings; received by the editors November 26, 1948.

[1] Bold face numerals refer to the bibliography at the end.

there are an infinite number of positive values of $n$ such that $T^n x \in E$ [18, 79]. An equivalent way of phrasing this assertion, in terms of the characteristic function $\chi_E$ of $E$, is to say that for every measurable set $E$ the series $\sum_{n=0}^{\infty} \chi_E(T^n x)$ diverges for almost every point $x$ in $E$. A related result, proved by Hopf [67], is that if $f$ is any positive measurable function, then the series $\sum_{n=0}^{\infty} f(T^n x)$ diverges for almost every point $x$ in $X$. From the Poincaré recurrence theorem it follows easily that if $E$ is any measurable set of positive measure, then there are an infinite number of positive values of $n$ such that $\mu(E \cap T^n E) > 0$, and, conversely, this formulation of Poincaré's theorem implies the original one. In this direction the result has been strengthened by Khintchine who showed that if $E$ is any measurable set of positive measure and if $\epsilon$ is any positive number, then the set of values of $n$ for which

$$\mu(E \cap T^n E) > (\mu(E))^2 - \epsilon$$

is relatively dense in the sense of Bohr [90, 126].

Birkhoff's ergodic theorem may also be viewed as a statement concerning the recurrence properties of $\{T^n\}$. In its most primitive form the ergodic theorem says that if $E$ is any measurable set and if, for each positive integer $n$ and for every point $x$ in $X$, $s_n(x)$ is the number of values of $i$ for which $T^i x \in E$, $0 \leqq i \leqq n-1$, then $\lim (1/n) s_n(x) = s^*(x)$ exists almost everywhere and, moreover, $\int s^*(x) d\mu(x) = \mu(E)$. If the discrete parameter $n$ is interpreted as time, then $s^*(x)$ is the relative amount of time (time of sojourn) that $x$ spends in $E$, and the assertion is that the mean sojourn time is almost everywhere defined and that its value depends in an obvious way on the size of $E$. In terms of characteristic functions the statement of the ergodic theorem concerns the existence almost everywhere of $\lim (1/n) \sum_{i=0}^{n-1} \chi_E(T^i x)$ and the value of the integral of the limit function. A more general statement asserts that, for any integrable function $f$, $\lim (1/n) \sum_{i=0}^{n-1} f(T^i x) = f^*(x)$ exists almost everywhere and that, moreover, $\int f^*(x) d\mu(x) = \int f(x) d\mu(x)$.

3. **Generalizations.** The first attempts to generalize the ergodic theorem consisted of replacing the underlying space $X$ by abstract measure spaces [88] and dropping the hypothesis $\mu(X) < \infty$ [57, 122]. A slightly more interesting generalization was given by Khintchine [89] who proved that if $f$ is any integrable function on $X$ and $g$ is any periodic function of $n$, then $\lim (1/n) \sum_{i=0}^{n-1} f(T^i x) g(i)$ exists almost everywhere. The most powerful result in this direction is due to Wiener and Wintner [129] who proved that, if $f$ is, as before, an integrable function, then not only does $\lim (1/n) \sum_{j=0}^{n-1} e^{2\pi i \lambda j} f(T^j x)$

exist almost everywhere for each real number $\lambda$ (here $i=(-1)^{1/2}$), but in fact the exceptional set of measure zero can be chosen to be independent of $\lambda$.

The ergodic theorem is a statement about a space, a function, and a transformation. In the preceding paragraph I mentioned the possibility of generalizing the space and the function; there remains the possibility of generalizing the transformation. What, for instance, can be said if the transformation is not necessarily one-to-one, but is still, in one sense or another, measure preserving? Suppose, for the sake of definiteness, that $T$ is a transformation of $X$ onto itself such that the inverse image of every measurable set is a measurable set of the same measure. (A nontrivial example of such a transformation $T$ is defined in the unit interval by $Tx=2x$ [mod 1].) If $T$ is such a transformation, if $f$ is any measurable function, and if $g_n(x)=f(T^n x)$, $n=0, 1, 2, \cdots$, then the sequence $\{g_n\}$ has the homogeneity property that the joint $k$-dimensional distribution of any $k$ of its terms depends only on the relative distances between the indices and not on their values. More precisely, it is true that if $A$ is any Borel set in $k$-dimensional space, and if $(n_1, \cdots, n_k)$ is any $k$-tuple of distinct non-negative integers, then

$$\mu(\{x: (g_{n_1+m}(x), \cdots, g_{n_k+m}(x)) \in A\})$$

is independent of $m$. It is a nontrivial extension of the ergodic theorem to say that for any sequence $\{g_n\}$ of integrable functions with this homogeneity property $\lim (1/n) \sum_{i=0}^{n-1} g_i(x)$ exists almost everywhere—this extension is due to Doob [**24**].

The final generalization that will be mentioned here is one in which the transformation is not required to be measure preserving. The first significant result here was obtained by Hurewicz [**73**]. I shall state a special case of Hurewicz's theorem—a special case which, however, is typical of the general case and from which, in fact, the general case can easily be derived. Suppose that the transformation $T$ is a one-to-one transformation of a measure space $X$ onto itself such that, for every measurable subset $E$ of $X$, both $TE$ and $T^{-1}E$ are measurable and such that if $\mu(E)=0$, then $\mu(TE)=\mu(T^{-1}E)=0$. Suppose moreover that $T$ is incompressible in the sense that, for every measurable set $E$, $E\subset TE$ implies $\mu(E-TE)=0$. It follows from the Radon-Nikodym theorem that, for each integer $n$, there exists a positive measurable function $\omega_n$ such that $\mu(T^nE)=\int_E \omega_n d\mu$ for every measurable set $E$. In the notation so established Hurewicz's ergodic theorem says that, for any integrable function $f$, the weighted averages

$$\left(\sum_{i=1}^{n-1} f(T^i x)\omega_i(x)\right) \Big/ \left(\sum_{i=0}^{n-1} \omega_i(x)\right)$$

converge to a finite limit almost everywhere [**42**].

4. **Indecomposability.** If the interval $X$ is the union of two disjoint measurable sets $E$ and $F$ of positive measure, each of which is invariant under the measure preserving transformation $T$, then the study of any property of $T$ on $X$ reduces to the separate studies of the corresponding property of $T$ on $E$ and $T$ on $F$. In such a situation the transformation $T$ may be called decomposable. The most significant transformations are the indecomposable ones—they are usually called metrically transitive [**7**] or ergodic. In the early days of the theory many special examples of indecomposable transformations were presented in the literature—they occur in fields as apparently diverse as geometry [**51, 52, 53, 54, 55, 56, 66, 69, 70, 71, 100, 120, 121, 123, 124, 125**], probability [**22, 74**], and topological groups [**39**].

For indecomposable transformations the statement of the ergodic theorem can be strengthened by adjoining to it a description of the limit function. Precisely speaking, if $f$ is an integrable function, if $f^*(x) = \lim\ (1/n) \sum_{i=0}^{n-1} f(T^i x)$, and if the measure preserving transformation $T$ is indecomposable, then $f^*(x)$ is equal almost everywhere to a constant, and the value of that constant is $\int f d\mu$. This assertion is at the basis of the celebrated and sometimes misunderstood interchangeability of time means and phase means. It is of interest to observe that the principle of interchangeability is in fact equivalent to indecomposability for measure-preserving transformations on a finite measure space; if, in other words, for each integrable function $f$, $f^*$ is equal almost everywhere to a finite constant, then $T$ is indecomposable [**103**].

A question might be raised as to the extent to which Birkhoff's theorem could be extended to functions which are not necessarily integrable. It is clear that if a transformation is sufficiently decomposable (meaning that there exists a disjoint infinite sequence of measurable invariant sets of positive measure), then the conclusion of Birkhoff's theorem is true for many non integrable functions. The identity transformation is from this point of view the extreme case—if $T$ is the identity, then the conclusion of Birkhoff's theorem is true for every function. For indecomposable transformations $T$, on the other hand, it can be proved that if $f$ is a non-negative measurable function, or, more generally, a measurable function with the property that either its positive part or its negative part is integrable, and if $\lim\ (1/n) \sum_{i=0}^{n-1} f(T^i x)$ exists and is finite almost everywhere, then $f$

must be integrable. M. Gerstenhaber has recently shown me an example which proves that for arbitrary measurable functions this result is not necessarily true.

5. **Decompositions.** A natural problem in connection with the concept of indecomposability is whether or not every measure preserving transformation may be decomposed into indecomposable components. In order to clarify the question and motivate the answer it is helpful to consider an example. Suppose that the measure space $X$ is the unit square in the Cartesian plane, and that the measure preserving transformation $T$ on $X$ has the property that it leaves unaltered the first coordinate of each point. This implies that every subset $E$ of $X$ which depends on the first coordinate alone, that is, every set $E$ which is the union of a class of vertical segments, is invariant under $T$. Suppose moreover that the invariant sets so obtained exhaust essentially all possibilities—that is, that every measurable invariant set is, modulo sets of measure zero, a union of vertical segments. Since each vertical segment may be considered as a measure space on its own right (with linear Lebesgue measure) and since, in the presence of the conditions described above, the transformation $T$ is indecomposable on each such segment, this situation is an example of a decomposable transformation which is in a certain intuitively obvious sense made up of many little indecomposable components.

The general situation exemplified by the preceding paragraph may be described as follows. Suppose that to each point $x$ of a measure space $X$ (with measure $\mu$) there corresponds a measure space $Y_x$ (with measure $\mu_x$) so that the spaces corresponding to distinct points are disjoint. Suppose that a concept of measurability and a numerical measure $\lambda$ are introduced into the set $Z$ of all those pairs $(x, y)$ for which $x \in X$ and $y \in Y_x$, in such a way that whenever $A$ is a measurable subset of $X$, then the set $A^* = \{(x, y) : x \in A\}$ is a measurable subset of $Z$. If, for every measurable subset $E$ of $Z$,

$$\lambda(E) = \int_X \mu_x(E \cap Y_x) d\mu(x),$$

then the measure space $Z$ (with measure $\lambda$) is called a direct sum of the measure spaces $Y_x$ with respect to the measure space $X$. The best possible theorem on the decomposability of a measure preserving transformation $T$ on a measure space $Z$ would presumably assert that $Z$ may be represented as a direct sum in such a way that the class of measurable invariant sets coincides, except possibly for sets of measure zero, with the class of all sets of the form $A^*$; this would imply that the transformation $T$ on $Y_x$ is indecomposable for almost every $x$ in $X$. The first theorem of this type (in case $Z$ is a complete metric space) was proved by von Neumann [**103**]; extensions to more general spaces were later given by Dieudonné [**21**] and myself [**36, 44**].

6. **Density and category theorems.** How likely is a measure preserving transformation to be indecomposable? Birkhoff conjectured that in some sense the indecomposable case is the general case. One possible way of establishing this conjecture was to introduce a suitable metric or topology into the set $\boldsymbol{T}$ of all measure preserving transformations (after identifying two transformations which differ only on a set of measure zero) and then to show that the subset $\boldsymbol{D}$ of decomposable transformations is of the first category. This was first done by Oxtoby and Ulam for measure preserving homeomorphisms of certain subsets of Euclidean spaces and later by me for arbitrary measure preserving transformations [**40, 107**]. These topological investigations had some interesting byproducts of which at least one is worth mentioning. A class of particularly simple measure preserving transformations of the interval is obtained by dividing the interval into a finite number of subintervals of equal length and sending each such interval into another such interval by translation. The assertion concerning the class $\boldsymbol{P}$ of transformations so obtained is that it is everywhere dense in $\boldsymbol{T}$—in other words every measure preserving transformation is the limit of a sequence of permutations of intervals.

7. **Strong mixing.** It is possible to define the concept of indecomposability of a transformation $T$ in terms of the asymptotic behavior of the sequence of powers of $T$. It is in fact an easy consequence of the ergodic theorem that if $T$ is indecomposable, then

$$\lim \frac{1}{n} \sum_{i=0}^{n-1} \mu(E \cap T^i F) = \mu(E)\mu(F) \tag{*}$$

for every pair of measurable sets $E$ and $F$ and that, conversely, the validity of this relation for every pair of measurable sets implies that $T$ is indecomposable. Since the condition $\mu(E \cap F) = \mu(E)\mu(F)$ is the usual requirement in the definition of independence in the sense of the theory of probability, the equation (*) may be viewed as asserting that asymptotically, in the sense of Cesaro convergence, any two measurable sets, of which one is held fixed and the other is allowed to move under the influence of the transformation $T$, tend to become independent of each other.

The equation (*) has another natural physical interpretation. Suppose that the transformation $T$ is visualized as a particular way of stirring a container (of total volume 1) full of an incompressible fluid which may be thought of as 99.44 per cent water and .56 per cent red ink. If $F$ is the region occupied by the red ink, then, for any part $E$ of the container, the relative amount of red ink in $E$, after $n$ repetitions of the act of stirring, is given by $\mu(E \cap T^n F)/\mu(E)$. The indecomposability of the transformation $T$ implies therefore that on the average this relative amount of redness is exactly equal to .56. In general, in physical situations like this one, one expects to be justified in making a much stronger statement, namely that, after the liquid has been stirred sufficiently often, every part $E$ of the container will contain approximately .56 per cent red ink. In mathematical language this pious hope amounts to replacing Cesaro convergence by ordinary convergence, that is, replacing (*) by

$$(**) \qquad \lim \mu(E \cap T^n F) = \mu(E)\mu(F).$$

Transformations $T$ satisfying (**) for every pair $E$ and $F$ of measurable sets are called strongly mixing.

It may be worth while to give an example of a strongly mixing transformation on the interval $X$. The definition of such a transformation is surprisingly simple. For each $x$ in $X$, let $x = \sum_{i=1}^{\infty} a_i/2^i$ be the binary expansion of $x$, and let $p$ be a cyclic permutation of the set of all positive integers; for instance $p$ may be defined by

$$p(1) = 2, \ p(2n) = 2n + 2, \text{ and } p(2n+1) = 2n - 1, \qquad n = 1, 2, \cdots.$$

If $S$ is defined by $Sx = \sum_{i=1}^{\infty} a_{p(i)}/2^i$, then, except for some easily rectifiable trouble caused by the non uniqueness of the binary expansion, it follows that $S$ is one-to-one from $X$ onto $X$ and measure preserving. Strongly mixing transformations similar to this one occur frequently in probability theory.

8. **Weak mixing.** Between indecomposable transformations and strongly mixing transformations there is room for another concept—the concept of a weakly mixing transformation. This apparently artificial concept is of great technical significance. A measure preserving transformation $T$ is, by definition, weakly mixing if

$$(***) \qquad \lim \frac{1}{n} \sum_{i=0}^{n-1} | \mu(E \cap T^i F) - \mu(E)\mu(F) | = 0$$

for every pair $E$ and $F$ of measurable sets. In mathematical language, the definition of weak mixing substitutes strong Cesaro convergence for the Cesaro convergence occurring in the definition of indecomposability and for the ordinary convergence occurring in the definition of strong mixing. It is an analytic exercise to show that (***) is satisfied if and only if there exists a set $N$ of positive integers such that $N$ has density zero and such that

$$\lim_{n \notin N} \mu(E \cap T^n F) = \mu(E)\mu(F).$$

If indecomposability is expressed by saying that on the average $E$ is .56 per cent red, and if strong mixing is expressed by saying that after a while $E$ will be .56 per cent red, then weak mixing can be expressed by saying that after a while $E$ will be .56 per cent red, with the exception of a few rare instants during which it may be either too scarlet or else too pale a pink.

For any two measure preserving transformations $S$ and $T$, the direct product $S \times T$ is defined as that transformation on the Cartesian product of the space $X$ with itself which sends each point $(x, y)$ into $(Sx, Ty)$. The first indication that weak mixing is more than an analytic artificiality is in the assertion that $T$ is weakly mixing if and only if its direct product with itself is indecomposable [**65, 83, 84**].

The physical intuition which motivated the conjecture that in general a measure preserving transformation is indecomposable seems also to indicate that in general a measure preserving transformation is weakly and even strongly mixing. With the usual (category) interpretation of the phrase "in general," I proved that this conjecture is right for weak mixing [**41**]. In a recent paper Rokhlin [**115**] showed, by a very simple and elegant argument, that for strong mixing on the other hand the conjecture is just as wrong as it can be, and that, in fact the set $\boldsymbol{S}$ of strongly mixing transformations is a set of the first category in the set $\boldsymbol{T}$ of all measure preserving transformations.

9. **Automorphisms and unitary operators.** A measure preserving transformation $T$ on a measure space $X$ induces in an obvious way an automorphism of the measure algebra of measurable sets modulo sets of measure zero, or, in other words, a set transformation which assigns to each class of sets, any two members of which differ only on a set of measure zero, another such class. The question of whether or not the converse is true has received some attention [**38, 102**]. The answer is yes except in pathological measure spaces. Since this pathology is not of very much interest from the point of view of this paper, I shall from now on require that the space $X$ be non patho-

logical in this sense. It has been proved that all the well known measure spaces satisfy this requirement, that is, that for them every automorphism of the measure algebra of measurable sets modulo sets of measure zero is indeed induced by a one-to-one measure preserving transformation of the space onto itself.

Another, very similar, problem is suggested by the consideration of the complex Hilbert space $L_2$. If, for each $f$ in $L_2$, an element $Uf$ in $L_2$ is defined by $(Uf)(x)=f(Tx)$, then the fact that $T$ is measure preserving implies by an easy and familiar argument (proceeding through finite linear combinations of characteristic functions of measurable sets) that

$$\|Uf\|^2 = \int |f(Tx)|^2 d\mu(x) = \int |f(x)|^2 d\mu(x) = \|f\|^2.$$

Since the transformation $U$ is a one-to-one linear transformation of $L_2$ onto itself, the last written relation means that $U$ is unitary [**93**, **133**]. In other words, to each measure preserving transformation $T$ on $X$ there corresponds a unitary operator $U$ on $L_2$; the first problem that has to be settled is the characterization of the unitary operators that can arise in this way. A more or less satisfactory answer is known: a unitary operator is induced by a measure preserving transformation $T$ in the way just now outlined if and only if it sends each bounded measurable function into a bounded measurable function and is such that, for any two bounded measurable functions $f$ and $g$, $U(fg) = Uf \cdot Ug$, where the indicated multiplications denote the pointwise product of the factors [**103**].

It is natural to hope that at least some of the measure theoretic properties of $T$ can be described in the language of Hilbert space and that, conversely, the ideas suggested by the Hilbert space point of view may have measure theoretic significance. There is a fact quite near the surface which seems at least partially to fulfill this hope: a necessary and sufficient condition that $T$ be indecomposable is that the complex number 1 be a simple proper value of $U$ [**103**]. (A proper function of proper value 1 is of course simply an invariant function. Each function in the one-dimensional family of constant functions is invariant; the theorem just stated characterizes indecomposability by the absence of any other invariant functions.) A considerably deeper fact, known as the mixing theorem, is that $T$ is weakly mixing if and only if 1 is a simple proper value of $U$ and moreover $U$ has no other proper values [**62**, **63**, **65**, **94**].

10. **Isomorphism and pure point spectrum.** A fundamental prob-

lem of the theory of measurable transformations is the problem of isomorphism. Two measure preserving transformations $T_1$ and $T_2$ are called isomorphic if there exists a measure preserving transformation $T$ which carries $T_1$ into $T_2$, that is, for which $TT_1T^{-1}=T_2$. An obvious solution of the isomorphism problem is suggested by the first result mentioned in the preceding section—two measure preserving transformations are isomorphic if and only if the automorphisms they induce are conjugate elements in the group $\boldsymbol{T}$ of all automorphisms.

One might make the more promising and less trivial conjecture that all the measure theoretic properties of a transformation $T$ are reflected by the operatorial properties of its unitary operator $U$, that is, that $T_1$ and $T_2$ are isomorphic if and only if the corresponding unitary operators $U_1$ and $U_2$ are spectrally equivalent, or, in other words, if and only if there exists a unitary operator $U$ for which $UU_1U^*=U_2$.

There is an interesting class of measure preserving transformations for which the conjecture expressed in the preceding paragraph is true. In order to motivate the introduction of this class, let the measure space $X$ be a compact abelian group (with Haar measure) and let $\alpha$ be any fixed element of $X$. If $T$ is defined, for every $x$ in $X$, by $Tx=x+\alpha$, then $T$ is a measure preserving transformation of $X$ onto itself. If $\phi$ is any character of the group $X$, then the equation $\phi(x+\alpha)=\phi(\alpha)\phi(x)$ shows that $\phi$ is a proper function of the unitary operator $U$ corresponding to $T$, with proper value $\phi(\alpha)$. Since the set of all characters is a complete orthonormal set in $L_2$, the transformation $T$ is said to have pure point spectrum—the general definition of a measure preserving transformation with pure point spectrum is that the set of proper functions of its induced unitary operator is large enough to contain a complete orthonormal set. The conjecture concerning the relation of isomorphism and spectral equivalence is true in this sense: two indecomposable measure preserving transformations with pure point spectrum are isomorphic if and only if their induced unitary operators are spectrally equivalent—this was proved by von Neumann [**103**]. In this connection it should also be mentioned that every indecomposable measure preserving transformation with pure point spectrum is known to be of the type described in the example above—that is, each such transformation is necessarily a translation by a suitable element in a suitable compact abelian group [**38**].

**11. Isomorphism and mixed spectrum.** The result of the preceding section is not true for arbitrary measure preserving transformations—it is possible, in other words, to construct an example of two measure preserving transformations $T_1$ and $T_2$ such that $T_1$ and $T_2$ are not isomorphic but such that the corresponding unitary operators are spectrally equivalent. This construction has not been published so far—it is the result of joint work by von Neumann and myself. I proceed to sketch the details of the theory behind the construction.

With each measure preserving transformation $T$ it is possible to associate a sequence $\{G_n\}$ of classes of bounded measurable functions as follows. The initial class $G_0$ contains only the function which is identically equal to 1; for $n \geqq 1$, the class $G_n$ consists of those functions $f$ which satisfy almost everywhere an equation of the form $f(Tx) = g(x)f(x)$, with $g \in G_{n-1}$. The class $G_1$ is then the class of invariant functions; if $T$ is indecomposable, so that $G_1$ is the class of constant functions, then $G_2$ is the class of proper functions; for positive values of $n$ the functions of $G_n$ may be viewed as generalized proper functions belonging to proper values which instead of being necessarily constants are elements of $G_{n-1}$. It may happen that, for a suitable positive integer $n$, $G_n = G_{n+1}$ (and hence $G_n = G_{n+p}$ for every positive integer $p$); the least positive integer for which this happens is then denoted by $n(T)$. If for instance $T$ is weakly mixing, then (since $T$ is a fortiori indecomposable) $G_1$ is the class of constant functions and therefore, by the mixing theorem, $n(T) = 1$.

An example which shows that isomorphism is not the same as spectral equivalence is now easy to construct. Let the measure space be the torus represented as all pairs $(x, y)$ of real numbers modulo 1. If $T_1(x, y) = (x+\alpha, y+x)$ [mod 1], and if $T_2(x, y) = (x+\alpha, Sy)$ [mod 1], where $\alpha$ is an irrational number and $S$ is the transformation described in §8 as an example of a strongly mixing transformation, then a reasonably straightforward calculation shows that the induced unitary operators are spectrally equivalent. The proof that, nevertheless, $T_1$ and $T_2$ are not isomorphic leans on the concepts introduced in the preceding paragraph; it can in fact be shown that $n(T_1) = 3$ and $n(T_2) = 2$.

12. **Invariant measures.** With the exception of the ergodic theorem, most of the considerations in the preceding sections deal with measure preserving transformations. How restrictive is the requirement that a transformation preserve some measure? One of the best precise and nontrivial formulations of this question is the following one. Suppose that $T$ is a one-to-one transformation of a measure space $X$ onto itself such that for every measurable subset $E$ of $X$ both $TE$ and $T^{-1}E$ are measurable and such that if $\mu(E) = 0$, then $\mu(TE) = \mu(T^{-1}E) = 0$; does there then exist a measure $\lambda$ on the class of all measurable subsets of $X$ such that $\lambda(E)$ vanishes if and only if $\mu(E)$

vanishes, such that $X$ is the union of countably many measurable sets on each of which $\lambda$ is finite, and such that $\lambda(TE)=\lambda(E)$ for every measurable set $E$?

An interesting necessary and sufficient condition for the existence of a finite $\lambda$ was given by Hopf [**64**]. In order to describe this condition it is necessary to introduce some new concepts. Two measurable sets $E$ and $F$ are primitively equivalent with respect to a transformation $T$ if there exists an integer $n$ such that $T^nE=F$. Two measurable sets $E$ and $F$ are equivalent with respect to a transformation $T$ if (after possibly omitting a set of measure zero from both of them) both $E$ and $F$ may be written as unions of disjoint sequences of measurable sets so that each term in the sequence representing $E$ is primitively equivalent to the corresponding term in the sequence representing $F$. In imitation of Dedekind's definition of finiteness, it is customary to say that a measurable set $E$ is bounded (or finite) with respect to $T$ if whenever $E$ is equivalent to a subset $F$ of itself, then $\mu(E-F)=0$. Hopf's condition for the existence of a finite invariant measure is that $X$ be bounded.

The concept of boundedness, while it gives quite a bit of insight into the structure of measurable transformations, is not very easy to apply. Until recently, for example, it was not known whether or not the condition was vacuously satisfied, that is, whether or not there existed any transformations at all which did not satisfy it. In 1947 I succeeded in showing that such a transformation does exist and hence that the search for invariant measures must be conducted among measures which are allowed to take infinite values [**43**]. The proof is based on the fact that it is sufficient to exhibit an indecomposable measure preserving transformation on a non atomic measure space of infinite measure; the first transformation with these properties was constructed by Oxtoby. The condition of Hopf has a natural extension to the infinite case, but is then even harder to apply [**43, 85**]. While it seems plausible to conjecture that an invariant measure need not always exist, the question is still open. Either an example or, if the conjecture is wrong, a general existence theorem would be of considerable interest. In the meantime the theory of measurable but not necessarily measure preserving transformations deserves a little investigation. The first steps of such an investigation were carried out by Rademacher in 1916 [**110**] and a few fragmentary results have been obtained since then [**43, 46**].

13. **Flows.** Virtually all the problems discussed above for a single transformation $T$ and its iterates $T^n$, $n=0, \pm 1, \pm 2, \cdots$, make sense for a one-parameter family of transformation $\{T_t\}$, $-\infty<t$

$<+\infty$, which have the group property $T_{t+s}=T_tT_s$ for all $t$ and $s$. Such a family of measure preserving transformations is called a flow; a flow is called measurable if $T_tx$ is a measurable function of $(x, t)$. If sums are replaced by integrals and the problems of measurability receive a modicum of attention, then the ergodic theorem, the mixing theorem, the decomposition into indecomposable parts, and many other essential results can be extended from transformations to flows without any new conceptual difficulties.

There is, however, at least one notion which is suggested by the study of flows and not by the automatic process of generalization. This is the concept of a flow built under a function. Suppose that $f$ is any positive measurable function on a measure space $X$ and that $T$ is any one-to-one measure preserving transformation of $X$ onto itself. Let $\overline{X}$ be the ordinate set of $f$, that is, $\overline{X}$ is a subset of the product space of $X$ with the real line and is defined by

$$\overline{X} = \{(x, t): 0 \leqq t < f(x)\}.$$

For convenience of language conceive $X$ as a horizontal interval, and the parameter $t$ as time; it will then make sense to speak of points of $\overline{X}$ going "up" with a certain "velocity." Let each point $(x, t)$ of $\overline{X}$ move up, in this sense, with a uniform unit velocity, until it hits the "ceiling" $f(x)$; let it then be put back at the place $(Tx, 0)$ and continue on up from there. This procedure defines a flow in $\overline{X}$; the flow so defined is called the flow built on $T$ under $f$. The study of such a flow may obviously be reduced to the study of the transformation $T$ and the function $f$; it is consequently a very useful thing to know that every measurable indecomposable flow is isomorphic to a flow built under a function. This result was proved by Ambrose [2] and later extended to all non pathological flows by Ambrose and Kakutani [3].

It is interesting to observe that the concept of a flow built under a function has a discrete analogue which promises to be of interest even in the study of a single transformation. Let $f$ be a measurable function from a measure space $X$ to the set of non-negative integers, and write $E_n=\{x: f(x)\geqq n\}$, $n=0, 1, 2, \cdots$. The set $\overline{X}$ of all pairs $(x, n)$, where $n=0, 1, 2, \cdots$ and $x\in E_n$, is an analogue of what in the continuous case was called the ordinate set of $f$. If $T$ is a transformation on $X$, then a transformation $\overline{T}$ may be defined on $\overline{X}$ by setting $\overline{T}(x, n)=(x, n+1)$ whenever $x\in E_{n+1}$ and $\overline{T}(x, n)=(Tx, 0)$ otherwise. The example (mentioned in the preceding section) of a transformation which does not preserve any suitable finite measure was constructed with the aid of such methods; a systematic study of the relation between $T$ and $\bar{T}$ was recently begun by Kakutani [**82**].

**14. Unsolved problems.** In connection with the discussion above I have already had occasion to mention two or three unsolved problems. The most important one among these is the problem of the existence of invariant measures. In this section I shall mention a few more directions in which further progress would be desirable.

Quite a few of the results discussed above, but by no means all, can be extended to measure spaces of not necessarily finite measure; a systematic investigation of measure preserving transformations on infinite measure spaces is still lacking. Another obvious problem is to extend the ergodic theorem (known for the additive group of the integers and the additive group of the real numbers) to more general groups of transformations; some results in this direction are known for finite dimensional vector groups [**109, 127, 128**].

In order to state the next problem, it is convenient to introduce one more concept. If, for each of a countable set of indices $i$, $X_i$ is a measure space with measure $\mu_i$ such that $\mu_i(X_i)=1$, then it is possible to make the Cartesian product $X$ of all the $X_i$ into a measure space in a way which is a natural extension of the concept of Cartesian product for a finite number of spaces. Such product spaces are well known in probability theory. Suppose in particular that the spaces $X_i$ are all equal to a fixed space $X_0$ and that the domain of the index $i$ is the set of all integers. A point $x$ of $X$ is in this case a sequence, $x=(\cdots, x_{-2}, x_{-1}, x_0, x_1, x_2, \cdots)$, $x_i \in X_0$. If a transformation $T_0$ is defined on $X$ by $T_0x=(\cdots, y_{-2}, y_{-1}, y_0, y_1, y_2, \cdots)$, $y_n=x_{n+1}$, then this coordinate shift $T_0$ is a one-to-one measure preserving transformation of $X$ onto itself. If $X_0$ is the real line (with a suitable measure $\mu_0$ satisfying $\mu_0(X_0)=1$) and if the function $f_0$ on $X$ is defined by $f_0(x)=x_0$, then results much more precise than the ergodic theorem are known about the asymptotic behavior of the sequence $\{f_0(T_0^n x)\}$. It would be of interest to obtain analogues of these results (for example, the law of the iterated logarithm and the central limit theorem) for a wider class of transformations and functions than the ones here described.

An attempt in this direction has been made by Izumi [**75**]. Motivated by the fact that (under a mild restriction on the measure $\mu_0$) not only do the averages $(1/n)\sum_{i=0}^{n-1} f_0(T_0^i x)$ converge to zero almost everywhere but even the series $\sum_{n=1}^{\infty}(1/n)f_0(T_0^n x)$ is convergent almost everywhere, Izumi formulated a set of conditions on the measure preserving transformation $T$ (and on the function $f$) sufficient to ensure the convergence almost everywhere of $\sum_{n=1}^{\infty}(1/n)f(T^n x)$. Unfortunately Izumi's requirements turned out to be so stringent that the only transformation which can satisfy them is the identity transformation on a measure space consisting of exactly one point [45].

In conclusion I emphasize again that the isomorphism problem is almost completely untouched in most nontrivial cases. It is, of course, hard to formulate precisely what is meant by the injunction: "find necessary and sufficient conditions for the isomorphism of two measure preserving transformations." There are, however, many concrete examples of measure preserving transformations concerning which it is not known whether or not any two are isomorphic—the vague task of finding a complete set of invariants can at least at the beginning be replaced by the specific task of finding sufficiently many invariants to sort out into isomorphism classes these known examples [39]. If, for instance, the coordinate shift on the Cartesian product of countably many $k$-point spaces (in which each point has measure $(1/k)$) is denoted by $T_k$, then is $T_2$ isomorphic to $T_3$ or is it not? Nobody knows.

## Bibliography

The following is a list, as nearly complete as I could make it, of the papers and books directly relevant to the theory of measurable and measure preserving transformations. It does *not* contain works concerning applications to physics or stochastic processes, generalizations to Markoff chains or mean ergodic theorems, and analytical or topological results about dynamical systems.


1. W. Ambrose, *Change of velocities in a continuous ergodic flow*, Duke Math. J. vol. 8 (1941) pp. 425–440.

2. ———, *Representation of ergodic flows*, Ann. of Math. vol. 42 (1941) pp. 723–739.

3. W. Ambrose and S. Kakutani, *Structure and continuity of measurable flows*, Duke Math. J. vol. 9 (1942) pp. 25–42.

4. W. Ambrose, P. R. Halmos, and S. Kakutani, *The decomposition of measures*, II, Duke Math. J. vol. 9 (1942) pp. 43–47.

5. M. Bebutoff and W. Stepanoff, *Sur le changement du temps dans les systèmes dynamiques possédant une mesure invariante*, C. R. (Doklady) Acad. Sci. URSS vol. 24 (1939) pp. 217–219.

6. ———, *Sur la mesure invariante dans les systèmes dynamiques qui ne diffèrent que par le temps*, Rec. Math. (Mat. Sbornik) N.S. vol. 7 (1940) pp. 143–166.

7. G. D. Birkhoff and P. A. Smith, *Structure analysis of surface transformations*, J. Math. Pures Appl. vol. 7 (1928) pp. 345–379.

8. G. D. Birkhoff, *Proof of a recurrence theorem for strongly transitive systems*, Proc. Nat. Acad. Sci. U.S.A. vol. 17 (1931) pp. 650–655.

9. ———, *Proof of the ergodic theorem*, Proc. Nat. Acad. Sci. U.S.A. vol. 17 (1931) pp. 656–660.

10. G. D. Birkhoff and B. O. Koopman, *Recent contributions to ergodic theory*, Proc. Nat. Acad. Sci. U.S.A. vol. 18 (1932) pp. 279–282.

11. G. D. Birkhoff, *Probability and physical systems*, Bull. Amer. Math. Soc. vol. 38 (1932) pp. 361–379.

12. ———, *Some unsolved problems of theoretical dynamics*, Science vol. 94 (1941) pp. 598–600.

13. ———, *What is the ergodic theorem?*, Amer. Math. Monthly vol. 49 (1942) pp. 222–226.

14. ———, *The ergodic theorems and their importance in statistical mechanics*, Revista de Ciencias vol. 44 (1942) p. 251.

15. N. Bogoliouboff and N. Kryloff, *Les mesures invariantes et la transitivité*, C. R. Acad. Sci. Paris vol. 201 (1935) pp. 1454–1456.

16. ———, *Les mesures invariantes et transitives dans la mécanique non linéaire*, Rec. Math. (Mat. Sbornik) N.S. vol. 1 (1936) pp. 707–711.

17. ———, *La théorie générale de la mesure dans son application à l'étude des systèmes dynamiques de la mécanique non linéaire*, Ann. of Math. vol. 38 (1937) pp. 65–113.

18. C. Carathéodory, *Über den Wiederkehrsatz von Poincaré*, Sitzungsberichte der Preussischen Akademie der Wissenschaften vol. 32 (1919) pp. 580–584.

19. ———, *Bemerkungen zum Riesz-Fischerschen Satz und zur Ergodentheorie*, Abh. Math. Sem. Hansischen Univ. vol. 14 (1941) pp. 351–389.

20. ———, *Bemerkungen zum Ergodensatz von G. Birkhoff*, Sitzungsberichte der Mathematisch-Naturwissenschaftlichen Klasse der Bayerischen Akademie der Wissenschaften zu München (1944) pp. 189–208.

21. J. Dieudonné, *Sur le théorème de Lebesgue-Nikodym* (III), Annales de l'Université de Grenoble vol. 23 (1948) pp. 25–53.

22. J. L. Doob, *Probability and statistics*, Trans. Amer. Math. Soc. vol. 36 (1934) pp. 759–775.

23. ———, *One-parameter families of transformations*, Duke Math. J. vol. 4 (1938) pp. 752–774.

24. ———, *The law of large numbers for continuous stochastic processes*, Duke Math. J. vol. 6 (1940) pp. 290–306.

25. J. L. Doob and R. A. Leibler, *On the spectral analysis of a certain transformation*, Amer. J. Math. vol. 65 (1943) pp. 263–272.

26. Y. N. Dowker, *Invariant measure and the ergodic theorems*, Duke Math. J. vol. 14 (1947) pp. 1051–1061.

27. N. Dunford, *Spectral theory*. I. *Convergence to projections*, Trans. Amer. Math. Soc. vol. 54 (1943) pp. 185–217.

28. ———, *Spectral theory*, Bull. Amer. Math. Soc. vol. 49 (1943) pp. 637–651.

29. N. Dunford and D. S. Miller, *On the ergodic theorem*, Trans. Amer. Math. Soc. vol. 60 (1946) pp. 538–549.

30. K. Fan, *Les fonctions asymptotiquement presque-périodiques d'une variable entière et leur application à l'étude de l'itération des transformations continues*, Math. Zeit. vol. 48 (1943) pp. 685–711.

31. S. Fomin, *Finite invariant measures in the flows*, Rec. Math. (Mat. Sbornik) N.S. vol. 12 (1943) pp. 99–108.

32. M. Fréchet, *Sur le théorème ergodique de Birkhoff*, C. R. Acad. Sci. Paris vol. 213 (1941) pp. 607–609.

33. ———, *Sur le problème ergodique*, La Revue Scientifique (Revue Rose Illustrée) vol. 81 (1943) pp. 155–157.

34. ———, *Une application des fonctions asymptotiquement presque-périodiques à l'étude des familles de transformations ponctuelles et au problème ergodique*, La Revue Scientifique (Revue Rose Illustrée) vol. 79 (1941) pp. 407–417.

**35.** M. Fukamiya, *On dominated ergodic theorems in* $L_p(p \geqq 1)$, Tôhoku Math. J. vol. 46 (1940) pp. 150–153.

**36.** P. R. Halmos, *The decomposition of measures*, Duke Math. J. vol. 8 (1941) pp. 386–392.

**37.** ———, *Square roots of measure preserving transformations*, Amer. J. Math. vol. 64 (1942) pp. 153–166.

**38.** P. R. Halmos and J. von Neumann, *Operator methods in classical mechanics*, II, Ann. of Math. vol. 43 (1942) pp. 332–350.

**39.** P. R. Halmos, *On automorphisms of compact groups*, Bull. Amer. Math. Soc. vol. 49 (1943) pp. 619–624.

**40.** ———, *Approximation theories for measure preserving transformations*, Trans. Amer. Math. Soc. vol. 55 (1944) pp. 1–18.

**41.** ———, *In general a measure preserving transformation is mixing*, Ann. of Math. vol. 45 (1944) pp. 786–792.

**42.** ———, *An ergodic theorem*, Proc. Nat. Acad. Sci. U.S.A. vol. 32 (1946) pp. 156–161.

**43.** ———, *Invariant measures*, Ann. of Math. vol. 48 (1947) pp. 735–754.

**44.** ———, *On a theorem of Dieudonné*, Proc. Nat. Acad. Sci. U.S.A. vol. 35 (1949) pp. 38–42.

**45.** ———, *A non homogeneous ergodic theorem*, Trans. Amer. Math. Soc. vol. 66 (1949) pp. 284–288.

**46.** ———, *Measure theory*, New York, 1949.

**47.** P. Hartman and A. Wintner, *Asymptotic distributions and the ergodic theorem*, Amer. J. Math. vol. 61 (1939) pp. 977–984.

**48.** ———, *Statistical independence and statistical equilibrium*, Amer. J. Math. vol. 62 (1940) pp. 646–654.

**49.** ———, *Integrability in the large and dynamical stability*, Amer. J. Math. vol. 65 (1943) pp. 273–278.

**50.** P. Hartman, *On the ergodic theorems*, Amer. J. Math. vol. 69 (1947) pp. 193–199.

**51.** G. A. Hedlund, *On the metrical transitivity of the geodesics on a surface of constant negative curvature*, Proc. Nat. Acad. Sci. U.S.A. vol. 20 (1934) pp. 136–140.

**52.** ———, *On the metrical transitivity of the geodesics on closed surfaces of constant negative curvature*, Ann. of Math. vol. 35 (1934) pp. 787–808.

**53.** ———, *A metrically transitive group defined by the modular group*, Amer. J. Math. vol. 57 (1935) pp. 668–678.

**54.** ———, *The dynamics of geodesic flows*, Bull. Amer. Math. Soc. vol. 45 (1939) pp. 241–260.

**55.** ———, *Fuchsian groups and mixtures*, Ann. of Math. vol. 40 (1939) pp. 370–383.

**56.** ———, *A new proof for a metrically transitive system*, Amer. J. Math. vol. 62 (1940) pp. 233–242.

**57.** H. Hilmy, *Sur le théorème ergodique*, C. R. (Doklady) Acad. Sci. URSS vol. 24 (1939) pp. 213–216.

**58.** ———, *Sur le récurrence ergodique dans les systèmes dynamiques*, Rec. Math. (Mat. Sbornik) N.S. vol. 7 (1940) pp. 101–109.

**59.** E. Hopf, *Zwei Sätze über den wahrscheinlichen Verlauf der Bewegungen dynamischer Systeme*, Math. Ann. vol. 103 (1930) pp. 710–719.

**60.** ———, *On the time average theorem in dynamics*, Proc. Nat. Acad. Sci. U.S.A. vol. 18 (1932) pp. 93–100.

**61.** ———, *Über lineare Gruppen unitärer Operatoren im Zusammenhange mit den*

*Bewegungen dynamischer Systeme*, Sitzungsberichte der Preussischen Akademie der Wissenschaften vol. 14 (1932) pp. 182–190.

**62.** ———, *Complete transitivity and the ergodic principle*, Proc. Nat. Acad. Sci. U.S.A. vol. 18 (1932) pp. 204–209.

**63.** ———, *Proof of Gibbs' hypothesis on the tendency toward statistical equilibrium*, Proc. Nat. Acad. Sci. U.S.A. vol. 18 (1932) pp. 333–340.

**64.** ———, *Theory of measure and invariant integrals*, Trans. Amer. Math. Soc. vol. 34 (1932) pp. 373–393.

**65.** ———, *On causality, statistics, and probability*, Journal of Mathematics and Physics vol. 13 (1934) pp. 51–102.

**66.** ———, *Fuchsian groups and ergodic theory*, Trans. Amer. Math. Soc. vol. 39 (1936) pp. 299–314.

**67.** ———, *Ergodentheorie*, Berlin, 1937.

**68.** ———, *Statistische Probleme und Ergebnisse in der klassischen Mechanik*, Actualités Scientifiques et Industrielles, no. 737, 1938, pp. 5–16.

**69.** ———, *Beweis des Mischungscharakters der geodätischen Strömung auf Flächen der Krümmung minus Eins und endlicher Oberfläche*, Sitzungsberichte der Preussischen Akademie der Wissenschaften (1938) pp. 333–344.

**70.** ———, *Statistik der geodätischen Linien in Mannigfaltigkeiten negativer Krümmung*, Berichte über die Verhandlungen der Sächsischen Akademie der Wissenschaften zu Leipzig vol. 91 (1939) pp. 261–304.

**71.** ———, *Statistik der Lösungen geodätischen Probleme vom unstabilen Typus*, II, Math. Ann. vol. 117 (1940) pp. 590–608.

**72.** ———, *Über eine Ungleichung der Ergodentheorie*, Sitzungsberichte der Mathematisch-Naturwissenschaftlichen Klasse der Bayerischen Akademie der Wissenschaften zu München (1944) pp. 171–176.

**73.** W. Hurewicz, *Ergodic theorem without invariant measure*, Ann. of Math. vol. 45 (1944) pp. 192–206.

**74.** K. Ito, *On the ergodicity of a certain stationary process*, Proc. Imp. Acad. Tokyo vol. 20 (1944) pp. 54–55.

**75.** S. Izumi, *A non-homogeneous ergodic theorem*, Proc. Imp. Acad. Tokyo vol. 15 (1939) pp. 189–192.

**76.** ———, *A remark on ergodic theorems*, Proc. Imp. Acad. Tokyo vol. 19 (1943) pp. 102–104.

**77.** B. Jessen, *Abstract theory of measure and integration*. IX, Matematisk Tidsskrift. B (1947) pp. 21–26.

**78.** ———, *Abstrakt Maal-og Integralteori*, Copenhagen, 1947.

**79.** M. Kac, *On the notion of recurrence in discrete stochastic processes*, Bull. Amer. Math. Soc. vol. 53 (1947) pp. 1002–1010.

**80.** S. Kakutani and K. Yosida, *Birkhoff's ergodic theorem and the maximal ergodic theorem*, Proc. Imp. Acad. Tokyo vol. 15 (1939) pp. 165–168.

**81.** S. Kakutani, *Representation of measurable flows in Euclidean 3-space*, Proc. Nat. Acad. Sci. U.S.A. vol. 28 (1942) pp. 16–21.

**82.** ———, *Induced measure preserving transformations*, Proc. Imp. Acad. Tokyo vol. 19 (1943) pp. 635–641.

**83.** Y. Kawada, *Über die masstreuen Abbildungen vom Mischungstypus im weiteren Sinne*, Proc. Imp. Acad. Tokyo vol. 19 (1943) pp. 520–524.

**84.** ———, *Über die masstreuen Abbildungen in Produkträumen*, Proc. Imp. Acad. Tokyo vol. 19 (1943) pp. 525–527.

**85.** ———, *Über die Existenz der invarianten Integrale*, Jap. J. Math. vol. 19 (1944) pp. 81–95.

**86.** A. Khintchine, *Zur mathematischen Begründung der statistischen Mechanik*, Zeitschrift für Angewandte Mathematik und Mechanik, vol. 13 (1933) pp. 101–103.

**87.** ———, *The method of spectral reduction in classical dynamics*, Proc. Nat. Acad. Sci. U.S.A. vol. 19 (1933) pp. 567–573.

**88.** ———, *Zu Birkhoffs Lösung des Ergodenproblems*, Math. Ann. vol. 107 (1933) pp. 485–488.

**89.** ———, *Fourierkoeffizienten längs einer Bahn im Phasenraum*, Rec. Math. (Mat. Sbornik) vol. 41 (1934) pp. 14–15.

**90.** ———, *Eine Verschärfung des Poincaréschen "Wiederkehrsatzes,"* Compositio Math. vol. 1 (1934) pp. 177–179.

**91.** ———, *Korrelationstheorie der stationären stochastischen Prozesse*, Math. Ann. vol. 109 (1934) pp. 604–615.

**92.** A. Kolmogoroff, *Ein vereinfachter Beweis des Birkhoff-Khintchineschen Ergodensatzes*, Rec. Math. (Mat. Sbornik) N.S. vol. 2 (1937) pp. 366–368.

**93.** B. O. Koopman, *Hamiltonian systems and transformations in Hilbert space*, Proc. Nat. Acad. Sci. U.S.A. vol. 17 (1931) pp. 315–318.

**94.** B. O. Koopman and J. von Neumann, *Dynamical systems of continuous spectra*, Proc. Nat. Acad. Sci. U.S.A. vol. 18 (1932) pp. 255–263.

**95.** T. Levi-Civita, *A general survey of the theory of adiabatic invariants*, Journal of Mathematics and Physics vol. 13 (1934) pp. 18–40.

**96.** F. Maeda, *Application of the theory of set functions to the mixing of fluids*, Journal of Science of the Hirosima University. Ser. A vol. 5 (1935) pp. 1–6.

**97.** ———, *Transitivities of conservative mechanism*, Journal of Science of the Hirosima University. Ser. A vol. 6 (1936) pp. 1–18.

**98.** D. Maharam, *On homogeneous measure algebras*, Proc. Nat. Acad. Sci. U.S.A. vol. 28 (1942) pp. 108–111.

**99.** P. T. Maker, *The ergodic theorem for a sequence of functions*, Duke Math. J. vol. 6 (1940) pp. 27–30.

**100.** M. H. Martin, *Metrically transitive point transformations*, Bull. Amer. Math. Soc. vol. 40 (1934) pp. 606–612.

**101.** J. von Neumann, *Proof of the quasi-ergodic hypothesis*, Proc. Nat. Acad. Sci. U.S.A. vol. 18 (1932) pp. 70–82.

**102.** ———, *Einige Sätze über messbare Abbildungen*, Ann. of Math. vol. 33 (1932) pp. 574–586.

**103.** ———, *Zur Operatorenmethode in der klassischen Mechanik*, Ann. of Math. vol. 33 (1932) pp. 587–642.

**104.** ———, *Zusätze zur Arbeit "Zur Operatorenmethode . . . ,"* Ann. of Math. vol. 33 (1932) pp. 789–791.

**105.** J. C. Oxtoby, *Note on transitive transformations*, Proc. Nat. Acad. Sci. U.S.A. vol. 23 (1937) pp. 443–446.

**106.** J. C. Oxtoby and S. M. Ulam, *On the existence of a measure invariant under a transformation*, Ann. of Math. vol. 40 (1939) pp. 560–566.

**107.** ———, *Measure-preserving homeomorphisms and metrical transitivity*, Ann. of Math. vol. 42 (1941) pp. 874–920.

**108.** J. C. Oxtoby, *On the ergodic theorem of Hurewicz*, Ann. of Math. vol. 49 (1948) pp. 872–884.

**109.** H. R. Pitt, *Some generalizations of the ergodic theorem*, Proc. Cambridge Philos. Soc. vol. 38 (1942) pp. 325–343.

**110.** H. Rademacher, *Eineindeutige Abbildungen und Messbarkeit*, Monatshefte für Mathematik und Physik vol. 27 (1916) pp. 183–290.

**111.** F. Riesz, *Sur quelques problèmes de la théorie ergodique*, Matematikai és Fizikai Lapok vol. 49 (1942) pp. 34–62.

**112.** ———, *Sur la théorie ergodique*, Comment. Math. Helv. vol. 17 (1945) pp. 221–239.

**113.** V. Rokhlin, *On the classification of measurable decompositions*, Doklady Akademii Nauk SSSR vol. 58 (1947) pp. 29–32.

**114.** ———, *On the problem of the classification of automorphisms of Lebesgue spaces*, Doklady Akademii Nauk SSSR vol. 48 (1947) pp. 189–191.

**115.** ———, *A "general" measure preserving transformation is not mixing*, Doklady Akademii Nauk SSSR vol. 60 (1948) pp. 349–351.

**116.** ———, *Unitary rings*, Doklady Akademii Nauk SSSR vol. 59 (1948) pp. 643–646.

**117.** G. Scorza Dragoni, *Sul teorema ergodico*, Rend. Circ. Mat. Palermo vol. 58 (1934) pp. 311–325.

**118.** ———, *Transitivita metrica e teoremi di media*, Rend. Circ. Mat. Palermo vol. 59 (1935) pp. 235–255.

**119.** ———, *Sul fondamento matematico della teoria degli invarianti adiabatici*, Annali di Mathematica Pura ed Applicata vol. 13 (1935) pp. 335–362.

**120.** W. Seidel, *Note on a metrically transitive system*, Proc. Nat. Acad. Sci. U.S.A. vol. 19 (1933) pp. 453–456.

**121.** ———, *On a metric property of Fuchsian groups*, Proc. Nat. Acad. Sci. U.S.A. vol. 21 (1935) pp. 475–478.

**122.** W. Stepanoff, *Sur une extension du théorème ergodique*, Compositio Math. vol. 3 (1936) pp. 239–253.

**123.** M. Tsuji, *On Hopf's ergodic theorem*, Proc. Imp. Acad. Tokyo vol. 20 (1944) pp. 640–647.

**124.** ———, *Some metrical theorems on Fuchsian groups*, Proc. Imp. Acad. Tokyo vol. 21 (1945) pp. 104–109.

**125.** ———, *On Hopf's ergodic theorem*, Jap. J. Math. vol. 19 (1945) pp. 259–284.

**126.** C. Visser, *On Poincaré's recurrence theorem*, Bull. Amer. Math. Soc. vol. 42 (1936) pp. 397–400.

**127.** N. Wiener, *The homogeneous chaos*, Amer. J. Math. vol. 60 (1938) pp. 897–936.

**128.** ———, *The ergodic theorem*, Duke Math. J. vol. 5 (1939) pp. 1–18.

**129.** N. Wiener and A. Wintner, *Harmonic analysis and ergodic theory*, Amer. J. Math. vol. 63 (1941) pp. 415–426.

**130.** ———, *On the ergodic dynamics of almost periodic systems*, Amer. J. Math. vol. 63 (1941) pp. 794–824.

**131.** ———, *The discrete chaos*, Amer. J. Math. vol. 65 (1943) pp. 279–298.

**132.** A. Wintner, *Remarks on the ergodic theorem of Birkhoff*, Proc. Nat. Acad. Sci. U.S.A. vol. 18 (1932) pp. 248–251.

**133.** ———, *Dynamische Systeme und unitäre Matrizen*, Math. Zeit. vol. 36 (1933) pp. 630–637.

**134.** ———, *On the ergodic analysis of the remainder term of mean motions*, Proc. Nat. Acad. Sci. U.S.A. vol. 26 (1940) pp. 126–129.

**135.** K. Yosida, *Ergodic theorems of Birkhoff-Khintchine's type*, Jap. J. Math. vol. 17 (1940) pp. 31–36.

**136.** ———, *An abstract treatment of the individual ergodic theorem*, Proc. Imp. Acad. Tokyo vol. 16 (1940) pp. 280–284.

UNIVERSITY OF CHICAGO

# ENTROPY IN ERGODIC THEORY[1]

PAUL R. HALMOS

## CONTENTS



[1]University of Chicago, September 1959.

## PREFACE

Shannon's theory of information appeared on the mathematical scene in 1948; in 1958 Kolmogorov applied the new subject to solve some relatively old problems of ergodic theory. Neither the general theory nor its special application is as well known among mathematicians as they both deserve to be; the reason, probably, is faulty communication. Most extant expositions of information theory are designed to make the subject palatable to non-mathematicians, with the result that they are full of words like "source" and "alphabet". Such words are presumed to be an aid to intuition; for the serious student, however, who is anxious to get at the root of the matter, they are more likely to be confusing than helpful. As for the recent ergodic application of the theory, the communication trouble there is that so far the work of Kolmogorov and his school exists in Doklady abstracts only, in Russian only. The purpose of these notes is to present a stop-gap exposition of some of the general theory and some of its applications. While a few of the proofs may appear slightly different from the corresponding ones in the literature, no claim is made for the novelty of the results. As a prerequisite, some familiarity with the ideas of the general theory of measure is assumed; Halmos's *Measure theory* (1950) is an adequate reference.

Chapter I begins with relatively well-known facts about conditional expectations; for the benefit of the reader who does not know this technical probabilistic concept, several standard proofs are reproduced. Standard reference: Doob, *Stochastic Processes* (1953). A special case of the martingale convergence theorem is proved by what is essentially Lévy's original method [Théorie de l'addition des variables aléatoires (1937)]. The reader who knows the martingale theorem can skip the whole chapter, except possibly Section 9, and, in particular, equation (9.1).

Chapter II motivates and defines information. Standard reference: Khinchin, *Mathematical foundations of information theory* (1957). The more recent book of Feinstein, *Foundations of information theory* (1958), is quite technical, but highly recommended. The chapter ends with a proof of McMillan's theorem (mean convergence); the reader who knows that theorem can skip the chapter after looking at it just long enough to absorb the notation. Almost everywhere convergence probably holds. A recent paper by Breiman [*Ann. Math. Stat.* **28**

(1957) 809–811] asserts it, but that paper has an error; at the time these lines were written, the correction has not appeared yet. In any case, for the ergodic application not even mean convergence is necessary; all that is needed is the convergence of the integrals, which is easy to prove directly.

Chapter III studies entropy (average amount of information); all the facts here are direct consequences of the definitions, via the machinery built up in the first two chapters.

Chapter IV contains the application to ergodic theory. In general terms, the idea is that information theory suggests a new invariant (entropy) of measure-preserving transformations. The new invariant is sharp enough to distinguish between some hitherto indistinguishable transformations (e.g., the 2-shift and the 3-shift). The original idea of using this invariant is due to Kolmogorov [*Doklady* **119** (1958) 861–864 and **124** (1959) 754–755]. An improved version of the definition is given by Sinai [*Doklady* **124** (1959) 768–771], who also computes the entropy of ergodic automorphisms of the torus. The new invariant is in some respects not so sharp as older ones. Thus for instance, Rokhlin [*Doklady* **124** (1959) 980–983] asserts that all translations (in compact abelian groups) have the same entropy (namely zero); he also begins the study of the connection between entropy and spectrum. Much remains to be done along all these lines.

## CHAPTER I. CONDITIONAL EXPECTATION

**Section 1. Definition.** We shall work, throughout what follows, with a fixed probability space

$$(X, \mathcal{S}, P).$$

Here $X$ is a non-empty set, $\mathcal{S}$ is a field of subsets of $X$, and $P$ is a probability measure on $\mathcal{S}$. The word "field" in these notes is an abbreviation for "collection of sets closed under the formation of complements and *countable* unions". A probability measure on a field of subsets of $X$ is a measure $P$ such that

$$P(X) = 1.$$

Suppose that $\mathcal{C}$ is a subfield of $\mathcal{S}$ and $f$ is an integrable real function on $X$. If

$$Q(C) = \int_C f\,dP$$

for each $C$ in $\mathcal{C}$, then $Q$ is a signed measure on $\mathcal{C}$, absolutely continuous with respect to $P$ (or, rather, with respect to the restriction of $P$ to $\mathcal{C}$). The Radon–Nikodym theorem implies the existence of an integrable function $f^*$, measurable $\mathcal{C}$, such that

$$Q(C) = \int_C f^*\,dP$$

for each $C$ in $\mathcal{C}$. The function $f^*$ is uniquely determined (to within a set of measure zero); its dependence on $f$ and $\mathcal{C}$ is indicated by writing

$$f^* = E(f/\mathcal{C}).$$

The function $E(f/\mathcal{C})$ is called "the conditional expectation of $f$ with respect to $\mathcal{C}$". It is worthwhile to repeat the characteristic properties of conditional expectation; they are that

$$E(f/\mathcal{C}) \text{ is measurable } \mathcal{C} \tag{1.1}$$

and

$$\int_C E(f/\mathcal{C})\, dP = \int_C f\, dP \tag{1.2}$$

for each $C$ in $\mathcal{C}$.

**Section 2. Examples.** If $\mathcal{C}$ is the largest subfield of $\mathcal{S}$, that is, $\mathcal{C} = \mathcal{S}$, then $f$ itself satisfies (1.1) and (1.2), so that

$$E(f/\mathcal{S}) = f.$$

This result has a trivial generalization: since $f$ always satisfies (1.2) ($\int_C f\, dP = \int_C f\, dP$), it follows that if the field $\mathcal{C}$ is such that $f$ is measurable $\mathcal{C}$, then

$$E(f/\mathcal{C}) = f. \tag{2.1}$$

To look at the other extreme, let $2$ be the smallest subfield of $\mathcal{S}$, that is the field whose only non-empty member is $X$. Since the only functions measurable $2$ are constants, and since the only constant [in the role of $E(f/\mathcal{C})$] that satisfies (1.2) is $\int_C f\, dP$, it follows that

$$E(f/2) = \int f\, dP. \tag{2.2}$$

The constant $\int_C f\, dP$ is sometimes called the absolute (as opposed to conditional) expectation of $f$, and, in that case, it is denoted by $E(f)$.

Here is an illuminating example. Suppose that $X$ is the unit square, with the collection of Borel sets in the role of $\mathcal{S}$ and Lebesgue measure in the role of $P$. We say that a set in $\mathcal{S}$ is "vertical" in case its intersection with each vertical line $L$ in the plane is either empty or else equal to $X \cap L$. The collection $\mathcal{C}$ of all vertical sets in $\mathcal{S}$ is a subfield of $\mathcal{S}$. A function $f$ is measurable $\mathcal{C}$ if and only if it does not depend on its second (vertical) argument; it follows easily that if $f$ is integrable, then

$$E(f/\mathcal{C})(x, y) = \int f(x, u)\, du.$$

**Section 3. Algebraic properties.** Conditional expectation is a generalized integral and in one form or another it has all the properties of an integral. Thus, for instance,

$$E(1/\mathcal{C}) = 1, \tag{3.1}$$

where this equation, as well as all other asserted equations and inequalities involving conditional expectations, holds almost everywhere. (To prove (3.1), apply (2.1).) If $f$ and $g$ are integrable functions and if $a$ and $b$ are constants, then

$$E(af + bg/\mathcal{C}) = aE(f/\mathcal{C}) + bE(g/\mathcal{C}). \tag{3.2}$$

(Proof. if $C$ is in $\mathcal{C}$, then the integrals over $C$ of the two sides of (3.2) are equal to each other.) If $f \geqq 0$, then

$$E(f/\mathcal{C}) \geqq 0. \tag{3.3}$$

(Proof. if $C = \{x\colon E(f/\mathcal{C})(x) < 0\}$, then $C$ is in $\mathcal{C}$ and $\int_C f\,dP = 0$; this implies that $P(C) = 0$.) It is a consequence of (3.3) that

$$|E(f/\mathcal{C})| \leqq E(|f|/\mathcal{C}). \tag{3.4}$$

(Proof. both $|f| - f \geqq 0$ and $|f| + f \geqq 0$, and therefore, by (3.2) and (3.3), both $E(-f/\mathcal{C}) \leqq E(|f|/\mathcal{C})$ and $E(f/\mathcal{C}) \leqq E(|f|/\mathcal{C})$.)

Conditional expectations also have the following multiplicative property: if $f$ is integrable and if $g$ is bounded and measurable $\mathcal{C}$, then

$$E(fg/\mathcal{C}) = E(f/\mathcal{C})g. \tag{3.5}$$

Since the right side of (3.5) is measurable $\mathcal{C}$, the thing to prove is that

$$\int_C E(f/\mathcal{C})g\,dP = \int_C fg\,dP \tag{3.6}$$

for each $C$ in $\mathcal{C}$. In case $g$ is the characteristic function of a set in $\mathcal{C}$, (3.6) follows from the defining equation (1.2) for conditional expectations. This implies that (3.6) holds whenever $g$ is a finite linear combination of such characteristic functions, and hence, by approximation, that (3.6) holds whenever $g$ is a bounded function measurable $\mathcal{C}$.

**Section 4. Dominated convergence.** The usual limit theorems for integrals also have their analogues for conditional expectations. Thus if $f$, $g$, and $f_n$ are integrable functions, if $|f_n| \leqq g$ and $f_n \to f$ almost everywhere, then

$$E(f_n/\mathcal{C}) \to E(f/\mathcal{C}) \tag{4.1}$$

almost everywhere and, also, in the mean. For the proof, write

$$g_n = \sup(|f_n - f|, |f_{n+1} - f|, |f_{n+2} - f|, \ldots);$$

observe that the sequence $\{g_n\}$ tends monotonely to 0 almost everywhere and that $g_n \leqq 2g$. It follows that the sequence $\{E(g_n/\mathcal{C})\}$ is monotone decreasing and, therefore, has a limit $h$ almost everywhere. Since

$$\int h\,dP \leqq \int E(g_n/\mathcal{C})\,dP = \int g_n\,dP, \tag{4.2}$$

and since $\int g_n\,dP \to 0$, this implies that $E(g_n/\mathcal{C}) \to 0$ almost everywhere. Since, finally,

(4.3) $$|E(f_n/\mathcal{C}) - E(f/\mathcal{C})| \leqq E(|f_n - f|/\mathcal{C}) \leqq E(g_n/\mathcal{C}),$$

the proof of almost everywhere convergence is complete. Mean convergence is implied by the inequality

(4.4) $$\int |E(f_n/\mathcal{C}) - E(f/\mathcal{C})| \, dP \leqq \int |f_n - f| \, dP$$

and the Lebesgue dominated convergence theorem.

**Section 5. Conditional probability.** If $A$ is a measurable set (that is $A$ is in $\mathcal{S}$) and if

$$f = c(A)$$

(where $c(A)$ is the characteristic function of $A$), we write

$$E(f/\mathcal{C}) = P(A/\mathcal{C}).$$

The function $P(A/\mathcal{C})$ is called "the conditional probability of $A$ with respect to $\mathcal{C}$". The characteristic properties of conditional probability are that

$$P(A/\mathcal{C}) \text{ is measurable } \mathcal{C}$$

and

$$\int_C P(A/\mathcal{C}) \, dP = P(A \cap C)$$

for each $C$ in $\mathcal{C}$. If $A$ is in $\mathcal{C}$, then

(5.1) $$P(A/\mathcal{C}) = c(A),$$

and, in any case,

(5.2) $$P(A/2) = P(A).$$

For this reason the constant $P(A)$ is sometimes called the absolute (as opposed to conditional) probability of $A$.

The converse of the conclusion (5.1) is true and sometimes useful. The assertion is that if $P(A/\mathcal{C})$ is the characteristic function of some set, say $B$, then

(5.3) $$A \text{ is in } \mathcal{C}$$

(and therefore $B = A$). To prove this, note that

$$\int_C c(B) \, dP = P(A \cap C),$$

and therefore $P(A \cap C) = P(B \cap C)$ for each $C$ in $\mathcal{C}$. Since $P(A/\mathcal{C})$ is measurable $\mathcal{C}$, the set $B$ itself belongs to $\mathcal{C}$. It is therefore permissible to put $C = B$ and to put $C = X - B$; it follows that $B \subset A$ and $A \subset B$ (almost), so that $B = A$ (almost).

Just as conditional expectation has the properties of an integral, conditional probability has the properties of a probability measure. Thus if $A$ is a measurable set, then

$$0 \leqq P(A/\mathcal{C}) \leqq 1,$$

and if $\{A_n\}$ is a disjoint sequence of measurable sets with union $A$, then

$$P(A/\mathcal{C}) = \sum_n P(A_n/\mathcal{C}).$$

**Section 6. Jensen's inequality.** A useful analytic property of integration is known as Jensen's inequality, which we now proceed to state and prove in its generalized (conditional) form.

A real-valued function $F$ defined on an interval of the real line is called *convex* if

$$F(ps+qt) \leqq pF(s) + qF(t)$$

whenever $s$ and $t$ are in the domain of $F$ and $p$ and $q$ are non-negative numbers with sum 1. It follows immediately, by induction, that if $t_1, \ldots, t_n$ are in the domain of $F$ and $p_1, \ldots, p_n$ are non-negative numbers with sum 1, then

$$F\left(\sum_{i=1}^{n} p_i t_i\right) \leqq \sum_{i-1}^{n} p_i F(t_i). \tag{6.1}$$

Suppose now that $F$ is a continuous convex function whose domain is a finite subinterval of $[0,\infty)$, suppose that $g$ is a measurable function on $X$ whose range is (almost) included in the domain of $F$, and suppose that $\mathcal{C}$ is an arbitrary subfield of $\mathcal{S}$. Jensen's inequality asserts that under these conditions

$$f(E(g/\mathcal{C})) \leqq E(F(g)/\mathcal{C}) \tag{6.2}$$

almost everywhere. Since $g$ is the limit of an increasing sequence of simple functions, and since $F$ is continuous, it is sufficient to prove (6.2) in case

$$g = \sum_A c(A) t_A,$$

where the summation extends over the atoms of some finite subfield of $\mathcal{S}$. If $g$ has this form, then

$$F(g) \leqq \sum_A c(A) F(t_A)$$

and

$$E(g/\mathcal{C}) = \sum_A P(A/\mathcal{C}) t_A.$$

Since $E(F(g)/\mathcal{C}) = \Sigma_A P(A/\mathcal{C}) F(t_A)$, the inequality (6.2) is in this case a special case of (6.1).

In the extreme case, $\mathcal{C} = \mathcal{S}$, the conditional form of Jensen's inequality reduces to a triviality ($F(g) \leqq F(g)$); in the other extreme case, $\mathcal{C} = 2$, it becomes the classical absolute Jensen's inequality

$$F\left(\int g \, dP\right) \leqq \int F(g) \, dP.$$

**Section 7. Transformations.** Later we shall need to know the effect of measure-preserving transformations on conditional expectations and probabilities. Suppose therefore that $T$ is a measure-preserving transformation on $X$; this means that if $A$ is measurable, then $T^{-1}A$ is measurable and

$$P(T^{-1}A) = P(A).$$

(For present purposes, $T$ need not be invertible.)

If $\mathcal{C}$ is a subfield of $\mathcal{S}$, then

$$T^{-1}\mathcal{C}$$

is the collection (field) of all sets of the form $T^{-1}C$ with $C$ in $\mathcal{C}$; if $f$ is a function on $X$, then $fT$ is the composite of $f$ and $T$. The basic change-of-variables result is that if $f$ is integrable, then

$$\int_C f\,dP = \int_{T^{-1}C} fT\,dP$$

for each measurable set $C$. If, in particular, $C$ is in $\mathcal{C}$, then

$$\begin{aligned}\int_{T^{-1}C} E(fT/T^{-1}\mathcal{C})\,dP &= \int_{T^{-1}C} fT\,dP\\ &= \int_C f\,dP = \int_C E(f/\mathcal{C})\,dP\\ &= \int_{T^{-1}C} E(f/\mathcal{C})T\,dP.\end{aligned}$$

Since both $E(fT/T^{-1}\mathcal{C})$ and $E(f/\mathcal{C})T$ are measurable $T^{-1}\mathcal{C}$, it follows that

$$E(fT/T^{-1}\mathcal{C}) = E(f/\mathcal{C})T. \tag{7.1}$$

Since $c(A)T = c(T^{-1}A)$, this implies that

$$P(T^{-1}A/T^{-1}\mathcal{C}) = P(A/\mathcal{C})T. \tag{7.2}$$

**Section 8. Lattice properties.** The next item of interest is the dependence of $E(f/\mathcal{C})$ on $\mathcal{C}$. The collection of all subfields of $\mathcal{S}$ has a reasonable amount of structure; it is partially ordered (by inclusion), and, in fact, it is a complete lattice. (The infimum of two subfields $\mathcal{B}$ and $\mathcal{C}$ is just their intersection, and their supremum $\mathcal{B} \vee \mathcal{C}$ is the field they generate; similar assertions hold for the infimum and supremum of any family of subfields.) It might therefore be hoped that the dependence of $E(f/\mathcal{C})$ on $\mathcal{C}$ exhibits some algebraically pleasant behavior, such as monotony or additivity. Nothing like this is true.

If, for instance, $\mathcal{B}$ and $\mathcal{C}$ are subfields of $\mathcal{S}$ with $\mathcal{B} \subset \mathcal{C}$, then the best that can be said is that

$$E(E(f/\mathcal{C})/\mathcal{B}) = E(f/\mathcal{B}) \tag{8.1}$$

and

$$\int |E(f/\mathcal{B})|\, dP \leqq \int |E(f/\mathcal{C})|\, dP. \tag{8.2}$$

To prove (8.1), observe that if $B$ is in $\mathcal{B}$, then $B$ is in $\mathcal{C}$, and therefore

$$\int_B E(E(f/\mathcal{C})/\mathcal{B})\, dP = \int_B E(f/\mathcal{C})\, dP$$

$$= \int_B f\, dP = \int_B E(f/\mathcal{B})\, dP.$$

To prove (8.2), write $B = \{x\colon\ E(f/\mathcal{B})(x) \geqq 0\}$. Since, clearly, $b$ is in $\mathcal{B}$, it follows that

$$\int_B E(f/\mathcal{B})\, dP = \int_B E(f/\mathcal{C})\, dP \leqq \int_B |E(f/\mathcal{C})|\, dP$$

and

$$-\int_{X-B} E(f/\mathcal{B})\, dP = -\int_{X-B} E(f/\mathcal{C})\, dP \leqq \int_{X-B} |E(f/\mathcal{C})|\, dP.$$

In case $f = c(A)$, the equation (8.1) becomes

$$E(P(A/\mathcal{C})/\mathcal{B}) = P(A/\mathcal{B}). \tag{8.3}$$

**Section 9. Finite fields.** Some insight can be gained by studying the finite subfields of $\mathcal{S}$. A finite subfield $\mathcal{B}$ of $\mathcal{S}$ is atomic. This means that $\mathcal{B}$ has a finite number of (necessarily disjoint) atoms whose union is (almost) equal to $X$; an atom is a set of positive measure in $\mathcal{B}$ that includes no subset of strictly smaller positive measure in $\mathcal{B}$. A function is measurable $\mathcal{B}$ if and only if it is a constant on each atom of $\mathcal{B}$. It follows easily that if $f$ is integrable, then on each atom $B$ of $\mathcal{B}$,

$$E(f/\mathcal{B}) = \frac{1}{P(B)} \int_B f\, dP.$$

A convenient way of expressing this fact is to write

$$E(f/\mathcal{B}) = \sum_B \frac{c(B)}{P(B)} \int_B f\, dP,$$

where the summation extends over all the atoms of $\mathcal{B}$. In the case $f = c(A)$, this becomes

$$P(A/\mathcal{B}) = \sum_B c(B) \frac{P(A \cap B)}{P(B)}.$$

Suppose next that $\mathcal{B}$ and $\mathcal{C}$ are two finite subfields of $\mathcal{S}$. It follows, of course, that $\mathcal{B} \vee \mathcal{C}$ is a finite subfield of $\mathcal{S}$, and the atoms of $\mathcal{B} \vee \mathcal{C}$ are the non-empty sets of the form $B \cap C$, where $B$ is an atom of $\mathcal{B}$ and $C$ is an atom of $\mathcal{C}$. More generally, if $\mathcal{B}$ and $\mathcal{C}$ are subfields of $\mathcal{S}$ such that $\mathcal{B}$ is finite, then every element of $\mathcal{B} \vee \mathcal{C}$ is obtained as follows: for each atom $B$ of $\mathcal{B}$ let $C_B$ be an element of $\mathcal{C}$ and form the union $\cup_B (B \cap C_B)$, extended over all atoms of $\mathcal{B}$. Something useful can be said about conditional expectations and probabilities in this case: the assertion is that if $f$ is integrable, then on each atom $B$ of $\mathcal{B}$

$$E(f/\mathcal{B} \vee \mathcal{C}) = \frac{1}{P(B/\mathcal{C})} E(c(B)f/\mathcal{C}).$$

Equivalently

$$E(f/\mathcal{B} \vee \mathcal{C}) = \sum_B \frac{c(B)}{P(B/\mathcal{C})} E(c(B)f/\mathcal{C}), \tag{9.1}$$

where the summation extends over all the atoms of $\mathcal{B}$. In case $f = c(A)$, this becomes

$$P(A/\mathcal{B} \vee \mathcal{C}) = \sum_B c(B) \frac{P(A \cap B/\mathcal{C})}{P(B/\mathcal{C})}. \tag{9.2}$$

To prove (9.1), observe that both sides are measurable $\mathcal{B} \vee \mathcal{C}$; it is therefore sufficient to prove that whenever $B$ is an atom of $\mathcal{B}$ and $C$ is in $\mathcal{C}$, then the integrals of the two sides of (9.1) over $B \cap C$ are equal. The proof is the following straightforward computation:

$$\begin{aligned}
\int_C c(B) \frac{E(c(B)f/\mathcal{C})}{P(B/\mathcal{C})} dP &= \int_C E\left( c(B) \frac{E(c(B)f/\mathcal{C})}{P(B/\mathcal{C})} \Big/ \mathcal{C} \right) dP \\
&= \int_C E(c(B)f/\mathcal{C})\, dP \qquad [\text{by } (3.5)] \\
&= \int_C c(B) f\, dP = \int_{B \cap C} f\, dP \\
&= \int_{B \cap C} E(f/\mathcal{B} \vee \mathcal{C})\, dP.
\end{aligned}$$

**Section 10. The martingale theorem.** The only thing along these lines that remains to be discussed is the continuity of $E(f/\mathcal{C})$ in $\mathcal{C}$. A typical and useful result is that if $\{\mathcal{C}_n\}$ is an increasing sequence of subfields of $\mathcal{S}$ (that is, $\mathcal{C}_n \subset \mathcal{C}_{n+1}$ for all $n$), and if $\mathcal{C}$ is the subfield they generate, then

$$E(f/\mathcal{C}_n) \to E(f/\mathcal{C}) \tag{10.1}$$

almost everywhere and in the mean. This is a non-trivial assertion; it is a special case of Doob's justly celebrated martingale convergence theorem. The special case $f = c(A)$ is all that we shall need; the assertion for that case is that

$$P(A/\mathcal{C}_n) \to P(A/\mathcal{C}) \tag{10.2}$$

almost everywhere. We shall give the proof for a statement of intermediate generality: we shall prove (10.1) in case $f$ is bounded (and, of course, measurable). Mean convergence is then a consequence of the Lebesgue bounded convergence theorem.

Assume for a moment that it is already known that

$$E(g/\mathcal{C}_n) \to g \tag{10.3}$$

almost everywhere whenever $g$ is measurable $\mathcal{C}$ (and bounded). Note that in this case $E(g/\mathcal{C}) = g$. If $f$ is an arbitrary bounded measurable function, write $g = E(f/\mathcal{C})$. It follows that $g$ is bounded and measurable $\mathcal{C}$. Since, by (8.1),

$$E(g/\mathcal{C}_n) = E(f/\mathcal{C}_n),$$

the desired result (10.1) is proved. Conclusion: it is sufficient to prove (10.3).

If $g$ is bounded and measurable $\mathcal{C}$, then it is the limit almost everywhere of a uniformly convergent sequence $\{g_k\}$ of simple functions measurable $\mathcal{C}$. Since, by (3.4),

$$|E(g/\mathcal{C}_n) - E(g_k/\mathcal{C}_n)| \leqq E(|g - g_k|/\mathcal{C}_n),$$

it follows that

$$|E(g/\mathcal{C}_n) - g| \leqq E(|g - g_k|/\mathcal{C}_n) + |E(g_k/\mathcal{C}_n) - g_k| + |g_k - g|. \tag{10.4}$$

The first and third terms on the right side of (10.4) tend to zero uniformly as $k \to \infty$. Conclusion: it is sufficient to prove (10.3) for simple functions measurable $\mathcal{C}$, and hence for the characteristic functions of sets in $\mathcal{C}$. The desired result has thus been reduced to

$$P(A/\mathcal{C}_n) \to c(A) \tag{10.5}$$

almost everywhere, whenever $A$ is in $\mathcal{C}$.

Let $\varepsilon$ and $\delta$ be arbitrary positive numbers less than 1. Since the union of the fields $\mathcal{C}_n$ is dense in $\mathcal{C}$, there exists a set $B$ in that union, say $B$ is in $\mathcal{C}_k$, such that $P(A + B) < \varepsilon\delta/2$ (where the plus sign denotes Boolean sum: $A + B = (A - B) \cup (B - A)$). Write

$$D_n = \{x\colon 1 - P(A/\mathcal{C}_n)(x) \geqslant \varepsilon\},$$

and write $\{F_n\}$ for the sequence obtained by disjointing the sequence $\{B \cap D_n\}$; that is, $F_1 = B \cap D_1$, $F_2 = (B \cap D_2) - F_1$, $F_3 = (B \cap D_3) - (F_1 \cup F_2)$, etc. Observe that $F_n$ is in $\mathcal{C}_n$ whenever $n \geqq k$. It follows that if $n \geqq k$, then

$$P(F_n - A) = \int_{F_n} [1 - P(A/\mathcal{C}_n)]\, dP \geqq P(F_n)\varepsilon,$$

and hence that if $F = \bigcup_{n \geqq k} F_n$, then

$$P(F - A) \geqq P(F)\varepsilon.$$

This implies that $P(B - A) \geqq P(F)\varepsilon$. Since $P(B - A) \leqq P(B + A) < \varepsilon\delta/2$, it follows that $P(F) < \delta/2$.

If $x$ is in $B - F$, then $P(A/\mathcal{C}_n)(x) \geqq 1 - \varepsilon$ whenever $n \geqq k$. This can be expressed as follows: if $n \geqq k$, then $P(A/\mathcal{C}_n) \geqq 1 - \varepsilon$ throughout $B$, except possibly in a set of measure less than $\delta$; recall that $\varepsilon\delta/2 + \delta/2 < \delta$. Since $\varepsilon$ is arbitrarily small, it follows that $P(A/\mathcal{C}_n)$ converges to 1 throughout $A$, except possibly in a set of measure less than $\delta$; since $\delta$ is arbitrarily small, it follows that $P(A/\mathcal{C}_n)$ converges to 1 almost everywhere in $A$. This result applied to $X-A$ in place of $A$ shows that $P(A/\mathcal{C}_n)$ converges to $c(A)$ almost everywhere.

# CHAPTER II. INFORMATION

**Section 11. Motivation.** What is a reasonable numerical measure of the amount of information conveyed by a statement? How much information, for instance, do we get about a point $x$ of $X$ when we are told that $x$ belongs to a subset $A$ of $X$? It seems reasonable to require that the answer should depend on the size of $A$ (that is, on $P(A)$) and on nothing else. In other words, the answer should be expressed in terms of a function $F$ on the unit interval; the amount of information conveyed by $A$ shall be $F(P(A))$. Two further reasonable requirements are that the function $F$ be non-negative and continuous.

Two experiments (or, alternatively, two statements) do not necessarily yield more information than one. If, however, two experiments are independent of each other, then it is reasonable to expect that the amount of information obtainable from the two together is the sum of the two separate amounts of information. If we perform two independent experiments, and if the result of the first tells us that $x$ is in $A$ and the result of the second tells us that $x$ is in $B$, then we know that $x$ belongs to a set (namely $A\cap B$) of measure $P(A)P(B)$ (by independence). The requirement of additivity implies, therefore, that the function $F$ should satisfy the functional equation

$$F(st)=F(s)+F(t)$$

throughout its domain.

It is well known, and it is easy to prove, that the conditions imposed on $F$ uniquely characterize $F$ in the interval $(0,1]$, to within a multiplicative constant. Indeed, two inductions (one for the numerator and one for the denominator) show that $F(t^r)=rF(t)$ whenever $r$ is a positive rational number. Continuity yields the same result whenever $r$ is a positive real number. Put $t=1/e$ to get $F(e^{-r})=rF(e^{-1})$, or, in other words,

$$F(t)=F(e^{-1})(-\log t).$$

Conclusion: except for a constant factor, $F(t)=-\log t$; the amount of information conveyed by the assertion that $x$ is in $A$ is satisfactorily measured by $-\log P(A)$.

**Section 12. Definition.** As the preceding discussion indicates, the mathematical model of an experiment with a finite number of possible outcomes is a finite partition of $X$ into measurable sets, or equivalently, and from the point of view of the intended applications more elegantly, a finite subfield of $\mathcal{S}$. The amount of information conveyed by one of the possible outcomes of an experiment is a number depending on that outcome; in other words, the amount of information is a function of outcomes (associated with some experiments). These considerations motivate the following definition. If $\mathcal{A}$ is any finite subfield of $\mathcal{S}$, the information function $I(\mathcal{A})$ associated with $\mathcal{A}$ is the function whose value at each point of each atom $A$ of $\mathcal{A}$ is $-\log P(A)$; equivalently

$$I(\mathcal{A}) = -\sum_{A} c(A)\log P(A). \tag{12.1}$$

Conditional information is a natural and useful generalization of this concept; it is obtained by using conditional probabilities instead of absolute probabilities. Explicitly, if $\mathcal{A}$ and $\mathcal{C}$ are subfields of $\mathcal{S}$, with $\mathcal{A}$ finite, then, by definition,

$$I(\mathcal{A}/\mathcal{C}) = -\sum_{A} c(A)\log P(A/\mathcal{C}). \tag{12.2}$$

Observe that $I(\mathcal{A}/\mathcal{C})$ is always non-negative. The connection between conditional information and absolute information is that

$$I(\mathcal{A}/2) = I(\mathcal{A}); \tag{12.3}$$

see (5.2). The function $I(\mathcal{A}/\mathcal{C})$ is measurable $\mathcal{A} \vee \mathcal{C}$, and, in particular, $I(\mathcal{A})$ is measurable $\mathcal{A}$.

**Section 13. Transformations.** If $T$ is a measure-preserving transformation on $X$, then

$$I(\mathcal{A}/\mathcal{C})T = I(T^{-1}\mathcal{A}/T^{-1}\mathcal{C}); \tag{13.1}$$

the proof is a straightforward application of (7.2). It follows in particular, that

$$I(\mathcal{A})T = I(T^{-1}\mathcal{A}). \tag{13.2}$$

**Section 14. Information zero.** If $\mathcal{A} \subset \mathcal{C}$, then

$$I(\mathcal{A}/\mathcal{C}) = 0. \tag{14.1}$$

The proof is based on (5.1) and the convention that if $t = 0$, then $t\log t = 0$. It follows in particular, with $\mathcal{A} = 2$ and $\mathcal{C} = 2$, that

$$I(2) = 0. \tag{14.2}$$

These equations are in harmony with the intuitive meaning of information. Thus, for example, (14.1) expresses that if before some particular experiment is performed more is already known than the experiment can possibly reveal, then the amount of information conveyed by the outcome is certainly zero.

The converse of the conclusion (14.1) is true; the assertion is that if $I(\mathcal{A}/\mathcal{C}) = 0$ almost everywhere, then

$$\mathcal{A} \subset \mathcal{C}. \tag{14.3}$$

Indeed, if $I(\mathcal{A}/\mathcal{C}) = 0$ almost everywhere, then $c(A)\log P(A/\mathcal{C}) = 0$ almost everywhere for each atom $A$ of $\mathcal{A}$. This implies that $P(A/\mathcal{C})$ is a characteristic function and hence (by (5.3)) that $A$ is in $\mathcal{C}$.

**Section 15. Additivity.** Conditional probability (of a set with respect to a field) is algebraically well-behaved as a function of its first argument and analytically well behaved as a function of its second argument. These facts are reflected by the behavior of conditional information.

If $\mathcal{A}$, $\mathcal{B}$, and $\mathcal{C}$ are fields such that $\mathcal{A}$ and $\mathcal{B}$ are finite, then

$$I(\mathcal{A} \vee \mathcal{B}/\mathcal{C}) = I(\mathcal{B}/\mathcal{C}) + I(\mathcal{A}/\mathcal{B} \vee \mathcal{C}). \tag{15.1}$$

The left side of (15.1) is symmetric in $\mathcal{A}$ and $\mathcal{B}$, whereas the right side is apparently not. This is no cause for alarm; (15.1) is merely one of two equations, which between them describe the whole (symmetric) truth. If $\mathcal{C} = 2$, then (15.1) becomes

$$I(\mathcal{A} \vee \mathcal{B}) = I(\mathcal{B}) + I(\mathcal{A}/\mathcal{B}); \tag{15.2}$$

if $\mathcal{A} \subset \mathcal{C}$, then (15.1) becomes

$$I(\mathcal{A} \vee \mathcal{B}/\mathcal{C}) = I(\mathcal{B}/\mathcal{C}). \tag{15.3}$$

If, in particular, $\mathcal{C} = \mathcal{A}$, then

$$I(\mathcal{A} \vee \mathcal{B}/\mathcal{A}) = I(\mathcal{B}/\mathcal{A}). \tag{15.4}$$

If, finally, $\mathcal{A} \subset \mathcal{B}$, then

$$I(\mathcal{A}/\mathcal{C}) \leqq I(\mathcal{B}/\mathcal{C}); \tag{15.5}$$

this follows from (15.1) and the fact that $I(\mathcal{B}/\mathcal{A} \vee \mathcal{C})$ is non-negative.

The proof of (15.1) is the following computation:

$$\begin{aligned}
I(\mathcal{A} \vee \mathcal{B}/\mathcal{C}) &= -\sum_A \sum_B c(A)c(B) \log P(A \cap B/\mathcal{C}) \\
&= -\sum_A \sum_B c(A)c(B) \log\left( P(B/\mathcal{C}) \frac{P(A \cap B/\mathcal{C})}{P(B/\mathcal{C})} \right) \\
&= -\sum_A \sum_B c(A)c(B) \log P(B/\mathcal{C}) \\
&\qquad - \sum_A \sum_B c(A)c(B) \log \frac{P(A \cap B/\mathcal{C})}{P(B/\mathcal{C})} \\
&= -\sum_A c(A) \sum_B c(B) \log P(B/\mathcal{C}) \\
&\qquad - \sum_A c(A) \log \sum_B c(B) \frac{P(A \cap B/\mathcal{C})}{P(B/\mathcal{C})} \\
&= \sum_A c(A) I(\mathcal{B}/\mathcal{C}) - \sum_A c(A) \log P(A/\mathcal{B} \vee \mathcal{C}) \\
&= I(\mathcal{B}/\mathcal{C}) + I(\mathcal{A}/\mathcal{B} \vee \mathcal{C}).
\end{aligned}$$

**Section 16. Finite additivity.** The following assertion is a useful corollary of (15.2). If $\mathcal{B}_0, \mathcal{B}_1, \ldots, \mathcal{B}_n$ are finite fields (with $\mathcal{B}_0 = 2$, and $n = 1, 2, 3, \ldots$), then

$$(16.1) \qquad I\left(\bigvee_{i=1}^{n} \mathcal{B}_i\right) = \sum_{k=1}^{n} I\left(\mathcal{B}_k / \bigvee_{i=0}^{k-1} \mathcal{B}_i\right).$$

The proof is inductive. For $n=1$, the assertion reduces to $I(\mathcal{B}_1) = I(\mathcal{B}_1/2)$. For the induction step, use (15.2) as follows:

$$\begin{aligned} I\left(\bigvee_{i=1}^{n+1} \mathcal{B}_i\right) &= I\left(\bigvee_{i=1}^{n} \mathcal{B}_i \vee \mathcal{B}_{n+1}\right) \\ &= I\left(\bigvee_{i=1}^{n} \mathcal{B}_i\right) + I\left(\mathcal{B}_{n+1} / \bigvee_{i=1}^{n} \mathcal{B}_i\right). \end{aligned}$$

An important special case of (16.1) is obtained as follows. Suppose that $\mathcal{A}$ is a finite subfield of $\mathcal{S}$ and that $T$ is a measure-preserving transformation on $X$. For each positive integer $n$, write $\mathcal{B}_i = T^{-(n-i)}\mathcal{A}$ $(i = 1, \ldots, n)$, and apply (16.1). The result is

$$I\left(\bigvee_{i=1}^{n} T^{-(n-i)}\mathcal{A}\right) = I(T^{-(n-1)}\mathcal{A}) + \sum_{k=2}^{n} I\left(T^{-(n-k)}\mathcal{A} / \bigvee_{i=1}^{k-1} T^{-(n-i)}\mathcal{A}\right)$$

for $n > 1$. This can be rewritten in the form

$$I\left(\bigvee_{i=0}^{n-1} T^{-i}\mathcal{A}\right) = I(T^{-(n-1)}\mathcal{A}) + \sum_{k=1}^{n-1} I\left(T^{-(n-k-1)}\mathcal{A} / \bigvee_{i=1}^{k} T^{-(n-i)}\mathcal{A}\right).$$

Since $T^{-(n-i)}\mathcal{A} = T^{-(n-k-1)}T^{-(k-i+1)}\mathcal{A}$, it follows that

$$I\left(\bigvee_{i=0}^{n-1} T^{-i}\mathcal{A}\right) = I(\mathcal{A})T^{n-1} + \sum_{k=1}^{n-1} I\left(\mathcal{A} / \bigvee_{i=1}^{k} T^{-(k-i+1)}\mathcal{A}\right)T^{n-k-1},$$

or, equivalently,

$$(16.2) \quad I\left(\bigvee_{i=0}^{n-1} T^{-i}\mathcal{A}\right) = I(\mathcal{A})T^{n-1} + \sum_{k=1}^{n-1} I\left(\mathcal{A} / \bigvee_{i=1}^{k} T^{-i}\mathcal{A}\right)T^{n-k-1}$$

whenever $n > 1$.

[The preceding computation made use of the fact that

$$(16.3) \qquad T^{-1}(\vee\, \mathcal{E}) = \vee\, (T^{-1}\mathcal{E})$$

for any collection $\mathcal{E}$ of sets; $\vee$ here denotes the generated field. To prove this, observe that $\mathcal{E} \subset \vee\, \mathcal{E}$, so that $T^{-1}\mathcal{E} \subset T^{-1}(\vee\, \mathcal{E})$; since $T^{-1}(\vee\, \mathcal{E})$ is a field, it follows that

$$\vee\, (T^{-1}\mathcal{E}) \subset T^{-1}(\vee\, \mathcal{E}).$$

For the reverse inclusion observe that $T^{-1}\mathcal{E} \subset \vee\, (T^{-1}\mathcal{E})$. Since the collection of those sets $A$ for which $T^{-1}A$ belongs to $\vee(T^{-1}\mathcal{E})$ is a field, it follows that

$$T^{-1}(\vee \mathcal{E}) \subset \vee(T^{-1}\mathcal{E}),$$

and the proof is complete.

If, in particular, $\mathcal{B}$ and $\mathcal{C}$ are fields, then put $\mathcal{E} = \mathcal{B} \cup \mathcal{C}$ and apply (16.3); the result is that

$$T^{-1}(\mathcal{A} \vee \mathcal{B}) = T^{-1}\mathcal{A} \vee T^{-1}\mathcal{B}.$$

The similar facts about infinite suprema are proved the same way.]

**Section 17. Convergence.** If $\{\mathcal{C}_n\}$ is an increasing sequence of subfields of $\mathcal{S}$, and if $\mathcal{C} = \vee_n \mathcal{C}_n$, then

$$(17.1) \qquad I(\mathcal{A}/\mathcal{C}_n) \to I(\mathcal{A}/\mathcal{C})$$

almost everywhere. The proof of this assertion is immediate from the definition (12.2) and the convergence theorem (10.2).

For some purposes, convergence in the mean is more useful than almost-everywhere convergence. Convergence in the mean is closely connected with uniform integrability. Recall that a sequence $\{f_n\}$ of measurable functions on $X$ is uniformly integrable if

$$\int_{\{x:\, |f_n(x)| \geqq t\}} |f_n| \, dP$$

tends to 0, as $t \to \infty$, uniformly in $n$. The facts are these. If $f$ is an integrable function and if $\{f_n\}$ is a sequence of integrable functions such that $f_n \to f$ in the mean, then $\{f_n\}$ is uniformly integrable. If, conversely, $f$ is a measurable function, $\{f_n\}$ is a uniformly integrable sequence, and if $f_n \to f$ almost everywhere, then $f$ is integrable and $f_n \to f$ in the mean. In view of these facts, the way to prove that (17.1) may also be interpreted in the sense of mean convergence is to prove uniform integrability. Note that even the integrability of one $I(\mathcal{A}/\mathcal{C})$ needs proof; there is no obvious boundedness to invoke.

**Uniform integrability theorem.** If $N$ is a positive integer, then the collection of all functions of the form $I(\mathcal{A}/\mathcal{C})$ (where $\mathcal{A}$ is a finite subfield of $\mathcal{S}$ with not more than $N$ atoms and $\mathcal{C}$ is an arbitrary subfield of $\mathcal{S}$) is uniformly integrable.

**Proof**. Take $N$, $\mathcal{A}$, and $\mathcal{C}$ as described, and let $r$ and $s$ be positive numbers, $r \leqq s$. Write

$$D = \{x: r \leqq I(\mathcal{A}/\mathcal{C})(x) \leqq s\}.$$

If $A$ is an atom of $\mathcal{A}$, then $I(\mathcal{A}/\mathcal{C}) = -\log P(A/\mathcal{C})$ on $A$. It follows that if

$$C_A = \{x: r \leqq -\log P(A/\mathcal{C})(x) \leqq s\},$$

then $A \cap D = A \cap C_A$. Since $P(A/\mathcal{C})$ is measurable $\mathcal{C}$, the set $C_A$ belongs to $\mathcal{C}$ for each $A$. Since on $C_A$

$$\log P(A/\mathcal{C}) \leqq -r$$

or

$$P(A/\mathcal{C}) \leqq e^{-r},$$

it follows that

$$P(A \cap D) = P(A \cap C_A) = \int_{C_A} P(A/\mathcal{C})\,dP$$
$$\leqq e^{-r}P(C_A) \leqq e^{-r}.$$

Since $I(\mathcal{A}/\mathcal{C}) \leqq s$ on $D$,

$$\int_{A \cap D} I(\mathcal{A}/\mathcal{C})\,dP \leqq sP(A \cap D) \leqq se^{-r}.$$

Sum over all the atoms $A$ of $\mathcal{A}$ to get

$$\int_D I(\mathcal{A}/\mathcal{C})\,dP \leqq Nse^{-r}.$$

Put $r = t + n$, $s = t + n - 1$ $(n = 0, 1, 2, \ldots)$; it follows that

$$(17.2) \qquad \int_{\{x:\, I(\mathcal{A}/\mathcal{C})(x) \geqq t\}} I(\mathcal{A}/\mathcal{C})\,dP \leqq N \sum_{n=0}^{\infty} \frac{t+n+1}{e^{t+n}}.$$

Since the sum in (17.2) tends to 0 as $t \to \infty$, the proof of uniform integrability is complete.

**Corollary.** *If $\{\mathcal{C}_n\}$ is an increasing sequence of subfields of $\mathcal{S}$ with $\bigvee_n \mathcal{C}_n = \mathcal{C}$, then $I(\mathcal{A}/\mathcal{C}_n) \to I(\mathcal{A}/\mathcal{C})$ in the mean for each finite field $\mathcal{A}$.*

**Section 18. McMillan's theorem.** The preceding corollary can be applied to the following important situation. Suppose that $\mathcal{A}$ is a finite subfield of $\mathcal{S}$ and that $T$ is a measure-preserving transformation on $X$. Write

$$f_n = I\left(\bigvee_{i=0}^{n-1} T^{-i}\mathcal{A}\right), \qquad n = 1, 2, 3, \ldots,$$
$$g_0 = I(\mathcal{A}),$$
$$g_n = I\left(\mathcal{A} / \bigvee_{i=1}^{n} T^{-i}\mathcal{A}\right), \qquad n = 1, 2, 3, \ldots.$$

It follows from (16.2) that

$$f_n = \sum_{k=0}^{n-1} g_k T^{n-k-1}, \qquad n = 1, 2, 3, \ldots.$$

If $g = I(\mathcal{A}/\bigvee_{i-1}^{\infty} T^{-i}\mathcal{A})$, then (17.1) implies that $g_n \to g$ almost everywhere, and the corollary of the uniform integrability theorem implies that $g_n \to g$ in the mean. The ergodic theorem applied to $g$ implies that the averages

$$\frac{1}{n} \sum_{k=0}^{n-1} gT^{n-k-1}$$

tend to some invariant integrable function $h$ both almost everywhere and in the mean. From these facts it is easy to deduce the following assertion, known as McMillan's theorem: $(1/n)f_n \to h$ in the mean. Indeed, if $\|f\| = \int |f|\, dP$, then

$$\left\|\frac{1}{n}f_n - h\right\| = \left\|\frac{1}{n}\sum_{k=0}^{n-1} g_k T^{n-k-1} - h\right\|$$
$$\leqq \left\|\frac{1}{n}\sum_{k=0}^{n-1}\left(g_k T^{n-k-1} - gT^{n-k-1}\right)\right\| + \left\|\frac{1}{n}\sum_{k=0}^{n-1} gT^{n-k-1} - h\right\|$$
$$\leqq \frac{1}{n}\sum_{k=0}^{n-1}\|g_k - g\| + \left\|\frac{1}{n}\sum_{k=0}^{n-1} gT^k - h\right\|,$$

and the proof of McMillan's theorem (for not necessarily invertible transformations) is complete. Presumably McMillan's theorem can be sharpened by adding the conclusion that $(1/n)f_n \to g$ almost everywhere.

## CHAPTER III. ENTROPY

**Section 19. Definition.** What we have been studying so far is the amount of information (absolute and conditional) furnished by an experiment. An associated concept of importance is the average amount of information so furnished. The best sense of "average" here is "expected value", where the expectation in question may itself be absolute or conditional. Suppose, to be specific, that $\mathcal{A}$, $\mathcal{C}$, and $\mathcal{D}$ are subfields of $\mathcal{S}$, with $\mathcal{A}$ finite, and write

$$H((\mathcal{A}/\mathcal{C})/\mathcal{D}) = E(I(\mathcal{A}/\mathcal{C})/\mathcal{D});$$

the function $H$ is called entropy, the entropy of $\mathcal{A}$, the conditional entropy of $\mathcal{A}$ with respect to $\mathcal{C}$, or the mean conditional entropy with respect to $\mathcal{D}$ of $\mathcal{A}$ with respect to $\mathcal{C}$.

If $\mathcal{D} = 2$, then the mean conditional entropy with respect to $\mathcal{D}$ is a constant $\overline{H}(\mathcal{A}/\mathcal{C})$; it follows from the elementary properties of conditional expectations that

$$\overline{H}(\mathcal{A}/\mathcal{C}) = \int I(\mathcal{A}/\mathcal{C})\, dP. \tag{19.1}$$

If $\mathcal{D} = \mathcal{C}$, the mean conditional entropy with respect to $\mathcal{D}$ is a function $H(\mathcal{A}/\mathcal{C})$,

$$H(\mathcal{A}/\mathcal{C}) = E(I(\mathcal{A}/\mathcal{C})/\mathcal{C}); \tag{19.2}$$

it follows from the elementary properties of conditional expectations that

$$\int H(\mathcal{A}/\mathcal{C})\, dP = \overline{H}(\mathcal{A}/\mathcal{C}). \tag{19.3}$$

If $H(\mathcal{A}) = H(\mathcal{A}/2)$ and $\overline{H}(\mathcal{A}) = \overline{H}(\mathcal{A}/2)$, then it follows from (19.2) that $H(\mathcal{A})$ is a constant, and it follows from (19.3) that the constant is $\overline{H}(\mathcal{A})$. Note also that

$$(19.4) \qquad H(\mathcal{A}) = \overline{H}(\mathcal{A}) = \int I(\mathcal{A})\, dP.$$

Conditional entropy can be computed directly from the definition (12.2) of information;

$$(19.5) \qquad H(\mathcal{A}/\mathcal{C}) = -\sum_A P(A/\mathcal{C}) \log P(A/\mathcal{C}).$$

If, in particular, $\mathcal{C} = 2$, then

$$(19.6) \qquad \overline{H}(\mathcal{A}) = -\sum_A P(A) \log P(A).$$

**Section 20. Transformations.** Most of the properties of entropies are relatively easy to derive from the corresponding properties of information. Thus, for instance, if $T$ is a measure-preserving transformation on $X$, then, by (13.1) and (7.1),

$$H(\mathcal{A}/\mathcal{C})T = H(T^{-1}\mathcal{A}/T^{-1}\mathcal{C}),$$

and, in particular,

$$\overline{H}(\mathcal{A}/\mathcal{C}) = \overline{H}(T^{-1}\mathcal{A}/T^{-1}\mathcal{C})$$

and

$$\overline{H}(\mathcal{A}) = \overline{H}(T^{-1}\mathcal{A}).$$

**Section 21. Entropy zero.** If $\mathcal{A} \subset \mathcal{C}$, then

$$(21.1) \qquad H(\mathcal{A}/\mathcal{C}) = 0,$$

and, in particular,

$$(21.2) \qquad \overline{H}(\mathcal{A}/\mathcal{C}) = 0$$

and

$$\overline{H}(2) = 0;$$

cf. (14.1) and (14.2).

Conversely if $H(\mathcal{A}/\mathcal{C}) = 0$ almost everywhere (or, equivalently, if $\overline{H}(\mathcal{A}/\mathcal{C}) = 0$), then $I(\mathcal{A}/\mathcal{C}) = 0$ almost everywhere (by (19.1)), and therefore $\mathcal{A} \subset \mathcal{C}$ (by (14.3)).

**Section 22. Concavity.** If $\mathcal{A}$, $\mathcal{B}$, and $\mathcal{C}$ are fields such that $\mathcal{A}$ is finite and $\mathcal{B} \subset \mathcal{C}$, then

$$(22.1) \qquad \overline{H}(\mathcal{A}/\mathcal{C}) \leqq \overline{H}(\mathcal{A}/\mathcal{B}).$$

The proof is an application of Jensen's inequality. If $F(t) = t \log t$ whenever $t > 0$ and $F(0) = 0$, then $F$ is a continuous convex function; it follows that

$$(22.2) \qquad F(E(P(A/\mathcal{C})/\mathcal{B})) \leqq E(F(P(A/\mathcal{C})/\mathcal{B}))$$

for each atom $A$ of $\mathcal{A}$. By (8.3) the left side of (22.2) is equal to $F(P(A/\mathcal{B}))$; this implies that

$$\sum_A P(A/\mathcal{B})\log P(A/\mathcal{B}) \leqq E\Big(\sum_A P(A/\mathcal{C})\log P(A/\mathcal{C})/\mathcal{B}\Big).$$

To get (22.1), change sign and integrate; cf. (19.5).

The special case $\mathcal{B} = 2$ is worthy of note:

$$\overline{H}(\mathcal{A}/\mathcal{C}) \leqq \overline{H}(\mathcal{A}). \tag{22.3}$$

**Section 23. Additivity.** If $\mathcal{A}$, $\mathcal{B}$, and $\mathcal{C}$ are fields such that $\mathcal{A}$ and $\mathcal{B}$ are finite, then, by (15.1),

$$H(\mathcal{A} \vee \mathcal{B}/\mathcal{C}) = H(\mathcal{B}/\mathcal{C}) + H((\mathcal{A}/\mathcal{B} \vee \mathcal{C})/\mathcal{C}). \tag{23.1}$$

It follows that

$$\overline{H}(\mathcal{A} \vee \mathcal{B}/\mathcal{C}) = \overline{H}(\mathcal{B}/\mathcal{C}) + \overline{H}(\mathcal{A}/\mathcal{B} \vee \mathcal{C}), \tag{23.2}$$

and, in particular,

$$\overline{H}(\mathcal{A} \vee \mathcal{B}) = \overline{H}(\mathcal{B}) + \overline{H}(\mathcal{A}/\mathcal{B}). \tag{23.4}$$

Further special cases: if $\mathcal{A} \subset \mathcal{C}$, then

$$H(\mathcal{A} \vee \mathcal{B}/\mathcal{C}) = H(\mathcal{B}/\mathcal{C}),$$
$$\overline{H}(\mathcal{A} \vee \mathcal{B}/\mathcal{C}) = \overline{H}(\mathcal{B}/\mathcal{C}),$$
$$\overline{H}(\mathcal{A} \vee \mathcal{B}/\mathcal{A}) = \overline{H}(\mathcal{B}/\mathcal{A}).$$

It follows from (23.2) and (22.1) that

$$\overline{H}(\mathcal{A} \vee \mathcal{B}/\mathcal{C}) \leqq \overline{H}(\mathcal{A}/\mathcal{C}) + \overline{H}(\mathcal{B}/\mathcal{C}), \tag{23.5}$$

and, hence, in particular (cf. (23.4)),

$$\overline{H}(\mathcal{A} \vee \mathcal{B}) \leqq \overline{H}(\mathcal{A}) + \overline{H}(\mathcal{B}). \tag{23.6}$$

If $\mathcal{A} \subset \mathcal{B}$, then (by (15.5))

$$H(\mathcal{A}/\mathcal{C}) \leqq H(\mathcal{B}/\mathcal{C})$$

and, therefore,

$$\overline{H}(\mathcal{A}/\mathcal{C}) \leqq \overline{H}(\mathcal{B}/\mathcal{C}).$$

It follows (put $\mathcal{C} = 2$) that if $\mathcal{A} \subset \mathcal{B}$, then

$$\overline{H}(\mathcal{A}) \leqq \overline{H}(\mathcal{B}).$$

**Section 24. Finite additivity.** If $\mathcal{B}_0, \mathcal{B}_1, \ldots, \mathcal{B}_n$ are finite fields (with $\mathcal{B}_0 = 2$, and $n = 1, 2, 3, \ldots$), then

$$\overline{H}\left(\vee_{i=1}^{n} \mathcal{B}_i\right) = \sum_{k=1}^{n} \overline{H}\left(\mathcal{B}_k / \vee_{i=0}^{k-1} \mathcal{B}_i\right); \tag{24.1}$$

this is an immediate consequence of (16.1). If $T$ is a measure-preserving transformation on $X$, then (16.2) applies; the conclusion is that

$$\overline{H}\left(\bigvee_{i=0}^{n-1} T^{-i}\mathcal{A}\right) = \overline{H}(\mathcal{A}) + \sum_{k=1}^{n-1} \overline{H}\left(\mathcal{A}/\bigvee_{i=1}^{k} T^{-i}\mathcal{A}\right) \tag{24.2}$$

whenever $n > 1$.

**Section 25. Convergence.** If $\{\mathcal{C}_n\}$ is an increasing sequence of subfields of $\mathcal{S}$, and if $\mathcal{C} = \bigvee_n \mathcal{C}_n$, then

$$H(\mathcal{A}/\mathcal{C}_n) \to H(\mathcal{A}/\mathcal{C}) \tag{25.1}$$

almost everywhere and in the mean. The proof is immediate from the convergence theorem (10.2) and equation (19.5); recall that since $t\log t$ is bounded for $t$ in $[0,1]$, the Lebesgue bounded convergence theorem is applicable. An immediate corollary is that

$$\overline{H}(\mathcal{A}/\mathcal{C}_n) \to \overline{H}(\mathcal{A}/\mathcal{C}). \tag{25.2}$$

# CHAPTER IV. APPLICATION

**Section 26. Relative entropy.** If $T$ is a measure-preserving transformation on $X$, then

$$\overline{H}\left(\mathcal{A}/\bigvee_{i=}^{k} T^{-i}\mathcal{A}\right) \to \overline{H}\left(\mathcal{A}/\bigvee_{i=1}^{\infty} T^{-i}\mathcal{A}\right), \tag{26.1}$$

and, therefore, by (24.2),

$$\frac{1}{n}\overline{H}\left(\bigvee_{i=0}^{n-1} T^{-i}\mathcal{A}\right) \to \overline{H}\left(\mathcal{A}/\bigvee_{i=1}^{\infty} T^{-i}\mathcal{A}\right); \tag{26.2}$$

alternatively, the same conclusion can be derived by applying term-by-term integration to McMillan's theorem.

The limit in (26.2) is an important one in the theory of measure-preserving transformations. If we write

$$h(\mathcal{A},T) = \overline{H}\left(\mathcal{A}/\bigvee_{i=1}^{\infty} T^{-i}\mathcal{A}\right), \tag{26.3}$$

then

$$h(\mathcal{A},T) = \lim \frac{1}{n}\overline{H}\left(\bigvee_{i=0}^{n-1} T^{-i}\mathcal{A}\right). \tag{26.4}$$

We may call $h(\mathcal{A},T)$ the entropy of $T$ relative to $\mathcal{A}$.

**Section 27. Elementary properties.** If $\mathcal{A}$ and $\mathcal{B}$ are finite fields, then, by (23.6) and (26.4),

$$h(\mathcal{A}\vee\mathcal{B},T) \leqq h(\mathcal{A},T) + h(\mathcal{B},T). \tag{27.1}$$

If $\mathcal{A}\subset\mathcal{B}$, then

$$h(\mathcal{A},T) \leqq h(\mathcal{B},T). \tag{27.2}$$

If $S$ and $T$ are commutative measure-preserving transformations, then

$$h(S^{-1}\mathcal{A}, T) = h(\mathcal{A}, T).$$

Observe, indeed, that

$$\vee_{i=1}^{\infty} T^{-i}(S^{-1}\mathcal{A}) = \vee_{i=1}^{\infty} S^{-1}(T^{-i}\mathcal{A}) = S^{-1}\left(\vee_{i=1}^{\infty} T^{-i}\mathcal{A}\right);$$

it follows that

$$\begin{aligned}\bar{H}\left(S^{-1}\mathcal{A}/\vee_{i=1}^{\infty} T^{-i}(S^{-1}\mathcal{A})\right) &= \bar{H}\left(S^{-1}\mathcal{A}/S^{-1}\vee_{i=1}^{\infty} T^{-i}\mathcal{A}\right)\\ &= \bar{H}\left(\mathcal{A}/\vee_{i=1}^{\infty} T^{-i}\mathcal{A}\right).\end{aligned}$$

A special case of value is

$$h(T^{-1}\mathcal{A}, T) = h(\mathcal{A}, T).$$

**Section 28. Strong monotony.** The next result is that if $T$ is invertible, then (27.2) can be given an infinitely sharper form. If, to be precise, $\mathcal{A}$ and $\mathcal{B}$ are finite fields such that

$$\mathcal{A} \subset \vee_{i=-\infty}^{+\infty} T^{i}\mathcal{B},$$

then

$$h(\mathcal{A}, T) \leqq h(\mathcal{B}, T). \tag{28.1}$$

The proof begins with the observation that

$$\vee_{i=0}^{k-1} T^{-i}\mathcal{A} \subset \vee_{i=0}^{k-1} T^{-i}\mathcal{A} \vee \vee_{j=-n-k+1}^{+n} T^{j}\mathcal{B}$$

for all positive integers $k$ and $n$. It follows that

$$\begin{aligned}\bar{H}\left(\vee_{i=0}^{k-1} T^{-i}\mathcal{A}\right) \leqq\ & \bar{H}\left(\vee_{j=-n-k+1}^{+n} T^{j}\mathcal{B}\right)\\ &+ \bar{H}\left(\vee_{i=0}^{k-1} T^{-i}\mathcal{A}/\vee_{j=-n-k+1}^{+n} T^{j}\mathcal{B}\right).\end{aligned} \tag{28.2}$$

The second summand on the right side of (28.2) is dominated by

$$\begin{aligned}&\sum_{i=0}^{k-1} \bar{H}\left(T^{-i}\mathcal{A}/T^{-i}\vee_{j=-n-k+1}^{+n} T^{j+i}\mathcal{B}\right)\\ &\qquad\leqq \sum_{i=0}^{k-1} \bar{H}\left(\mathcal{A}/\vee_{j=-n-i}^{+n-i} T^{j+i}\mathcal{B}\right)\\ &\qquad= \sum_{i=0}^{k-1} \bar{H}\left(\mathcal{A}/\vee_{j=-n}^{+n} T^{j}\mathcal{B}\right),\end{aligned}$$

and therefore

$$\bar{H}\left(\vee_{i=0}^{k-1} T^{-i}\mathcal{A}\right) \leqq \bar{H}\left(T^{n}\vee_{j=0}^{2n+k-1} T^{-j}\mathcal{B}\right) + k\bar{H}\left(\mathcal{A}/\vee_{j=-n}^{+n} T^{j}\mathcal{B}\right).$$

This implies that

$$\begin{aligned}\frac{1}{k}\bar{H}\left(\vee_{i=0}^{k-1} T^{-i}\mathcal{A}\right) \leqq\ & \frac{2n+k}{k}\,\frac{1}{2n+k}\bar{H}\left(\vee_{j=0}^{2n+k-1} T^{-j}\mathcal{B}\right)\\ &+ \bar{H}\left(\mathcal{A}/\vee_{j=-n}^{+n} T^{j}\mathcal{B}\right).\end{aligned} \tag{28.3}$$

Since $\mathcal{A} \subset \bigvee_{j=-\infty}^{+\infty} T^j\mathcal{B}$, it follows from (25.2) and (21.2) that the second summand on the right side of (28.3) tends to 0 as $n \to \infty$. Choose $n$ large and then, for fixed $n$, let $k \to \infty$; the result is that $h(\mathcal{A}, T)$ is dominated by $h(\mathcal{B}, T)$ plus an arbitrarily small positive number. This completes the proof of (28.1).

**Section 29. Algebraic properties.** The preceding results give some idea of how $h(\mathcal{A}, T)$ depends on $\mathcal{A}$; we turn now to study the way it depends on $T$. The principal result along these lines is that

$$h(\mathcal{A}, T^k) \leqq kh(\mathcal{A}, T) \tag{29.1}$$

for each positive integer $k$.

To prove this, write $\mathcal{B} = \bigvee_{i=0}^{k-1} T^{-i}\mathcal{A}$; then

$$\bigvee_{j=0}^{n-1} (T^k)^{-j}\mathcal{B} = \bigvee_{j=0}^{n-1} T^{-jk} \bigvee_{i=0}^{k-1} T^{-i}\mathcal{A} = \bigvee_{i=0}^{kn-1} T^{-i}\mathcal{A}.$$

It follows that

$$\frac{1}{n}\overline{H}\left(\bigvee_{j=0}^{n-1} (T^k)^{-j}\mathcal{B}\right) = k\frac{1}{kn}\overline{H}\left(\bigvee_{i=0}^{kn-1} T^{-i}\mathcal{A}\right)$$

and hence that

$$h(\mathcal{B}, T^k) = kh(\mathcal{A}, T). \tag{29.2}$$

Since $\mathcal{A} \subset \mathcal{B}$, assertion (29.1) follows from (27.2) (for $T^k$).

If $T$ is invertible, then

$$h(\mathcal{A}, T^{-1}) = h(\mathcal{A}, T). \tag{29.3}$$

This follows immediately from the equations

$$\overline{H}\left(\bigvee_{i=0}^{n-1} T^i\mathcal{A}\right) = \overline{H}\left(T^{n-1}\bigvee_{i=0}^{n-1} T^{-i}\mathcal{A}\right) = \overline{H}\left(\bigvee_{i=0}^{n-1} T^{-i}\mathcal{A}\right).$$

Combining (29.1) and (29.3), we obtain

$$h(\mathcal{A}, T^k) \leqq |k|h(\mathcal{A}, T)$$

for every integer $k$.

**Section 30. Entropy.** The entropy $h(T)$ of a measure-preserving transformation $T$ is defined by

$$h(T) = \sup h(\mathcal{A}, T), \tag{30.1}$$

where the supremum extends over all finite subfields of $\mathcal{S}$.

Since, by (29.2),

$$kh(\mathcal{A}, T) = h\left(\bigvee_{i=0}^{k-1} T^{-i}\mathcal{A}, T^k\right) \leqq h(T^k),$$

and since, by (29.1),

$$h(\mathcal{A}, T^k) \leqq kh(\mathcal{A}, T) \leqq kh(T),$$

it follows that

$$h(T^k) = kh(T) \tag{30.2}$$

for every positive integer $k$.

If $T$ is invertible, then, by (29.3),

$$h(T^{-1}) = h(T),$$

and therefore

$$h(T^k) = |k| h(T)$$

for every integer $k$.

**Section 31. Generated fields.** The following result is an efficient tool for calculating the entropy of some transformations. If $T$ is invertible and if $\mathfrak{B}$ is a finite subfield of $\mathfrak{S}$ such that

$$\bigvee_{i=-\infty}^{+\infty} T^i\mathfrak{B} = \mathfrak{S},$$

then

$$h(T) = h(\mathfrak{B}, T). \tag{31.1}$$

Indeed, (28.1) implies that

$$h(\mathfrak{A}, T) \leqq h(\mathfrak{B}, T)$$

for all finite subfields $\mathfrak{A}$ of $\mathfrak{S}$, and hence that

$$h(T) \leqq h(\mathfrak{B}, T).$$

The reverse inequality follows from the definition of $h(T)$.

A curious corollary of the preceding result is the assertion that if $T$ is invertible and if $\mathfrak{B}$ is a finite subfield of $\mathfrak{S}$ such that

$$\bigvee_{i=0}^{\infty} T^{-i}\mathfrak{B} = \mathfrak{S},$$

then

$$h(T) = 0.$$

Indeed, (31.1) implies that $h(T) = h(\mathfrak{B}, T)$. The invertibility of $T$ implies that $T^{-1}\mathfrak{S} = \mathfrak{S}$ and therefore, since

$$T^{-1}\mathfrak{S} = \bigvee_{i=1}^{\infty} T^{-i}\mathfrak{B},$$

that $\mathfrak{B} \subset \bigvee_{i=1}^{\infty} T^{-i}\mathfrak{B}$. The conclusion now follows from (26.3) and (21.2).

**Section 32. Examples.** We are now in a position to compute the entropy of some transformations.

Suppose, to begin with, that $T$ is the identity. In that case, $T^{-i}\mathfrak{A} = \mathfrak{A}$ for all $\mathfrak{A}$, so that $\bigvee_{i=0}^{n-1} T^{-i}\mathfrak{A} = \mathfrak{A}$ for all $\mathfrak{A}$ and all $n$. This implies that $h(\mathfrak{A}, T) = \lim(1/n)\overline{H}(\mathfrak{A}) = 0$ for all $\mathfrak{A}$, and, hence, that $h(T) = 0$.

If $T$ has finite order, i.e., $T^k$ is the identity for some positive integer $k$, then $h(T^k) = 0$ by the preceding paragraph, and it follows (from (30.2)) that $h(T) = 0$.

Suppose next that $X$ is the circle group and that $T$ is a rotation, $Tx = cx$, where $c$ is not a root of unity. Let $\mathfrak{B}$ be the field consisting of the half-open top arc and its complement (and, of course, the empty set and $X$). For suitable values of $i$ the result of applying $T^{-i}$ to the top arc yields arbitrarily small arcs, which can then be rotated around to nearly arbitrary positions. (All rotations needed are by $T^{-i}$, $i = 0, 1, 2, \ldots$.) It follows that every half-open arc is approximable arbitrarily closely by sets in $\bigvee_{i=0}^{\infty} T^{-i}\mathfrak{B}$, and hence that every such arc belongs to $\bigvee_{i=0}^{\infty} T^{-i}\mathfrak{B}$. This implies, by (31.2), that $h(T) = 0$.

Consider now the measure space with the points $0, \ldots, k-1$ ($k = 1, 2, 3, \ldots$) bearing the measures $p_0, \ldots, p_{k-1}$. Let $X$ be the bilateral sequence space based on that space, let $P$ be the usual product measure in $X$, and let $T$ be the bilateral shift on $X$. If $\mathfrak{B}$ is the field generated by the sets $\{x\colon x_0 = i\}$, $i = 0, \ldots, k-1$, then $\bigvee_{i=-\infty}^{+\infty} T^{i}\mathfrak{B} = \mathfrak{S}$ and therefore, by (31.1), $h(T) = h(\mathfrak{B}, T)$. Since the fields $T^{-i}\mathfrak{B}$, $i = 0, \ldots, n-1$ are independent, it follows that

$$\overline{H}\left(\bigvee_{i=0}^{n-1} T^{-i}\mathfrak{B}\right) = n\overline{H}(\mathfrak{B}),$$

and, hence, that

$$h(\mathfrak{B}, T) = \overline{H}(\mathfrak{B}).$$

This implies, finally, that

$$h(T) = -\sum_{i-0}^{k-1} p_i \log p_i .$$

If, in particular, $p_0 = \cdots = p_{k-1} = 1/k$, then $h(T) = \log k$. This proves that the 2-shift is not conjugate to the 3-shift.

Reprinted from the
BULLETIN OF THE AMERICAN MATHEMATICAL SOCIETY
Vol. 67, No. 1, pp. 70–80, Jan. 1961

# RECENT PROGRESS IN ERGODIC THEORY[1]

PAUL R. HALMOS

**Prologue.** In 1948, at the November meeting of the Society in Chicago, I delivered an address entitled *Measurable transformations.* In the twelve years that have elapsed since then, ergodic theory (of which the theory of measurable transformations is the greatest part) has been spectacularly active. The purpose of today's address is to report some of the developments of those twelve years; its title might well have been *Measurable transformations revisited.* The subjects I chose for this purpose are: some new ergodic theorems, information theory and its connection with ergodic theory, and the problem of invariant measure.

The stage on which most ergodic performances take place is a measure space consisting of a set $X$ and of a measure $\mu$ defined on a specified $\sigma$-field of measurable subsets of $X$. At the most trivial level $X$ consists of a finite number of points, every subset of $X$ is measurable, and $\mu$ is a mass distribution on $X$ (which may or may not be uniform). At a more useful and typical level $X$ is the real line $(-\infty, +\infty)$, or the unit interval $[0, 1]$, measurability in either case is interpreted in the sense of Borel, and $\mu$ is Lebesgue measure. Another possibility is to consider a measure space having a finite number of points with total measure 1 and to let $X$ be the Cartesian product of a countably infinite number of copies of that space with itself; measurability and measure in this case are interpreted in the customary sense appropriate to product spaces. This latter example is easily seen to be measure-theoretically isomorphic to the unit interval, as also are most of the normalized measure spaces (measure spaces with total measure 1) that ever occur in honest analysis. The only measure spaces I shall consider in this report are the ones isomorphic to one of the spaces already mentioned. The expert will know just how little generality is lost thereby, and the casual passer-by, quite properly, will not care.

A transformation $T$ from a measure space $X$ into a measure space $Y$ is called *measurable* if the inverse image $T^{-1}E$ (in $X$) of each measurable set $E$ (in $Y$) is again a measurable set. A measurable transformation $T$ is *measure-preserving* if, for every measurable set $E$, the sets $E$ and $T^{-1}E$ have the same measure. A measurable (but not

An address delivered before the Summer Meeting of the Society in East Lansing on September 1, 1960, by invitation of the Committee to Select Hour Speakers for Summer and Annual Meetings; received by the editors September 1, 1960.

[1] The work was done while the author had a National Science Foundation grant.

necessarily measure-preserving) transformation $T: X \to Y$ is *invertible* if there exists a (necessarily unique) measurable transformation $T^{-1}: Y \to X$ such that each of the composites $T^{-1}T$ and $TT^{-1}$ is equal to the identity on its domain. If $T$ is measure-preserving and invertible, then $T^{-1}$ is measure-preserving also. Most of the transformations to be considered in this report are transformations from a measure space $X$ into itself, or, in the customary phrase, transformations on $X$.

A measure-preserving transformation on a finite set with uniform distribution is simply a permutation. A familiar example of a measure-preserving transformation on the real line is given by a translation, $Tx = x + c$; the same equation, interpreted modulo 1, gives an interesting example of a measure-preserving transformation on the unit interval. An important transformation on the product of countably many copies of a finite normalized measure space is obtained as follows. Let the index set used in the formation of the Cartesian product be the set of all integers (positive, negative, or zero), so that a point $x$ of the space $X$ under consideration is a two-way infinite sequence $(x_n)$; for each $x$ in $X$ let $Tx$ be the sequence $y$ such that $y_n = x_{n+1}$. Any transformation defined this way is called a *shift*, or, more precisely, the shift based on the given finite measure space. In case that space has $k$ points and the original measure on it is the uniform distribution, the resulting shift is completely determined by the number $k$. The shift so determined is called the *k-shift*; in what follows it will be denoted by $S_k$.

**Ergodic theorems.** Every transformation $T$ of a set $X$ into itself induces a functional operator $U$ that acts on functions whose domain is $X$; by definition $(Uf)(x) = f(Tx)$. The transformations that always have been and still are of central interest in ergodic theory are the measure-preserving transformations. The restriction to any of the Lebesgue spaces $L_p$ of the functional operator $U$ associated with a measure-preserving transformation $T$ turns out to be an isometry; the preservation of norm is an easy consequence of the preservation of measure. If $T$ is invertible, then $U$ is invertible. On $L_2$, the pleasantest of all function spaces, an invertible isometry is a unitary operator, to which the extensive and powerful spectral theory can be applied. The first result of such an application is von Neumann's mean ergodic theorem; it studies the norm convergence in $L_2$ of averages such as $(1/n)\sum_{i=0}^{n-1} U^i f$. The methods of the study extend immediately to all unitary operators on $L_2$, including the ones that are not induced by measure-preserving transformations. This fact motivated the subsequent extension of the theory to wider and wider classes of transformations on wider and wider classes of Banach spaces.

The finite special case of von Neumann's mean ergodic theorem is amusing and instructive. If, for instance, $X=\{1, 2, 3\}$, then $L_2$ (over $X$) is three-dimensional Euclidean space; if $T$ is the cyclic permutation (1, 2, 3), then the induced unitary operator $U$ is the one whose matrix (with respect to the usual and obvious basis) is

$$\begin{pmatrix} 0 & 1 & 0 \\ 0 & 0 & 1 \\ 1 & 0 & 0 \end{pmatrix}.$$

The original mean ergodic theorem asserts the convergence of the averages of the powers of such permutation matrices. The first and most natural generalization is to the convergence of the averages of the powers of arbitrary unitary matrices.

The so-called individual ergodic theorem of G. D. Birkhoff has the function space $L_1$ for its natural habitat; the assertion is that for every integrable function $f$ the averages $(1/n)\sum_{i=0}^{n-1} U^i f$ converge almost everywhere. It is natural to try to extend this result to functional operators that may not be induced by measure-preserving transformations. The spirit of the general mean ergodic theorems and the spirit of the generalization desired here are quite different. The former have a wider domain of applicability (Banach spaces) with a necessarily weaker conclusion (norm convergence); the latter has a sharp conclusion (almost everywhere convergence) for a necessarily more special structure (the function space $L_1$). Recent progress of the latter kind (based on the pioneering work of Doob, Kakutani, and others) was made by Eberhard Hopf. He assumes that $\mu(X)<\infty$, and that $U$ is a positive linear operator on $L_1$ subject to the following two conditions:

$$\int (Uf)d\mu = \int fd\mu$$

for all $f$ in $L_1$, and

$$U1 = 1.$$

(Positiveness means, of course, that if $f$ is in $L_1$ and $f \geqq 0$ a.e., then $Uf \geqq 0$ a.e.) The conclusion is that for each $f$ in $L_1$ the averages $(1/n)\sum_{i=0}^{n-1} U^i f$ converge almost everywhere. It is obvious that if $U$ is the functional operator induced by a measure-preserving transformation, then Hopf's conditions are satisfied; the assertion is, in other words, a bona fide generalization of the classical individual ergodic theorem.

A glance at the finite special case will clarify the meaning of Hopf's conditions. In the finite case, the linear transformation $U$ corresponding to a matrix $(u_{ij})$ is positive if and only if the entries of the matrix are positive $(u_{ij} \geqq 0)$. It is a simple and pleasant exercise to prove that, in the finite case, the first Hopf condition requires exactly that each column sum $(\sum_i u_{ij})$ be equal to 1, and the second one that each row sum $(\sum_j u_{ij})$ be equal to 1. The extent to which Hopf generalizes Birkhoff is now clear. Birkhoff treats matrices with exactly one 1 in each row and in each column, all other entries being 0, and Hopf treats matrices with positive entries whose row sums and column sums are equal to unity.

The theorem of Hopf was generalized by Dunford and Schwartz in one direction and by Chacon and Ornstein in another. The assumption that $\mu(X) < \infty$ is absent from both generalizations. The Dunford-Schwartz assumptions are that

$$\|U\|_1 \leqq 1 \quad \text{and} \quad \|U\|_\infty \leqq 1.$$

The first of these conditions is clear. It requires that the norm of the operator $U$ on $L_1$ do not exceed 1, i.e., that $\int |Uf| d\mu \leqq \int |f| d\mu$ for every integrable function $f$. The second condition requires that $U$ map each essentially bounded function in $L_1$ onto an essentially bounded function, without increasing the essential supremum. It is noteworthy that Dunford and Schwartz do not assume that $U$ is positive. The conclusion is, as before, that for each $f$ in $L_1$ (or, for that matter, in any $L_p$, with $1 \leqq p < \infty$) the averages $(1/n)\sum_{i=0}^{n-1} U^i f$ are almost everywhere convergent.

The $L_1$ norm of a finite matrix $(u_{ij})$ is the largest of the absolute column sums $(\sum_i |u_{ij}|)$ and the $L_\infty$ norm is the largest of the absolute row sums $(\sum_j |u_{ij}|)$. This observation indicates the extent to which the Dunford-Schwartz result generalizes Hopf's: positiveness is dropped and the equations involving column sums and row sums become inequalities.

Chacon and Ornstein assume, as Hopf does, that $U$ is a positive linear operator on $L_1$, and they assume, as Dunford and Schwartz do, that $\|U\|_1 \leqq 1$. Instead of adding further assumptions, they alter the form of the conclusion. The new conclusion asserts that if $f$ and $g$ are in $L_1$ and if $g \geqq 0$ almost everywhere, then the ratios

$$\sum_{i=0}^{n-1} U^i f \Big/ \sum_{i=0}^{n-1} U^i g$$

converge at almost every one of the points at which the denominator is ultimately different from zero. The Chacon-Ornstein result implies, in particular, that if for some $g$ (for instance $g=1$) the averages $(1/n)\sum_{i=0}^{n-1} U^i g$ converge to 1 almost everywhere, then the averages $(1/n)\sum_{i=0}^{n-1} U^i f$ converge almost everywhere; this exhibits the extent to which their result generalizes Hopf's.

In the finite case, the coordinates of the vector $U^n 1$ are the row sums of the matrix corresponding to $U^n$. Accordingly, a special case of the Chacon-Ornstein theorem applies to matrices with positive entries, column sums equal to 1, and row sums converging to 1 in the sense of Cesàro. The last qualification replaces Hopf's requirement that the column sums be equal to 1.

**Information theory.** If $\mu(X)=1$, the measurable subsets of the space $X$ may be viewed as the possible outcomes of some random process, and, from this point of view, the measure of a measurable set $E$ becomes the probability that a randomly chosen point of $X$ belong to $E$. It is well known that all the intuitive concepts of probability theory can be given a rigorous description in measure-theoretic language. Thus, for instance, an experiment with a finite number of possible outcomes is (corresponds to) a finite measurable partition $\mathfrak{A}$. (This means, of course, a finite disjoint collection of measurable sets whose union is $X$.)

How much information does an experimenter obtain from an experiment? If, to be slightly more precise, the random point of $X$ that the experiment specifies belongs to the measurable set $E$, how much more is known after that specification than before? There are several heuristic arguments that tend to show that a reasonable measure of the quantity of information is $-\log \mu(E)$. It follows that associated with each finite measurable partition $\mathfrak{A}$ there is a simple function $I(\mathfrak{A})$, which may be called the information conveyed by the performance of the experiment $\mathfrak{A}$; the value of $I(\mathfrak{A})$ throughout each set $E$ of $\mathfrak{A}$ is $-\log \mu(E)$.

The formation of the successive iterates of some particular measure-preserving transformation $T$ may be regarded as the action of the passage of time observed at equally spaced intervals, say once a day. Thus if "$x \in E$" is read as "$x$ belongs to $E$ today," then "$Tx \in E$" may be read as "$x$ will belong to $E$ tomorrow." Since "$Tx \in E$" means the same as "$x \in T^{-1}E$," it follows that if the mathematical model of some event today is the measurable set $E$, then the model of the same event tomorrow will be the measurable set $T^{-1}E$.

Suppose now that $\mathfrak{A}_0$ is an experiment (with a finite number of possible outcomes) and suppose that the experiment $\mathfrak{A}_1$ consists of the performance of $\mathfrak{A}_0$ both today and tomorrow. In mathematical language, the partition $\mathfrak{A}_1$ is the least common refinement of $\mathfrak{A}_0$ and $T^{-1}\mathfrak{A}_0$, where the sets of $T^{-1}\mathfrak{A}_0$ are the sets of the form $T^{-1}E$ with $E$

in $\mathfrak{A}_0$. In the standard concise lattice-theoretic notation $\mathfrak{A}_1 = \mathfrak{A}_0 \vee T^{-1}\mathfrak{A}_0$. If, more generally,

$$\mathfrak{A}_n = \mathfrak{A}_0 \vee \cdots \vee T^{-(n-1)}\mathfrak{A}_0, \qquad n = 1, 2, 3, \cdots,$$

then $\mathfrak{A}_n$ is (corresponds to) the experiment that consists of performing $\mathfrak{A}_0$ for $n$ successive days starting today.

What can be said about the average amount of information obtained in a day by repeated performances of $\mathfrak{A}_0$? In mathematical terms the question is one about the asymptotic behavior of the sequence of functions $(1/n)I(\mathfrak{A}_n)$. The question appears to call for, and the answer is, a theorem of ergodic type; the proof, indeed, makes use of the classical individual ergodic theorem. Motivated by Shannon's work on Markov chains, McMillan proved that the sequence $(1/n)I(\mathfrak{A}_n)$ converges in the mean of order 1, and, later, Breiman asserted that it converges almost everywhere. McMillan and Breiman treat ergodic transformations only; an elegant simple proof of the general theorem (whose spirit is substantially the same as that of Breiman's proof) was recently given by Alexandra Ionescu-Tulcea. (Recall that $T$ is ergodic in case, for each measurable set $E$ that is invariant under $T$, either $\mu(E) = 0$ or $\mu(X - E) = 0$.)

In the present state of science the information ergodic theorem of Shannon-McMillan-Breiman looks more like a special consequence of ergodic methods and results than like a statement of comparable generality with the ergodic theorem. The special theorem has, however, some powerful applications, and, I think, a promising future.

**The conjugacy problem.** The main outstanding problem of ergodic theory was (and is) to find usable necessary and sufficient conditions for the conjugacy of two transformations. To say that two measure-preserving transformations $T_1$ and $T_2$ are conjugate means that there exists an invertible measure-preserving transformation $S$ such that $ST_1 = T_2S$ modulo sets of measure zero. A long outstanding test problem was the conjugacy of the 2-shift and the 3-shift (defined above, in the prologue). The first decisive recent step along these lines was taken by Kolmogorov and the next one by Sinai. The conclusion is that if $n \neq m$, then $S_n$ and $S_m$ are not conjugate.

The Kolmogorov-Sinai method is to introduce a new conjugacy invariant $h^*(T)$ (called *entropy*) associated with a measure-preserving transformation $T$ on a normalized measure space. The invariant $h^*(T)$ can be computed in many cases, and, in particular, it turns out that $h^*(S_n) = \log n$.

The function $h^*$ is defined in two steps, via two other functions $H$ and $h$ (both of which are also called entropy). The domain of the first auxiliary function $H$ consists of finite measurable partitions. If $\mathfrak{A}$ is such a partition, then, by definition, $H(\mathfrak{A}) = -\sum_{E \in \mathfrak{A}} \mu(E) \log \mu(E)$, or, in terms of the notation used before, $H(\mathfrak{A}) = \int I(\mathfrak{A}) d\mu$. The number $H(\mathfrak{A})$ is, intuitively speaking, the average amount of information obtainable from one performance of the experiment $\mathfrak{A}$.

It is a trivial consequence of McMillan's theorem that, for each finite measurable partition $\mathfrak{A}$ the sequence of numbers

$$\frac{1}{n} H(\mathfrak{A} \vee \cdots \vee T^{-(n-1)}\mathfrak{A})$$

converges to a finite limit, say $h(T, \mathfrak{A})$. The number $h(T, \mathfrak{A})$ is, intuitively speaking, the average amount of information obtainable *per day* from repeated performances of the experiment.

Now that $h$ has been defined, the definition of $h^*$ is a simple matter; write $h^*(T) = \sup_{\mathfrak{A}} h(T, \mathfrak{A})$, where the supremum is extended over all finite measurable partitions $\mathfrak{A}$. Intuitively speaking, the entropy $h^*(T)$ is the greatest quantity of information obtainable about the universe per day by repeated performances of experiments with a finite (but possibly unbounded) number of possible outcomes. The value of $h^*(T)$ is a real number between 0 and $\infty$; both extremes can occur.

The computation of the supremum $h^*(T)$ is often made easy by a penetrating result of Sinai. The assertion is that if $T$ is invertible and if $\mathfrak{A}$ is a finite measurable partition such that every measurable set belongs to the $\sigma$-field generated by the elements of the partitions $T^i\mathfrak{A}$ $(i = 0, \pm 1, \pm 2, \cdots)$, then the supremum $h^*(T)$ is attained, and, in fact, it is equal to $h(T, \mathfrak{A})$. In intuitive language: if knowledge of the entire history of performances of the experiment $\mathfrak{A}$ (past, present, and future) entails knowledge of the exact state of the universe, then the average amount of information obtainable per day by repeated performances of $\mathfrak{A}$ is as large as it can be for any experiment.

Having shown that the concept of entropy can be used to solve a problem of ergodic theory, the Russian school is proceeding with enthusiasm to build a theory around the theorem. Sample question: what is the relation between the entropy of $T$ and the spectral properties of the induced unitary operator $U$? Partial answer by Rokhlin: if $U$ has pure point spectrum, then $h^*(T) = 0$. Sample question: if $T$ is an automorphism of a compact group, what can be said about the entropy of $T$? Partial answer by Sinai: if $T$ is an automorphism of the torus, then $T$ is in a sense the dual of a unimodular matrix $M$; if $\alpha$ and $\beta$ are the proper values of $M$, with $|\alpha| > |\beta|$, then $h^*(T) = \log |\alpha|$. Partial answer by Abramov: if $T$ is an automorphism of the character group of the additive group of rational numbers, then $T$ is in a sense the dual of a rational multiplier $r$; if $\alpha$ and $\beta$ are (not necessarily respectively) the numerator and denominator of $r$ in lowest form, with $|\alpha| > |\beta|$, then $h^*(T) = \log |\alpha|$.

In the early days of ergodic theory it was possible to conjecture that two invertible measure-preserving transformations on normalized measure spaces are conjugate if and only if the functional operators they induce are unitarily equivalent. That conjecture died long ago; the Kolmogorov-Sinai result buries it forever. The answer to the following question is, however, not at all obvious. If two invertible measure-preserving transformations on normalized measure spaces induce unitarily equivalent functional operators and if they have the same entropy, does it follow that they are conjugate? (This question was raised by Fomin.) It is instructive to examine it in the light of the following specific subquestion. If two shifts based on finite sets with not necessarily uniform distributions have the same entropy, does it follow that they are conjugate? (The subquestion was raised by Billingsley.) An affirmative answer to Fomin's question would imply an affirmative answer to Billingsley's. (The functional operators induced by any two shifts are unitarily equivalent.) Following Billingsley, I conjecture that the truth is the other way around; the answer to Billingsley's question is probably no.

**Non-invariant measures.** The preceding considerations had to do mostly with measure-preserving transformations. How much loss of generality does this involve? How likely is a transformation to preserve some measure? There are several possible ways to formulate a precise question along these lines; experimentation has revealed that the most fruitful formulation runs as follows. Suppose that $T$ is an invertible measurable transformation on, say, the unit interval (there is no essential loss of generality so far) such that if $E$ is a measurable set of measure zero, then both $T^{-1}E$ and $TE$ have measure zero; does there exist a finite or possibly $\sigma$-finite measure $\nu$ equivalent to $\mu$ and invariant under $T$? (Equivalence here means that $\mu$ and $\nu$ vanish on the same sets.) Several solutions have been offered for the problem of finite invariant measure in the course of the years, and, incidentally, it has been known for some time that the solution of that problem is not always affirmative.

The most recent solution of the problem of finite invariant measure is that of Hajian and Kakutani. The condition in terms of which that solution is stated is an elegant generalization of a well known condition. A set $E$ is a *wandering set* (for a transformation $T$) if the sets $E, T^{-1}E, T^{-2}E, \cdots$ are pairwise disjoint, and the transformation $T$

is *conservative* if every measurable wandering set has measure zero. These definitions are old; the new concepts of Hajian and Kakutani are defined as follows. A set $E$ is a *weakly wandering set* (for $T$) if the sequence of sets $E, T^{-1}E, T^{-2}E, \cdots$ has an infinite pairwise disjoint subsequence, and $T$ is *weakly conservative* if every measurable weakly wandering set has measure zero. (Hajian and Kakutani do not actually use the latter term, but it seems natural to do so.) The new result is that a transformation $T$ (satisfying the conditions stated before) preserves some finite invariant measure equivalent to the given one if and only if it is weakly conservative.

The general problem of ($\sigma$-finite) invariant measure has stood unsolved for a long time. There were, to be sure, some known necessary and sufficient conditions, but none of them was usable. In particular, for all anyone knew, the solution of the problem was always affirmative. The solution, when it finally came, turned out to be negative; Ornstein constructed an ingenious example of a transformation that satisfies all the stated conditions but does not preserve any equivalent $\sigma$-finite measure.

Ornstein defines his transformation on the unit interval; there is some value in an alternative approach via another measure space, which, however, is easily shown to be isomorphic to the interval. The construction uses a sequence of integers $m_0, m_1, m_2, \cdots$, such that $m_k > 1$ for all $k$, and such that the quotients $m_{k+1}/m_0 \cdots m_k$ are unbounded. (Example: $m_k = 2^{3^k}$. In this case the quotients even tend to $\infty$ with $k$.) These conditions are sufficient to guarantee the truth of the assertions that follow. Since, however, those assertions are not proved here, it is not necessary to keep the exact form of the conditions in mind.

Let $X_k$ be the additive group of integers modulo $m_k$ ($k = 0, 1, 2, \cdots$), and let $\mu_k$ be the (non-uniform) measure in $X_k$ that assigns the weight 1/2 to 0 and distributes the weight 1/2 uniformly among the remaining points. Let $X$ be the Cartesian product of the spaces $X_k$ and let $\mu$ be the product of the measures $\mu_k$. For an intuitive insight into the role that the elements of $X$ play, think of them as generalized (infinite) integers written in a generalized (infinite) decimal system. An ordinary non-negative integer $x$ has an expansion

$$x = x_0 + x_1 \cdot 10 + x_2 \cdot 10^2 + x_3 \cdot 10^3 + \cdots,$$

where the digits $x_k$ can take the values $0, 1, \cdots, 9$, and where $x_k \neq 0$ for finitely many values of $k$ only. The generalization replaces the $k$th 10 by $m_k$, and, at the same time, removes the finitely-non-zero restriction. The result is the consideration of formally infinite expansions of the form

$$x_0 + x_1 \cdot m_0 + x_2 \cdot m_0 m_1 + x_3 \cdot m_0 m_1 m_2 + \cdots,$$

where the digits $x_k$ can take the values $0, 1, \cdots, m_k - 1$.

What happens to the sequence $(x_0, x_1, x_2, \cdots)$ of decimal digits of an integer $x$ when $x$ is replaced by $x+1$? The answer is that, usually, $x_0$ is replaced by $x_0+1$. If, however, $x_0=9$, then $x_0$ is replaced by 0, and $x_1$ is replaced by $x_1+1$ ("carry one")—unless, that is, $x_1=9$ also, in which case both $x_0$ and $x_1$ are replaced by 0, and $x_2$ is replaced by $x_2+1$—unless, etc., etc. In these terms it is easy to define the promised transformation $T$ on $X$: it is the obvious generalization of adding 1. If

$$x = (x_0, x_1, x_2, \cdots)$$

is an element of $x$, and if $x_0 \neq m_0 - 1$, then

$$Tx = (x_0 + 1, x_1, x_2 \cdots).$$

If, however, $x_0 = m_0 - 1$, and if, in fact, $k$ is the smallest index such that $x_{k+1} \neq m_{k+1} - 1$, then

$$Tx = (0, \cdots, 0, x_{k+1} + 1, x_{k+2}, \cdots),$$

where the number of initial zeros is exactly $k+1$ (one each for $x_0, \cdots, x_k$). If $x_k = m_k - 1$ for all $k$, then

$$Tx = (0, 0, 0, \cdots).$$

Intuitively $X$ may be thought of as an infinite adding machine; the action of $T$ on the numbers in $X$ is simply to add 1. The proof that $T$ is not even potentially measure-preserving is a complicated combinatorial argument.

**Epilogue.** Ergodic theory is very much alive these days; there are new results and there are new problems. Now that it is known, for instance, that there exist measure-theoretically interesting transformations with no invariant measure of the proper sort, the conjugacy problem for such transformations and the applicability of the concept of entropy to them become worthy of consideration. The same thing is true for the operators on $L_1$ that enter into the generalized ergodic theorems discussed above. Other old results and techniques also deserve to be extended to the newly important transformations and operators; this is especially true of the topological studies sometimes pursued under the name of approximation theorems.

Even according to the narrowest known definition of ergodic theory there are some parts of the subject that were not mentioned in this report. There has been, for instance, some recent progress in the theory of conservative transformations and in the spectral theory of measure-preserving transformations. The latter subject, especially, constitutes a far from closed chapter. Since, in particular, asymptotic results (ergodic theorems) are usually true for both finite and infinite measure spaces, it is sometimes tacitly assumed that the same is true for spectral results. In fact, however, knowledge of the possible spectral behavior of measure-preserving transformations is meager for finite measure spaces and almost nil for infinite ones.

I hope that in the near future, in the course of the next twelve years, say, humanity learns sufficiently many new answers to these fascinating old questions to warrant another Society address on the subject. I should like to hear that address so that I may discover how everything came out and who did it.

## REFERENCES

1. L. M. Abramov, *The entropy of an automorphism of a solenoidal group*, Teor. Veroyatnost. i Primenen. vol. 4 (1959) pp. 249–254.

2. L. Breiman, [a] *The individual ergodic theorem of information theory*, Ann. Math. Statist. vol. 28 (1957) pp. 809–811. [b] *A correction to the individual ergodic theorem of information theory*, Ann. Math. Statist. To appear.

3. R. V. Chacon and D. S. Ornstein, *A general ergodic theorem*, Illinois J. Math. vol. 4 (1960) pp. 153–160.

4. N. Dunford and J. T. Schwartz, *Convergence almost everywhere of operator averages*, J. Rat. Mech. Anal. vol. 5 (1956) pp. 129–178.

5. A. Hajian and S. Kakutani, *Weakly wandering sets and invariant measures*. To appear.

6. E. Hopf, *The general temporally discrete Markoff process*, J. Rat. Mech. Anal. vol. 3 (1954) pp. 13–45.

7. A. Ionescu-Tulcea, *Contributions to information theory for abstract alphabets*. To appear.

8. A. N. Kolmogorov, [a] *A new metric invariant of transitive dynamical systems and automorphisms of Lebesgue spaces*, Dokl. Akad. Nauk SSSR vol. 119 (1958) pp. 861–864. [b] *Entropy per unit time as a metric invariant of automorphisms*, Dokl. Akad. Nauk SSSR vol. 124 (1959) pp. 754–755.

9. D. S. Ornstein, *On invariant measures*. Bull. Amer. Math. Soc. vol. 66 (1960) pp. 297–300.

10. B. McMillan, *The basic theorems of information theory*, Ann. Math. Statist. vol. 24 (1953) pp. 196–219.

11. V. A. Rokhlin, *Entropy of metric automorphism*, Dokl. Akad. Nauk SSSR vol. 124 (1959) pp. 980–983.

12. Y. Sinai, *On the concept of entropy for a dynamic system*, Dokl. Akad. Nauk SSSR vol. 124 (1959) pp. 768–771.

UNIVERSITY OF CHICAGO

Reprinted from the
AMERICAN MATHEMATICAL MONTHLY
Vol. 70, No. 3, pp. 241–247, Mar. 1963

# WHAT DOES THE SPECTRAL THEOREM SAY?

P. R. HALMOS, University of Michigan

Most students of mathematics learn quite early and most mathematicians remember till quite late that every Hermitian matrix (and, in particular, every real symmetric matrix) may be put into diagonal form. A more precise statement of the result is that every Hermitian matrix is unitarily equivalent to a diagonal one. The spectral theorem is widely and correctly regarded as the generalization of this assertion to operators on Hilbert space. It is unfortunate therefore that even the bare statement of the spectral theorem is widely regarded as somewhat mysterious and deep, and probably inaccessible to the nonspecialist. The purpose of this paper is to try to dispel some of the mystery.

Probably the main reason the general operator theorem frightens most people is that it does not obviously include the special matrix theorem. To see the relation between the two, the description of the finite-dimensional situation has to be distorted almost beyond recognition. The result is not intuitive in any language; neither Stieltjes integrals with unorthodox multiplicative properties, nor bounded operator representations of function algebras, are in the daily toolkit of every working mathematician. In contrast, the formulation of the spectral theorem given below uses only the relatively elementary concepts of measure theory. This formulation has been part of the oral tradition of Hilbert space for quite some time (for an explicit treatment see [6]), but it has not been called the spectral theorem; it usually occurs in the much deeper "multiplicity theory." Since the statement uses simple concepts only, this aspect of the present formulation is an advantage, not a drawback; its effect is to make the spirit of one of the harder parts of the subject accessible to the student of the easier parts.

Another reason the spectral theorem is thought to be hard is that its proof is hard. An assessment of difficulty is, of course, a subjective matter, but, in any case, there is no magic new technique in the pages that follow. It is the statement of the spectral theorem that is the main concern of the exposition, not the proof. The proof is essentially the same as it always was; most of the standard methods used to establish the spectral theorem can be adapted to the present formulation.

Let $\phi$ be a complex-valued bounded measurable function on a measure space $X$ with measure $\mu$. (All measure-theoretic statements, equations, and relations, e.g., "$\phi$ is bounded," are to be interpreted in the "almost everywhere" sense.) An operator $A$ is defined on the Hilbert space $\mathfrak{L}^2(\mu)$ by

$$(Af)(x) = \phi(x)f(x), \qquad x \in X;$$

the operator $A$ is called the *multiplication* induced by $\phi$. The study of the relation between $A$ and $\phi$ is an instructive exercise. It turns out, for instance, that the adjoint $A^*$ of $A$ is the multiplication induced by the complex conjugate $\bar{\phi}$ of $\phi$. If $\psi$ also is a bounded measurable function on $X$, with induced multiplication $B$, then the multiplication induced by the product function $\phi\psi$ is the product operator $AB$. It follows that a multiplication is always normal; it is Hermitian if and only if the function that induces it is real. (For the elementary concepts of operator theory, such as Hermitian operators, normal operators, projections, and spectra, see [**3**]. For present purposes a concept is called elementary if it is discussed in [**3**] before the spectral theorem, i.e., before p. 56.)

As a special case let $X$ be a finite set (with $n$ points, say), and let $\mu$ be the "counting measure" in $X$ (so that $\mu(\{x\})=1$ for each $x$ in $X$). In this case $\mathfrak{L}^2(\mu)$ is $n$-dimensional complex Euclidean space; it is customary and convenient to indicate the values of a function in $\mathfrak{L}^2(\mu)$ by indices instead of parenthetical arguments. With this notation the action on $f$ of the multiplication $A$ induced by $\phi$ can be described by

$$A\langle f_1, \cdots, f_n\rangle = \langle \phi_1 f_1, \cdots, \phi_n f_n\rangle.$$

To say this with matrices, note that the characteristic functions of the singletons in $X$ form an orthonormal basis in $\mathfrak{L}^2(\mu)$; the assertion is that the matrix of $A$ with respect to that basis is diag $\langle \phi_1, \cdots, \phi_n\rangle$.

The general notation is now established and the special role of the finite-dimensional situation within it is clear; everything is ready for the principal statement.

SPECTRAL THEOREM. *Every Hermitian operator is unitarily equivalent to a multiplication.*

In complete detail the theorem says that if $A$ is a Hermitian operator on a Hilbert space $\mathcal{H}$, then there exists a (real-valued) bounded measurable function $\phi$ on some measure space $X$ with measure $\mu$, and there exists an isometry $U$ from $\mathfrak{L}^2(\mu)$ onto $\mathcal{H}$, such that

$$(U^{-1}AUf)(x) = \phi(x)f(x), \qquad x \in X,$$

for each $f$ in $\mathfrak{L}^2(\mu)$. What follows is an outline of a proof of the spectral theorem, a brief discussion of its relation to the version involving spectral measures, and an illustration of its application.

Three tools are needed for the proof of the spectral theorem.

(1) *The equality of norm and spectral radius.* If the spectrum of $A$ is $\Lambda(A)$, then the *spectral radius* $r(A)$ is defined by

$$r(A) = \sup\{|\lambda| : \lambda \in \Lambda(A)\}.$$

It is always true that $r(A) \leqq \|A\|$ ([**3**, Theorem 2, p. 52]); the useful fact here is that if $A$ is Hermitian, then $r(A)=\|A\|$ ([**3**, Theorem 2, p. 55]).

(2) *The Riesz representation theorem for compact sets in the line.* If $L$ is a positive linear functional defined for all real-valued continuous functions on a com-

pact subset $X$ of the real line, then there exists a unique finite measure $\mu$ on the Borel sets of $X$ such that

$$L(f) = \int f d\mu$$

for all $f$ in the domain of $L$. (To say that $L$ is linear means of course that

$$L(\alpha f + \beta g) = \alpha L(f) + \beta L(g),$$

whenever $f$ and $g$ are in the domain of $L$ and $\alpha$ and $\beta$ are real scalars; to say that $L$ is positive means that $L(f) \geqq 0$ whenever $f$ is in the domain of $L$ and $f \geqq 0$.) For a proof, see [**4**, Theorem D, p. 247].

(3) *The Weierstrass approximation theorem for compact sets in the line.* Each real-valued continuous function on a compact subset of the real line is the uniform limit of polynomials. For a pleasant elementary discussion and proof see [**1**, p. 102].

Consider now a Hermitian operator $A$ on a Hilbert space $\mathcal{H}$. A vector $\xi$ in $\mathcal{H}$ is a *cyclic vector* for $A$ if the set of all vectors of the form $q(A)\xi$, where $q$ runs over polynomials with *complex* coefficients, is dense in $\mathcal{H}$. Cyclic vectors may not exist, but an easy transfinite argument shows that $\mathcal{H}$ is always the direct sum of a family of subspaces, each of which reduces $A$, such that the restriction of $A$ to each of them does have a cyclic vector. Once the spectral theorem is known for each such restriction, it follows easily for $A$ itself; the measure spaces that serve for the direct summands of $\mathcal{H}$ have a natural direct sum, which serves for $\mathcal{H}$ itself. Conclusion: there is no loss of generality in assuming that $A$ has a cyclic vector, say $\xi$.

For each real polynomial $p$ write

$$L(p) = (p(A)\xi, \xi).$$

Clearly $L$ is a linear functional; since

$$\begin{aligned} |L(p)| &\leqq \|p(A)\| \cdot \|\xi\|^2 = r(p(A)) \cdot \|\xi\|^2 \\ &= \sup \{ |\lambda| : \lambda \in \Lambda(p(A))\} \cdot \|\xi\|^2 \\ &= \sup \{ |p(\lambda)| : \lambda \in \Lambda(A)\} \cdot \|\xi\|^2, \end{aligned}$$

the functional $L$ is bounded for polynomials. (The last step uses the spectral mapping theorem; cf. [**3**, Theorem 3, p. 55].) It follows (by the Weierstrass theorem) that $L$ has a bounded extension to all real-valued continuous functions on $\Lambda(A)$. To prove that $L$ is positive, observe first that if $p$ is a real polynomial, then

$$((p(A))^2\xi, \xi) = \|p(A)\xi\|^2 \geqq 0.$$

If $f$ is an arbitrary positive continuous function on $\Lambda(A)$, then approximate $\sqrt{f}$ uniformly by real polynomials; the inequality just proved implies that $L(f) \geqq 0$

(since $f$ is then uniformly approximated by squares of real polynomials). The Riesz theorem now yields the existence of a finite measure $\mu$ such that

$$(p(A)\xi, \xi) = \int p d\mu$$

for every real polynomial $p$.

For each (possibly complex) polynomial $q$ write

$$Uq = q(A)\xi.$$

Since $A$ is Hermitian, $(q(A))^*(=\bar{q}(A))$ is a polynomial in $A$, and so is $(q(A))^*q(A)$ $(=|q|^2(A))$; it follows that

$$\int |q|^2 d\mu = (\bar{q}(A)q(A)\xi, \xi) = ((q(A))^*q(A)\xi, \xi) = \|q(A)\xi\|^2 = \|Uq\|^2.$$

This means that the linear transformation $U$ from a dense subset of $\mathfrak{L}^2(\mu)$ into $\mathfrak{H}$ is an isometry, and hence that it has a unique isometric extension that maps $\mathfrak{L}^2(\mu)$ into $\mathfrak{H}$. The assumption that $\xi$ is a cyclic vector implies that the range of $U$ is in fact dense in, and hence equal to, the entire space $\mathfrak{H}$.

It remains only to prove that the transform of $A$ by $U$ is a multiplication. Write $\phi(\lambda)=\lambda$ for all $\lambda$ in $\Lambda(A)$. Given a complex polynomial $q$, write $\tilde{q}(\lambda)$ $=\lambda q(\lambda)=\phi(\lambda)q(\lambda)$; then

$$U^{-1}AUq = U^{-1}Aq(A)\xi = U^{-1}\tilde{q}(A)\xi = U^{-1}U\tilde{q} = \tilde{q}.$$

In other words $U^{-1}AU$ agrees, on polynomials, with the multiplication induced by $\phi$, and that is enough to conclude that $U^{-1}AU$ is equal to that multiplication. This completes the outline of the proof of the spectral theorem for Hermitian operators.

The formulation of the spectral theorem given above yields fairly easily all the information contained in the more common versions. Thus if $A$ is the multiplication on $\mathfrak{L}^2(\mu)$ induced by the real function $\phi$ on $X$, and if $F$ is a (complex) Borel measurable function that is bounded on $\Lambda(A)$, then $F(A)$ can be defined as the multiplication induced by the composite function $F \circ \phi$. The mapping $F \to F(A)$ is the homomorphism that is frequently known by the impressive name of "the functional calculus." If, in particular, $F=F_M$ is the characteristic function of a Borel set $M$ in the real line, and if $E(M)$ is the multiplication induced by $F_M \circ \phi$, then $E$ is the spectral measure of $A$. The verification that $E$ is indeed a spectral measure is easy. To prove that it belongs to $A$ (i.e., that $A=\int \lambda dE(\lambda)$), proceed as follows. Fix $f$ and $g$ in $\mathfrak{L}^2(\mu)$ and write

$$\nu(M) = (E(M)f, g)$$

for each Borel set $M$; it is to be proved that

$$(Af, g) = \int \lambda d\nu(\lambda).$$

Since $(E(M)f, g) = \int (F_M \circ \phi) f\bar{g} d\mu$ and $\nu(M) = \int F_M d\nu$, it follows that

$$\int (F_M \circ \phi) f\bar{g} d\mu = \int F_M d\nu$$

for all Borel sets $M$. This implies that

$$\int (F \circ \phi) f\bar{g} d\mu = \int F d\nu$$

whenever $F$ is a simple function, and hence, by approximation, whenever $F$ is a bounded Borel measurable function. This conclusion (for $F(\lambda) \equiv \lambda$) is just what was wanted.

The multiplication version of the spectral theorem implies the spectral measure version, but the latter is canonical ($E$ is uniquely determined by $A$) whereas the former is not. Consider, for instance, the identity operator on a separable infinite-dimensional Hilbert space in the role of $A$. It is unitarily equivalent to multiplication by the constant function 1 on, say, the unit interval (with Lebesgue measure); it is also unitarily equivalent to multiplication by the constant function 1 on the set of positive integers (with the counting measure).

There is a spectral theorem for normal operators also; its statement can be obtained from the one given above by substituting "normal" for "Hermitian." It is a well-known technical nuisance that the proof of the spectral theorem for normal operators involves some difficulties that do not arise in the Hermitian case. The source of the trouble is that it is not enough just to replace polynomials in a real variable by polynomials in a complex variable; the Weierstrass theorem demands the consideration of polynomials in two real variables. There is a consequent difficulty in extending the spectral mapping theorem to the kind of functions (polynomials in the real and imaginary parts of a complex variable) that arise in the imitation of the proof above. Even the equality of norm and spectral radius, while true for normal operators, requires a proof quite a bit deeper than in the real case. One way around all this is not to imitate the proof but to use the result. In [**3**, p. 72], for instance, the spectral theorem for normal operators (spectral measure version) is derived from the Hermitian theorem (spectral measure version); the only additional tool needed is an essentially classical extension theorem for measures in the plane.

In any case, all this talk about proof is somewhat beside the point in this paper. The reason a proof is outlined above is not so much to induce belief in the result as to clarify it. The emphasis here is not on *how* but on *what*, not on *proof* but on *statement*, not on *How does the spectral theorem come about*? but on *What does the spectral theorem say*?

To see how the multiplication point of view can be used, consider the Fuglede commutativity theorem [**2**]. A possible statement is this: if $A$ is normal and if $B$ is an operator that commutes with $A$, then $B$ commutes with $F(A)$ for each Borel measurable function bounded on $\Lambda(A)$. (An alternative statement, only apparently weaker, is that if $B$ commutes with $A$, then $B$ commutes with $A^*$; for a recent elegant proof see [5].) The spectral theorem shows that there is no loss of generality in assuming that $A$ is the multiplication induced by $\phi$, say, on a measure space $X$ with measure $\mu$. If $F_M$ is, for each Borel set $M$ in the complex plane, the characteristic function of $M$, and if $E(M)$ is the multiplication induced by $F_M \circ \phi$, then it is sufficient to prove that $B$ commutes with each $E(M)$. (Approximate the general $F$ by simple functions, as before.) If $\mathcal{E}(M)$ is the range of the projection $E(M)$, then the desired result is that $\mathcal{E}(M)$ reduces $B$, but it is, in fact, sufficient to prove that $\mathcal{E}(M)$ is invariant under $B$. Reason: apply the invariance conclusion, once obtained, to the complement of $M$, and infer that both $\mathcal{E}(M)$ and $(\mathcal{E}(M))^\perp$ are invariant under $B$.

Observe now that $\mathcal{E}(M)$ is the set of all those functions in $\mathcal{L}^2(\mu)$ that vanish outside $\phi^{-1}(M)$, and consider first the case of the closed unit disc,

$$M = \{\lambda : |\lambda| \leqq 1\};$$

then

$$\phi^{-1}(M) = \{x : |\phi(x)| \leqq 1\}.$$

Assertion: $\mathcal{E}(M)$ consists of all $f$ in $\mathcal{L}^2(\mu)$ for which the sequence $\{\|Af\|, \|A^2f\|, \|A^3f\|, \cdots\}$ is bounded. Indeed, if $f$ vanishes outside $\phi^{-1}(M)$, then

$$\|A^n f\|^2 = \int |\phi^n f|^2 d\mu = \int_{\phi^{-1}(M)} |\phi^n|^2 \cdot |f|^2 d\mu \leqq \int |f|^2 d\mu.$$

If, on the other hand, there is a set $S$ of positive measure on which $f \neq 0$ and $|\phi| > 1$, then

$$\|A^n f\|^2 = \int |\phi^n f|^2 d\mu \geqq \int_S |\phi^2|^n |f|^2 d\mu \rightarrow \infty.$$

The assertion is proved, and the invariance of $\mathcal{E}(M)$ under $B$ follows: if $\|A^n f\| \leqq c$ for all $n$, then $\|A^n B f\| = \|BA^n f\| \leqq \|B\| \cdot \|A^n f\| \leqq \|B\| \cdot c$ for all $n$.

If $M$ is any closed disc, $M = \{\lambda : |\lambda - \lambda_0| \leqq r\}$, then

$$\phi^{-1}(M) = \{x : |\phi(x) - \lambda_0| \leqq r\} = \left\{x : \left|\left(\frac{\phi - \lambda_0}{r}\right)(x)\right| \leqq 1\right\}.$$

Since $B$ commutes with multiplication by $\phi$, it commutes with multiplication by $(\phi - \lambda_0)/r$ also, and it follows from the preceding paragraph that $\mathcal{E}(M)$ is invariant under $B$.

The rest of the proof is easy measure theory; from this point of view spectral measures behave even better than numerical measures. Since $\mathcal{E}(M)$ is invariant under $B$ whenever $M$ is a disc, the same is true whenever $M$ is the union of countably many discs. This implies that $\mathcal{E}(M)$ is invariant under $B$ whenever $M$ is open, and hence (regularity) for arbitrary Borel sets $M$.

## References

1. R. P. Boas, A primer of real functions, Math. Assoc. of America, Carus Monograph no. 13, 1960.

2. B. Fuglede, A commutativity theorem for normal operators, Proc. Nat. Acad. Sci. U.S.A., 36 (1950) 35–40.

3. P. R. Halmos, Introduction to Hilbert space, Chelsea, New York, 1951.

4. ———, Measure theory, Van Nostrand, New York, 1950.

5. M. Rosenblum, On a theorem of Fuglede and Putnam, J. London Math. Soc., 33 (1958) 376–377.

6. I. E. Segal, Decomposition of operator algebras, II, Memoirs, Amer. Math. Soc., 1951.

Reprinted from
LECTURES ON MODERN MATHEMATICS, VOL. I
T. Saaty (Ed.), pp. 1–22, J. Wiley & Sons Inc., 1963

# 1

# A Glimpse into Hilbert Space

*P. R. Halmos*

## PROLOGUE

The purpose of these lectures is to describe the current status of research in Hilbert space theory. This is impossible; it cannot be done in two hours and I doubt whether it could be done in two weeks. Since the subject began, in the early years of the twentieth century, the theory of Hilbert space has come to have ramifications in many parts of pure and applied mathematics. This makes the impossible a little harder; I doubt whether any one man could tell about it all in any finite time. I shall talk about some small fragments of the theory. The choice of these particular fragments was, of course, dictated by personal taste; I can talk only about what I have studied, and the parts of the subject that I studied most were, naturally, the parts that I liked best to begin with. If it turns out that I do not even mention something that you were especially hoping to hear about, please accept these remarks as my apology; the reason is that both my knowledge and my time are finite.

Hilbert space theory is still very much alive and growing. In an attempt to describe its vitality, in an attempt to indicate the direction of its growth, I should like to talk less about what is known and more about what is not. A field of mathematics is alive only so long as it continues to ask interesting and worth-while questions. To prove that Hilbert space theory is alive, I plan to push the description of several of its fragments far enough to be able to state some of their challenging unsolved problems.

A Hilbert space is usually defined as a complete complex inner-product space. In other words, a Hilbert space is a vector space $\mathcal{H}$ over the field of complex numbers, endowed with an inner product $(x, y)$ and the corresponding norm $\|x\| = \sqrt{(x, x)}$, such that with respect to the metric $\|x - y\|$ the space $\mathcal{H}$ is complete. It should be stated right away that the subject known as Hilbert space theory pays very little attention to Hilbert spaces; by far the largest part of it is, or should be, better known as operator theory. Some attention has been, and continues to be, paid to the geometric structure of Hilbert space; to illustrate what I am not going to talk about I mention two striking and famous results. The first one singles out Hilbert spaces from the more inclusive Banach spaces by the quadratic character of their norm. The precise statement is the theorem of von Neumann and Jordan [30]: a Banach space is a Hilbert space if and only if the parallelogram law $\|x - y\|^2 + \|x + y\|^2 = 2\|x\|^2 + 2\|y\|^2$ is valid. The second asserts that the topology of a general Hilbert space can be radically different from that of its Euclidean special cases. The result is due to Klee [24]; the assertion is that every infinite-dimensional Hilbert space is homeomorphic to its unit sphere (i.e., to the set $\{x: \|x\| = 1\}$).

As for operator theory, it can be done in several versions. Operators can be discussed one at a time or in bunches (algebras, rings, groups, semigroups); they come in several sizes (completely continuous, bounded, semibounded, unbounded); and their motivation and methods of study can come from either finite-dimensional matrix algebra or classical analysis and its applications (differential equations, integral equations, quantum mechanics). I believe that most of the depth and most of the interest of the entire subject is present (and perhaps more clearly visible) in its simplest and most concrete aspects, and, for that reason, I shall concentrate attention on some algebraic aspects of the study of a single bounded operator.

NUMERICAL RANGES

An easy and historically important concept to begin with is that of a quadratic form. An operator (here always bounded) is a linear transformation $A$ of a Hilbert space $\mathcal{H}$ into itself; the corresponding quadratic form $Q_A$ is the function on the unit sphere of $\mathcal{H}$ defined by $Q_A(x) = (Ax, x)$. The *numerical range* of $A$, in symbols $W(A)$, is the range of the quadratic form $Q_A$; explicitly, $W(A)$ is the set of all complex numbers of the form $(Ax, x)$, where $x$ varies over all unit vectors. Here are some examples. The numerical range of 1 (the identity operator) consists of a single point, namely 1 (the complex number); the numerical range of the matrix $\begin{pmatrix} 1 & 0 \\ 0 & 0 \end{pmatrix}$ (regarded as an operator on a two-dimensional Hilbert space) is the closed unit interval; the numerical range of $\begin{pmatrix} 0 & 0 \\ 1 & 0 \end{pmatrix}$ is the closed disc with center 0 and radius $\frac{1}{2}$; the numerical range of $\begin{pmatrix} 0 & 0 \\ 1 & 1 \end{pmatrix}$ is the ellipse (interior and boundary) with foci at 0 and 1, minor axis of length 1 and major axis of length $\sqrt{2}$. These examples illustrate everything that can happen in a two-dimensional Hilbert space (see [9, 42]); the numerical range of every two-by-two matrix is an ellipse (possibly degenerate). In spaces of higher dimension more things can happen. The numerical range of $\begin{pmatrix} 0 & 0 & \lambda \\ 1 & 0 & 0 \\ 0 & 1 & 0 \end{pmatrix}$, where $\lambda$ is a complex number of modulus 1, is the equilateral triangle (interior and boundary) whose vertices are the cube roots of $\lambda$; the numerical range of the operator on $\mathcal{L}^2[0, 1]$ defined by $f(t) \to e^{2\pi i t} f(t)$ is the open [!] unit disc.

An elegant general result about numerical ranges is the Toeplitz-Hausdorff theorem [20, 41]; the assertion is simply that the numerical range of every operator is convex. The proof is elementary but annoying. It usually proceeds by reducing the general theorem to the special case of two-by-two matrices and then establishing the special case by explicit computations. The reduction makes contact with another interesting topic, and I shall indicate how it goes; as for the computations, I leave you to carry them out (or to look them up) at your leisure.

Suppose that the Hilbert space $\mathcal{H}$ is a subspace of (meaning a closed linear manifold in) another Hilbert space $\mathcal{K}$, and suppose that $A$ and $B$ are operators on $\mathcal{H}$ and on $\mathcal{K}$ respectively. If $Q_A(x) = Q_B(x)$ whenever $x$ is a unit vector in $\mathcal{H}$, that is, if the function $Q_B$ is an extension of the function $Q_A$, then I shall say [12] that the operator $B$ is a *dilation* of the operator $A$. Equivalently, $B$ is a dilation of $A$ if, for every vector $x$ in $\mathcal{H}$, the image $Ax$ is the projection of the image $Bx$ into $\mathcal{H}$. Still another equivalent definition, perhaps the most convenient one, is this: $B$ is a dilation of $A$ if and only if its action on $\mathcal{K}$ can be represented by an operator matrix of the form $\begin{pmatrix} A & X \\ Y & Z \end{pmatrix}$, where $X$ maps $\mathcal{H}^{\perp}$ (in $\mathcal{K}$) into $\mathcal{H}$, $Y$ goes in the other direction, and $Z$ is an operator on $\mathcal{H}^{\perp}$. The idea of the reduction of the Toeplitz-Hausdorff theorem to the two-dimensional case is now easy to describe. Since convexity is a condition on only two vectors at a time, the general theorem about an arbitrary operator $A$ on an arbitrary Hilbert space $\mathcal{H}$ can be proved by restricting $Q_A$ to all possible two-dimensional subspaces of $\mathcal{H}$ and applying the two-dimensional theorem to each one.

For normal operators, and, in particular, for normal matrices (two-by-two, or any size), the proof of the Toeplitz-Hausdorff theorem is easy and natural. The reason is that normality is the same as diagonability. If $A = \operatorname{diag} \langle \lambda_1, \cdots, \lambda_n \rangle$, and if $x = \langle \xi_1, \cdots, \xi_n \rangle$, then $(Ax, x) = \Sigma_{i=1}^{n} \lambda_i |\xi_i|^2$; as $x$ varies over all vectors with $\Sigma_{i=1}^{n} |\xi_i|^2 = 1$, the values of $(Ax, x)$ vary over all convex linear combinations of the $\lambda$'s.

It is clear that if $B$ is a dilation of $A$, then $W(A) \subset W(B)$. It follows that $W(A)$ is included in $\bigcap_{B \in \mathfrak{N}(A)} W(B)$, where $\mathfrak{N}(A)$ is the set of all normal dilations of $A$. A pertinent recent result [17] is that for finite matrices the inclusion is in fact an equality: the numerical range of every matrix is the intersection of the numerical ranges of its normal dilations.† The proof makes use of the Toeplitz-Hausdorff theorem; a direct proof (if one could be found) would yield a clean conceptual proof of that theorem.

The examples $\begin{pmatrix} 0 & 0 \\ 1 & 0 \end{pmatrix}$ and $\begin{pmatrix} 0 & 0 & \lambda \\ 1 & 0 & 0 \\ 0 & 1 & 0 \end{pmatrix}$, mentioned above, illustrate

† Note added May 2, 1963. This theorem has recently been generalized by C. A. Berger. His result is that it is sufficient to consider normal dilations with the property that each of their positive powers is a dilation of the corresponding power of the given operator; cf. the concept of "strong" unitary dilation defined below.

the normal dilation theorem just mentioned, and a little more. The assumption $|\lambda| = 1$ implies that the larger matrix is not only normal but even unitary. A trivial necessary condition that an operator $A$ have a unitary dilation is that it be a contraction (i.e., that $\|A\| \leqq 1$); it is known (and not especially difficult to prove) that this condition is sufficient also. It may not follow that if $A$ is a contraction, then $A$ has many unitary dilations—sufficiently many to close down on its numerical range. The experimental evidence is good—recall $\begin{pmatrix} 0 & 0 \\ 1 & 0 \end{pmatrix}$—and the desired conclusion is permissible in the presence of various additional assumptions (e.g., in case $\|A\| \leqq \frac{1}{3}$, or in case $A$ itself is normal), but the general case is open. For the record, I formulate it explicitly.

*Problem 1. Is the numerical range of every contraction the intersection of the numerical ranges of its unitary dilations?*

## COMMUTATORS

I turn next to an algebraic topic (with roots in physics) that has engaged the attention of many authors, but in which, nevertheless, the number of questions is far greater than the number of answers. A mathematical formulation of the Heisenberg uncertainty principle is that a certain pair of operators $P$ and $Q$ satisfies (after suitable normalizations) the relation $PQ - QP = 1$. It is easy enough to produce a concrete example of this behavior; consider $\mathfrak{L}^2(-\infty, +\infty)$, and let $P$ and $Q$ be the differentiation operator and the position operator respectively (that is, $(Pf)(x) = f'(x)$ and $(Qf)(x) = xf(x)$). These are not bounded operators, of course, and their domains are far from being the whole space, and they misbehave in many other ways. For quite a few years (in the 1930's and 1940's) it was not known whether a commutator of bounded operators (that is, one of the form $PQ - QP$ with bounded $P$ and $Q$) could be equal to the identity. The finite-dimensional case is easy to settle. The reason is that in that case the concept of trace is available; the trace of a commutator is always zero, and, consequently, $PQ - QP$ can never be equal to something with a non-zero trace. The infinite-dimensional case turned out to have a negative solution too. By two independent beautiful arguments Wintner [46] and Wielandt [44] showed that $PQ - QP = 1$ has no solutions in bounded operators (or, for that matter, in any Banach algebra).

At this stage the problem may be settled for a physicist, but for an algebraist this is just where the fun begins. The general question is this: which operators are and which operators are not commutators? In finite-dimensional spaces every commutator must have trace 0, and it turns out to be an easy exercise to prove the converse: every matrix with trace 0 is a commutator. The non-trivial questions concern infinite-dimensional spaces.

Suppose that $\mathcal{H}$ is an infinite-dimensional Hilbert space and consider the infinite direct sum $\mathcal{H} \oplus \mathcal{H} \oplus \mathcal{H} \oplus \cdots$. Operators on this large space can be represented as infinite matrices whose entries are operators on $\mathcal{H}$. If, in particular, $A$ is an arbitrary operator on $\mathcal{H}$ (for instance, the identity), then the matrix

$$P = \begin{pmatrix} 0 & A & 0 & 0 & & & \\ 0 & 0 & A & 0 & & & \\ 0 & 0 & 0 & A & & & \\ 0 & 0 & 0 & 0 & & & \\ & & & & \cdot & & \\ & & & & & \cdot & \\ & & & & & & \cdot \end{pmatrix}$$

defines an operator; if

$$Q = \begin{pmatrix} 0 & 0 & 0 & 0 & & & \\ 1 & 0 & 0 & 0 & & & \\ 0 & 1 & 0 & 0 & & & \\ 0 & 0 & 1 & 0 & & & \\ & & & & \cdot & & \\ & & & & & \cdot & \\ & & & & & & \cdot \end{pmatrix}$$

then it can be painlessly verified that

$$PQ - QP = \begin{pmatrix} A & 0 & 0 & 0 & & & \\ 0 & 0 & 0 & 0 & & & \\ 0 & 0 & 0 & 0 & & & \\ 0 & 0 & 0 & 0 & & & \\ & & & & \cdot & & \\ & & & & & \cdot & \\ & & & & & & \cdot \end{pmatrix}$$

Since the direct sum of infinitely many copies of $\mathcal{H}$ is the direct sum of $\mathcal{H}$, the first copy, and infinitely many other copies, and since the latter direct sum itself is isomorphic (unitarily equivalent) to $\mathcal{H}$, it follows that every two-by-two operator matrix of the form $\begin{pmatrix} A & 0 \\ 0 & 0 \end{pmatrix}$ is a commutator [14, 15]. This assertion is the principal affirmative result in commutator theory; most other results that assert the existence of commutators of various sorts (on an infinite-dimensional space) either follow from the statement or imitate the proof of this one. The most striking general result is that every operator is the sum of two commutators. Two special matrices that can be proved to be commutators are $\begin{pmatrix} 1 & 0 \\ 1 & 0 \end{pmatrix}$ and $\begin{pmatrix} 1 & 1 \\ 1 & 1 \end{pmatrix}$. (All the entries in the matrices mentioned in this context are to be regarded as operators on an infinite-dimensional Hilbert space.) A few other special examples could be mentioned, but they constitute a minority; for most operators it is very difficult to decide whether or not they are commutators.

*Problem 2. Is the operator matrix* $\begin{pmatrix} 1 & 0 \\ 1 & 1 \end{pmatrix}$ *a commutator?*

A question of interest is the topological character of the set of commutators. Thus, for instance, it is not completely satisfying just to know that a commutator $C$ cannot be equal to 1; the question still remains, how near can $C$ come to 1? Easy arguments (based on traces again) show that in the finite-dimensional case $\|1 - C\| \geqq 1$. In 1953 I laboriously proved that the infinite-dimensional case is different; in that case there exists a commutator $C$ with $\|1 - C\|^2 \leqq 0.97$. I keep returning to the subject, but nothing happens. A painstaking re-examination of the same 1953 example has finally shown that, in fact, for that example, $\|1 - C\|^2 = \frac{8}{9} = 0.888 \ldots$, but the question of how small $\|1 - C\|$ can be made by other $C$'s is open.

*Problem 3. Is the identity operator on an infinite-dimensional Hilbert space the limit (in the norm) of commutators?*

It should be mentioned that in other, more pathological, senses of limit the answer is known and pathological: the set of commutators on an infinite-dimensional Hilbert space is dense in the strong operator topology.

What can be said about the spectrum of a commutator? The answer is: not much. It is known that every compact subset of the complex plane that contains 0 is the spectrum of some commutator; can the phrase "that contains 0" be dropped? The answer is not known, even in extreme special cases.

*Problem 4. Does there exist a commutator whose spectrum consists of the number 1 alone?*

Let me return briefly to the negative questions, that is, to the questions that do not ask for an example of a commutator with prescribed properties, but ask for a proof that certain prescribed operators cannot be commutators. There is, for instance, a certain amount of algebraic interest in commutators $C = PQ - QP$ for which $C$ commutes with $P$. Generalizing a result of Jacobson's, Kleinecke [25] proved that if $C$ is such a commutator, then $C$ is quasinilpotent (that is, $\|C^n\|^{1/n} \to 0$ as $n \to \infty$). This includes, in particular, the Wintner-Wielandt negative result; since 1 commutes with every $P$ and since 1 is not quasinilpotent, 1 cannot be commutator.

The identity is a projection; it is the unique projection whose null-space has dimension 0. What about a projection (on an infinite-dimensional Hilbert space) whose null-space has dimension 1; can it be a commutator? Intuition cries out for a negative answer, and, for once, intuition is right. The argument goes as follows. Consider the Banach algebra of all operators and in it the ideal of compact (completely continuous) operators. The quotient algebra is a Banach algebra [8, 38]. In that algebra the identity element is not a commutator (by Wintner and Wielandt); this means that the identity operator is not equal to a commutator plus a completely continuous operator. Since a projection whose null-space has dimension 1 is a very special example of such a sum, the proof is complete. The proof proves much more, of course, but, even so, it fails to say anything at all about most operators. Thus, for instance, the following absurd question remains unsettled.

*Problem 5. Is every operator that is not a scalar modulo the completely continuous operators a commutator?*

## NORMAL DILATIONS AND EXTENSIONS

The theory of dilations was mentioned above, and it deserves to be mentioned again. There are some beautiful results about normal dilations and extensions of operators, and I am sure that many more are waiting to be discovered. To be sure there do not seem to be any conspicuous and challenging yes-or-no questions that serve to indicate the direction in which the search for new results might begin, but I have faith. There is depth in the subject; the trouble is that the surface has not been explored enough to show where the deepest parts lie.

The theorem that every contraction has a unitary dilation is a special case of a much more useful result. If $||A|| \leqq 1$, then not only does there exist a unitary operator matrix $U$ that has $A$ in a specified diagonal position, but $U$ can even be constructed so that each positive integral power of $U$ has in that same diagonal position the corresponding power of $A$. A unitary dilation of this kind is called a *strong* unitary dilation; its existence was first proved by Nagy [28]. A remarkable proof of Nagy's theorem was given by Schäffer [37]. The proof is maximally constructive and almost wordless; it proves the existence of $U$ by writing it down as an infinite operator matrix, as follows:

$$U = \begin{pmatrix} \cdot & & & & & & & & \cdot \\ & \cdot & & & & & & \cdot & \\ & & \cdot & & & & \cdot & & \\ & & 0 & 0 & 0 & 0 & 0 & 0 & 0 \\ & & 1 & 0 & 0 & 0 & 0 & 0 & 0 \\ & & 0 & 1 & 0 & 0 & 0 & 0 & 0 \\ & & 0 & 0 & -S & (A) & 0 & 0 & 0 \\ & & 0 & 0 & A^* & T & 0 & 0 & 0 \\ & & 0 & 0 & 0 & 0 & 1 & 0 & 0 \\ & & 0 & 0 & 0 & 0 & 0 & 1 & 0 \\ & & \cdot & & & & \cdot & & \\ & \cdot & & & & & & \cdot & \\ \cdot & & & & & & & & \cdot \end{pmatrix}.$$

The operators $S$ and $T$ are the positive square roots of $1 - AA^*$ and $1 - A^*A$ respectively. The proofs (that $U$ is unitary and that the powers of $U$ have the corresponding powers of $A$ in the parenthetically indicated position) are simple matrix calculations.

A normal dilation $B$ of an operator $A$ on a space $\mathcal{H}$ to a larger space $\mathcal{K}$ is called *minimal* if the smallest subspace of $\mathcal{K}$ that includes $\mathcal{H}$ and reduces $B$ is $\mathcal{K}$ itself. It is a part of Nagy's formulation of his theorem that a minimal strong unitary dilation of $A$ is uniquely determined by $A$ to within unitary equivalence. For ordinary (not necessarily strong) normal dilations this is not true; the operator 0, for instance, on a one-dimensional space, has each of the matrices

$$\begin{pmatrix} 0 & 1 \\ 1 & 0 \end{pmatrix}, \quad \begin{pmatrix} 0 & 0 & 1 \\ 1 & 0 & 0 \\ 0 & 1 & 0 \end{pmatrix}, \quad \begin{pmatrix} 0 & 0 & 0 & 1 \\ 1 & 0 & 0 & 0 \\ 0 & 1 & 0 & 0 \\ 0 & 0 & 1 & 0 \end{pmatrix}, \quad \cdots$$

as a minimal unitary dilation. The set of all minimal normal (or unitary) dilations of a fixed operator might have a structure that gives valuable information about the operator; the study of this subject has not even begun. The minimal strong unitary dilation of an operator $A$ is not, by itself, a sufficiently delicate instrument for the study of $A$. It turns out for instance that every *proper* contraction has essentially the same minimal strong unitary dilation; the only free parameter is the dimension of the underlying Hilbert space. This remarkable result is due to Schreiber [39] (see also Nagy [29]). The precise statement is as follows: if $A$ and $B$ are operators on Hilbert spaces with the same separability character ($= \aleph_0$ times the dimension), such that $\|A\| < 1$ and $\|B\| < 1$, then their minimal strong unitary dilations are unitarily equivalent.

Some operators can be not only dilated but, in fact, extended so as to become normal. An operator $A$ on $\mathcal{H}$ is *subnormal* [12] if there exists a normal operator $B$ on a larger Hilbert space $\mathcal{K}$ such that $\mathcal{H}$ is invariant under $B$ and such that the restriction of $B$ to $\mathcal{H}$ coincides with $A$; in other words, $A$ is subnormal if it has a normal extension. Equivalently, $A$ is subnormal if it has a normal dilation of the form $\begin{pmatrix} A & X \\ Y & Z \end{pmatrix}$ with $Y = 0$.

A little of the basic theory of subnormal operators is already known; much of it is still mysterious. The first problem that was solved was that of finding an intrinsic characterization of subnormality; soon thereafter came some information about the spectrum of a subnormal operator and its relation to the spectrum of its (unique) minimal normal extension [4, 13]. The spectral result is that if $B$ is the minimal normal extension of $A$, then the spectrum of $B$ is included in the spectrum of $A$; in fact, the spectrum of $A$ is obtained from that of $B$ by filling in some of the holes. (*A* "hole" of a compact set is a bounded component of its complement.) Some topological results are known; perhaps the pleasantest among them is the theorem that the strong closure of the set of all normal operators is the set of all subnormal operators [2, 40].

An apparently deep and still unanswered question concerns the invariant subspaces of a subnormal operator. The spectral theorem shows that normal operators have many invariant subspaces; what about subnormal operators? The question is important because whenever a satisfactory geometric structure theory for operators does exist, it is always based on the existence of an ample supply of small subspaces invariant under those operators. The question is a stepping stone on the way to the more general question for arbitrary operators. For the record, here it is.

*Problem 6. Does every subnormal operator on a Hilbert space* $\mathcal{H}$ *leave invariant at least one (closed) subspace (other than* $\{0\}$ *and* $\mathcal{H}$*)?*

SHIFTS

There is a particular subnormal operator, the so-called unilateral shift, that plays a fundamental role in many studies of infinite-dimensional Hilbert spaces. (It enters naturally into studies of numerical ranges, commutators, and normal dilations and extensions; it took a valiant effort to refrain from mentioning it till now.) The simplest definition of the unilateral shift, the definition that explains its name, is this: let $\mathcal{H}$ be a Hilbert space with an orthonormal basis $\{e_n: n = 0, 1, 2, \cdots\}$, and let $U$ be the unique bounded operator such that $Ue_n = e_{n+1}$, $n = 0, 1, 2, \cdots$. The unilateral shift is an isometric mapping of the whole space $\mathcal{H}$ onto a proper subspace; this sort of behavior can only happen in, and is characteristic of, infinite-dimensional spaces. To see that $U$ is subnormal, let $\mathcal{K}$ be a Hilbert space with a bilaterally infinite orthonormal basis $\{e_n: n = 0, \pm 1, \pm 2, \cdots\}$, and let $V$ be the unique bounded operator such that $Ve_n = e_{n+1}$, $n = 0, \pm 1, \pm 2, \cdots$. The operator $V$ is called the *bilateral shift.* It is an isometric mapping of the whole space $\mathcal{K}$ onto itself; it is, consequently, unitary, and therefore normal. Clearly $\mathcal{H}$ is invariant under $V$ and the restriction of $V$ to $\mathcal{H}$ is $U$; this proves that $U$ is subnormal.

The bilateral shift (and hence also the unilateral one) has a natural functional representation. If $\mathcal{K} = \mathcal{L}^2[0, 1]$, and if $e_n(t) =$

$e^{2\pi int}(n = 0, \pm 1, \pm 2, \cdots)$, then the $e_n$'s constitute an orthonormal basis for $\mathcal{K}$. The multiplication operator $f(t) \to e^{2\pi it}f(t)$ shifts each basis vector onto the next one; it is unitarily equivalent to (I might as well say that it is the same as) the bilateral shift. The span of the $e_n$'s with $n \geqq 0$ is the domain of the functional representation of the unilateral shift. That span is known as the Hardy space of index 2, $\mathcal{H}^2$ for short; in a slightly different notation it plays an important role in several parts of analysis.

The answers to all questions about the bilateral shift $V$ are either trivial consequences of the spectral theorem or else reduce to delicate analytic problems about the unilateral shift $U$ on $\mathcal{H}^2$. The first published systematic study of $U$ is due to Beurling [2]; for subsequent simplification and generalizations see [16, 22, 26, 27]. The most useful part of all this work is a complete determination of all subspaces invariant under $U$ (and its generalization to direct sums of copies of $U$). The point is not to prove that $U$ does have invariant subspaces (that is trivial), but to exhibit the structure of the set of all invariant subspaces in some effective manner. Ever since Beurling showed the way, work of this kind has been going on. The hope is that if we know all about all invariant subspaces of many operators, we might get an insight into either the construction of an operator without any or the proof that all operators have one.

What can a subspace invariant under $U$ look like? The easiest way to obtain one is to fix a positive integer $k$ and consider the span $\mathfrak{M}_k$ of the $e_n$'s with $n \geqq k$. After this elementary observation most students of the subject must stop and think; it is not at all obvious that any other invariant subspaces exist. A careful study of the spectrum of $U$ ultimately reveals some more. Indeed, it turns out that each complex number $\lambda$ with $|\lambda| < 1$ is a simple proper value of the adjoint operator $U^*$ (with proper vector $x_\lambda = e_0 + \lambda e_1 + \lambda^2 e_2 + \cdots$), and it follows that the orthogonal complement of $\{x_\lambda\}$ is a non-trivial subspace invariant under $U$. More generally, if $\mathfrak{M}_k(\lambda)$ is the orthogonal complement of $\{x_\lambda, \cdots, U^{k-1}x_\lambda\}$, then $\mathfrak{M}_k(\lambda)$ is invariant under $U$; the spaces $\mathfrak{M}_k$ considered first are the same as the spaces $\mathfrak{M}_k(0)$. The lattice operations (intersection and span) applied to the $\mathfrak{M}_k(\lambda)$'s yield some not particularly startling new examples; after that the well seems to run dry again.

The final result, Beurling's theorem characterizing all subspaces invariant under $U$, is simple to state, and, although Beurling's own proof was quite involved, it is by now simple to prove; it depends on hardly anything more than the geometry of Hilbert space. The profitable point of view is not sequential but functional. If $\phi$ is a bounded function in $\mathcal{H}^2$ (better: if $\phi \in \mathcal{H}^\infty$), then, for every $f$ in $\mathcal{H}^2$, the product $\phi \cdot f$ belongs to $\mathcal{H}^2$, and the mapping $M_\phi$ defined by $M_\phi f = \phi \cdot f$ is a (bounded) operator on $\mathcal{H}^2$. If $\phi$ is *rigid* (this means that $|\phi(t)| = 1$ almost everywhere), then $M_\phi$ is an isometry, and, consequently, its range (which may be denoted by $\phi \cdot \mathcal{H}^2$) is a (closed) subspace of $\mathcal{H}^2$. The unilateral shift $U$ itself is of the form $M_\phi$ (with $\phi = e_1$); since any two multiplication operators commute, it follows that $\phi \cdot \mathcal{H}^2$ is invariant under $U$ for every rigid $\phi$. Beurling's theorem asserts the converse: every non-zero subspace invariant under $U$ is of the form $\phi \cdot \mathcal{H}^2$ for some rigid $\phi$. The correspondence between invariant subspaces and rigid functions is essentially one-to-one; the subspace determines the function to within a scalar factor of modulus 1. A small puzzle remains: what is the representation of the original invariant subspaces (the elements of the lattice generated by the $\mathfrak{M}_k(\lambda)$'s) in the Beurling form just mentioned? The solution of this puzzle leads to the easy part of some well known classical analysis, having to do with Blaschke products, and I am leaving it to you.

It is always a pleasure to see a piece of current mathematics reach half a century into the past to illuminate and simplify some of the work of the founding fathers; it is a pleasure to report that Beurling's theorem can do this too. The elements of $\mathcal{H}^2$ are related to certain analytic functions on the unit disc, and, although they themselves are defined on the unit interval only, and only almost everywhere at that, they tend to imitate the behavior of analytic functions. A crucial property of an analytic function is that it cannot vanish very often without vanishing everywhere. An important theorem of F. and M. Riesz [32] asserts that the elements of $\mathcal{H}^2$ exhibit the same kind of behavior: if $f \in \mathcal{H}^2$ and if the set $\{t: f(t) = 0\}$ has positive Lebesgue measure, then $f(t) = 0$ almost everywhere. Here is how the proof goes via Beurling's theorem. Let $\mathfrak{M}$ be the smallest subspace of $\mathcal{H}^2$ that contains $f$ and is invariant under $U$; equivalently, $\mathfrak{M}$ is the closure in $\mathcal{H}^2$ of the functions of the form $p \cdot f$, where $p$ ranges over all polynomials in $e_1$. Assume that $f$ is not the zero element of $\mathcal{H}^2$; in that case Beurling's theorem asserts the existence of a rigid function $\phi$ such that $\mathfrak{M} = \phi \cdot \mathcal{H}^2$. This implies in particular (since $\mathcal{H}^2$ contains the constant functions) that $\phi$ is a limit of products such as $p \cdot f$. Conclusion: $\phi$ vanishes whenever $f$ does, and, since $|\phi(t)| = 1$ almost everywhere, it follows that $f(t) \neq 0$ almost everywhere.

The F. and M. Riesz theorem has a generalization to analytic measures; the generalization is due to F. and M. Riesz also. An *analytic measure* is a finite complex measure $\mu$ on the Borel sets of the unit interval such that $\int_0^1 e_n(t)\, d\mu(t) = 0$ for $n = 1, 2, 3, \cdots$. The assertion of the generalized F. and M. Riesz theorem is that every such measure is absolutely continuous with respect to Lebesgue measure. This generalization has a simple proof too, based on some known geometric facts about the invariant subspaces of unitary operators; the proof will be a part of the doctor's thesis of D. E. Sarason [36].

## TOEPLITZ OPERATORS

In the functional representation the unilateral shift is a multiplication operator (or rather the restriction of a multiplication operator to $\mathcal{H}^2$); the multiplier is the basis vector $e_1$. A mild generalization of the study of the shift is the study of polynomials in the shift. These operators are multiplications also; the multiplier is a polynomial in $e_1$. It might seem that generalizations in this direction can be pushed as far as multiplications by bounded functions in $\mathcal{H}^2$ (better: functions in $\mathcal{H}^\infty$) and no further; any other kind of multiplier might multiply $\mathcal{H}^2$ out of itself. In fact, a moderately successful generalization exists for a much wider class of functions; this is what the theory of Toeplitz operators is about. Toeplitz operators act on $\mathcal{H}^2$; the most general one is obtained as follows. Let $\phi$ be an arbitrary bounded measurable function on the unit interval; the corresponding Toeplitz operator $T_\phi$ maps each element $f$ of $\mathcal{H}^2$ onto $P(\phi \cdot f)$, where $\phi \cdot f$ is the ordinary product of $\phi$ and $f$, and $P$ is the projection from $\mathcal{L}^2$ onto $\mathcal{H}^2$. If $\phi$ is in $\mathcal{H}^\infty$ (the case first considered), the projection is unnecessary; in this case $T_\phi$ is subnormal (with normal extension $M_\phi$) and I shall say that $T_\phi$ is *analytic.* If the complex conjugate of $\phi$ is in $\mathcal{H}^\infty$, I shall say that the Toeplitz operator $T_\phi$ is *co-analytic.* For an arbitrary $\phi$, the Toeplitz operator $T_\phi$ (on $\mathcal{H}^2$) has the multiplication operator $M_\phi$ (on $\mathcal{L}^2$) for a dilation.

The beginning, the algebraic part, of the theory of Toeplitz operators is relatively easy and quite a few things are known. The correspondence $\phi \to T_\phi$ is one-to-one, linear, and commutes with adjunction. (The latter assertion means that if $\psi = \bar{\phi}$, then $T_\psi = T_\phi{}^*$. This implies, for instance, that there are many Hermitian Toeplitz operators; they are just the ones that correspond to real multipliers.) Hartman and Wintner [19] proved that the spectrum of $M_\phi$ is included in the approximate point spectrum of $T_\phi$. From this it follows easily that $\|T_\phi\| = \|\phi\|_\infty$; in other words, the correspondence $\phi \to T_\phi$ is even norm-preserving. The chief defect of that correspondence (and the source of the depth of the theory of Toeplitz operators) is that it is not multiplicative: if $\phi = \chi \cdot \psi$, it does not follow that $T_\phi = T_\chi T_\psi$.

The space $\mathcal{H}^2$ has the natural orthonormal basis $\{e_0, e_1, e_2, \cdots\}$; what does the matrix (with respect to that basis) of a Toeplitz operator look like? The answer is easy (see Hartman and Wintner [18]); the matrix of $T_\phi$ is

$$\begin{pmatrix} \alpha_0 & \alpha_{-1} & \alpha_{-2} & & & & \\ \alpha_1 & \alpha_0 & \alpha_{-1} & \alpha_{-2} & & & \\ \alpha_2 & \alpha_1 & \alpha_0 & \alpha_{-1} & \alpha_{-2} & & \\ & \alpha_2 & \alpha_1 & \alpha_0 & \cdot & \cdot & \\ & & \alpha_2 & \cdot & \cdot & \cdot & \cdot \\ & & & \cdot & \cdot & \cdot & \cdot & \cdot \\ & & & & \cdot & \cdot & \cdot \end{pmatrix},$$

where the $\alpha_n$'s are just the Fourier coefficients of the function $\phi$ (that is, $\phi = \Sigma_n \alpha_n e_n$). A moderately careful examination [7] of such Toeplitz matrices yields some interesting information about the multiplicative behavior of Toeplitz operators. Here are some samples. An operator $A$ on $\mathcal{H}^2$ is a Toeplitz operator (that is, $A = T_\phi$ for some $\phi$) if and only if $U^*AU = A$ (where $U$ is, as before, the unilateral shift); an operator $A$ is an analytic Toeplitz operator if and only if $AU = UA$. The product of two Toeplitz operators is a Toeplitz operator if and only if either the left factor is co-analytic or the right factor is analytic. An invertible Toeplitz operator has a Toeplitz inverse if and only if it is either analytic or co-analytic. Conclusion: satisfactory multiplicative behavior can be expected only if something in sight is analytic.

The first major difficulty in the study of Toeplitz operators (far from completely overcome yet) is the determination of their spectra. Widom [43] has a necessary and sufficient condition on $\phi$ for the invertibility of $T_\phi$. The condition is the existence of a factorization $\phi = \chi \cdot \psi$, where $T_\chi$ is co-analytic, $T_\psi$ is analytic, and $\chi$ and $\psi$ are subject to some other rather severe restrictions. The trouble is that a factorization condition does not behave well under translation; the passage from $T_\phi$ to $T_\phi - \lambda$ $(= T_{\phi-\lambda})$ is not an easy one. Widom has succeeded nevertheless in applying his condition to several interesting special cases, and his result is sufficiently powerful to recapture the earlier results of Hartman and Wintner for analytic and for Hermitian Toeplitz operators.

If $\phi \in \mathcal{H}^2$, then $\phi$ has a Fourier expansion of the form $\Sigma_{n=0}^{\infty} \alpha_n e_n$. Since $\Sigma_{n=0}^{\infty} |\alpha_n|^2 < \infty$, the power series $\Sigma_{n=0}^{\infty} \alpha_n z^n$ converges in the open unit disc and defines there an analytic function $\tilde{\phi}$. The result of Wintner [45] on the spectrum of $T_\phi$ is easy to express in terms of $\tilde{\phi}$: the spectrum of $T_\phi$ is the closure of the range of $\tilde{\phi}$. If $\phi$ is a real-valued bounded measurable function on the unit interval (so that $T_\phi$ is a Hermitian Toeplitz operator), let $\alpha$ and $\beta$ be the essential infimum and supremum of $\phi$ respectively; Hartman and Wintner [19] have shown that the spectrum of $T_\phi$ is the closed interval $[\alpha, \beta]$. In the real case one more thing is known, namely that $T_\phi$ can have no proper values (except when $\phi$ is a constant, that is, when $T_\phi$ is a scalar).

If $\phi$ is an arbitrary complex-valued bounded measurable function, most questions about the spectrum of $T_\phi$ are still unanswered. It is known [7] that a non-scalar $T_\phi$ can never have any isolated proper values, but it is not known whether or not its spectrum can be finite. To illustrate the extent of our ignorance, I record here a test question; the answer to it does not promise to be of great intrinsic interest, but the method yielding the answer may be.

*Problem 7. Is the spectrum of every Toeplitz operator connected?*†

There are many unsolved problems about Toeplitz operators. The task of determining their invariant subspaces has just begun (see Duren [11]); in the general case not even the existence of non-trivial invariant subspaces is known. In the Hermitian case the spectrum of $T_\phi$ is known (in terms of $\phi$), but the full spectral structure (spectral measure, multiplicity) is not. Some special cases were known to Hellinger [21], and current studies are being made by

† Note added May 2, 1963. This question has recently been answered affirmatively by H. Widom in a forthcoming paper entitled "On the Spectrum of a Toeplitz Operator."

Putnam [31] and Rosenblum [33]. A usable complete set of invariants for the unitary equivalence of two Toeplitz operators is still far in the future. To conclude this part of the discussion, I mention another test question.

*Problem 8. Are two similar Toeplitz operators necessarily unitarily equivalent?*†

## INVARIANT SUBSPACES

The problem of the existence of invariant subspaces was raised above for subnormal operators (where there is some hope of an affirmative solution), but it might just as well be raised for any other class of operators; only in exceptional cases is a solution known. Thus, for instance, the spectral theorem implies that every normal operator has a large supply of invariant subspaces; this is classical and can be considered elementary by now. For completely continuous operators the answer is much more recent; it is the work of Aronszajn and Smith [1]. To show how delicate the techniques are, I raise explicitly one more problem about invariant subspaces; it is due to K. T. Smith.

*Problem 9. Does every operator with a completely continuous square on a Hilbert space $\mathcal{H}$ leave invariant at least one (closed) subspace (other than $\{0\}$ and $\mathcal{H}$)?*

Some of the current work on invariant subspaces asks not about their existence but about their structure. More precisely, in a typical work of this kind, the problem is to take some particular operator (or a reasonably small class of operators) and, granted that it has invariant subspaces, to study the lattice of all its invariant subspaces. Beurling's work on the unilateral shift is of this kind. A related general problem is this: given an abstractly described lattice, construct an operator (on a prescribed, preferably infinite-dimensional, Hilbert space) whose lattice of invariant subspaces is isomorphic to it. If the lattice has exactly two elements, 0 and 1, then this is just the problem of invariant subspaces. The three

† Note added May 2, 1963. This question was recently answered negatively by D. E. Sarason; his counterexample is based on Duren's work on tri-diagonal operators.

most valuable examples known are the unilateral shift, a certain weighted shift, and a Volterra integral operator.

The weighted shift was described by Donoghue [10]. In the notation used for the unilateral shift, let $A$ be the operator such that $Ae_n = \frac{1}{2^n} e_{n+1}$, $n = 0, 1, 2, \cdots$. If $\mathfrak{M}_k$ is, as before, the span of the $e_n$'s with $n \geqq k$, then $\mathfrak{M}_k$ is invariant under $A$, $k = 0, 1, 2, \cdots$, and so also is $\mathfrak{M}_\infty$ $(= \{0\})$; Donoghue's result is that every subspace invariant under $A$ is one of these $\mathfrak{M}_k$'s. The lattice of invariant subspaces is anti-isomorphic to the ordinal numbers less than or equal to $\omega$: "anti-" because the $\mathfrak{M}_k$'s decrease as $k$ increases.

The Volterra integral operator $V$ on $\mathfrak{L}^2[0, 1]$, defined by $(Vf)(t) = \int_0^t f(s)\, ds$, has been the subject of quite a bit of attention recently, both in the U.S. and in the U.S.S.R. [5, 6, 10, 23, 35]. Of the many facts known about it, I mention only the structure of its lattice of invariant subspaces. For each number $\alpha$, $0 \leqq \alpha \leqq 1$, let $\mathfrak{M}_\alpha$ be the set of all those functions in $\mathfrak{L}^2[0, 1]$ that vanish almost everywhere on the interval $[0, \alpha]$. The subspace $\mathfrak{M}_\alpha$ is invariant under $V$; the principal assertion is that every subspace invariant under $V$ is one of these $\mathfrak{M}_\alpha$'s. The lattice of invariant subspaces is isomorphic to the closed unit interval.

Weighted shifts are worth more attention. In the notation used for the unilateral shift once more, let $\{\alpha_n\}$ be a bounded sequence of positive numbers and let $A$ be the operator such that $Ae_n = \alpha_n e_{n+1}$. If $\alpha_n = 1$ for all $n$, the lattice of $A$ is the one described by Beurling; if $\alpha_n = \frac{1}{2^n}$, the lattice is the one obtained by Donoghue. Some current experimental work by my colleague A. L. Shields indicates that other possibilities can be realized, but it is not yet known what they are or how many different ones there can be.

## UNITARY EQUIVALENCE AND SIMILARITY

From a certain point of view the main problem of group theory is to decide when two groups are isomorphic, and the main problem of topology is to decide when two spaces are homeomorphic. From this point of view the main problem of operator theory is to decide when two operators are unitarily equivalent. Such problems are usually too broad (and too vague) ever to arrive at a satisfactory solution; their use is to indicate an important direction that no theory should ever lose sight of.

Special cases of the problem of unitary equivalence can sometimes turn out to be rewarding; it is usually hard to tell in advance whether they will or not. The unitary equivalence problem for weighted shifts looks relatively easy; the unitary equivalence problem for quasinilpotent operators is recalcitrant. You might be amused to hear about the recently observed spectacular failure of an attack on a special unitary equivalence problem; the attack was made jointly by A. Brown, J. E. McLaughlin, and myself.

The class of operators in question was the class of partial isometries. An operator $A$ on a Hilbert space $\mathcal{H}$ is a *partial isometry* if there exists a subspace $\mathfrak{M}$ of $\mathcal{H}$ such that the restriction of $A$ to $\mathfrak{M}$ is an isometry and the restriction of $A$ to $\mathfrak{M}^\perp$ is 0. An isometry is a partial isometry, and so, in particular, is a unitary operator; every projection is a partial isometry. In all these special cases the problem of unitary equivalence is solved; from this point of view the problem for partial isometries in general looks promising. Despite this plausibility argument, it turns out that the problem is hopeless; the reason is that any solution of it would immediately yield a solution of the problem of unitary equivalence for completely arbitrary operators.

The reduction goes as follows. The most general problem of unitary equivalence is equivalent to the problem for invertible contractions; this step involves nothing more than translation by a scalar and change of scale. If $S$ is an invertible contraction, let $T$ be the positive square root of $1 - S^*S$, and let $M$ be the operator matrix $\begin{pmatrix} S & 0 \\ T & 0 \end{pmatrix}$. It is quite easy to verify that $M$ is a partial isometry; for this step the invertibility of $S$ is not needed. (Stop and think about this a moment; it is quite shocking really. One consequence is that if a compact subset of the unit disc contains 0, as, for instance, the unit interval does, then it is the spectrum of some partial isometry.) It is slightly more complicated, but still quite elementary, to verify that two partial isometries $M$ obtained in this way are unitarily equivalent if and only if the $S$'s that induced them are unitarily equivalent. Conclusion: the unitary equivalence problem is no more solvable for $M$'s than it is for $S$'s.

Similarity is of less interest in Hilbert space than unitary equivalence, but it is still good to know when it holds. In order not to ignore it altogether, I shall at least mention a remarkable general theorem and an interesting special problem about similarity.

The general theorem is due to Rota [34]; it goes as follows. A *part* of an operator may be defined as the restriction of the operator to an invariant subspace. The operators whose parts are involved in Rota's theorem are (possibly infinite) direct sums of copies of $U^*$, where $U$ is the unilateral shift. For temporary convenience, call any such direct sum a *reverse shift*. The theorem says that every operator whose spectrum lies in the interior of the unit disc is similar to a part of a reverse shift. One consequence of the theorem is that if we knew all about the invariant subspaces of reverse shifts (a small class of operators), we could then solve the problem of invariant subspaces for all operators.

The special problem is due to Nagy; it has to do with the sequence of powers of an operator. Consider, to begin with, an operator on a one-dimensional Hilbert space. Such an operator is just multiplication by a scalar $\lambda$; its norm is $|\lambda|$. If the sequence of powers of the operator is bounded, that is, if the sequence $\{1, \lambda, \lambda^2, \cdots\}$ is bounded, then $|\lambda| \leqq 1$; the operator must be a contraction. This conclusion is not valid if the dimension of the space is greater than 1. Thus, for instance, if $T = \begin{pmatrix} 0 & 0 \\ 2 & 0 \end{pmatrix}$, then $T^2$ is 0, and so are all higher powers of $T$, but $T$ is not a contraction. All that can be said in this case is that $T$ is similar to a contraction (for example, to $\begin{pmatrix} 0 & 0 \\ \lambda & 0 \end{pmatrix}$ for any $\lambda$ with $0 < |\lambda| \leqq 1$). Nagy asks whether this conclusion is universally valid.

*Problem 10. Is every operator whose sequence of powers is bounded similar to a contraction?*†

EPILOGUE

That is all. I have told most of what I know and I have asked most of what I do not know; I hope that the process succeeded in giving you the glimpse that the title promised.

† Note added May 2, 1963. This question was recently answered negatively by S. R. Foguel; his counterexample will be published in the *Proceedings of the American Mathematical Society*.

## REFERENCES


1. N. Aronszajn and K. T. Smith, "Invariant subspaces of completely continuous operators," *Ann. Math.* **60** (1954) 345–350.
2. A. Beurling, "On two problems concerning linear transformations in Hilbert space," *Acta Math.* **81** (1949) 239–255.
3. Errett Bishop, "Spectral theory for operators on a Banach space," *Trans. Amer. Math. Soc.* **86** (1957) 414–445.
4. J. Bram, "Subnormal operators," *Duke Math. J.* **22** (1955) 75–93.
5. M. S. Brodskii, "On a problem of I. M. Gelfand," *Uspekhi Mat. Nauk* **12** (1957) 129–132.
6. M. S. Brodskii and M. S. Livshitz, "Spectral analysis of non-selfadjoint operators and intermediate systems," *Uspekhi Mat. Nauk* **13** (1958) 3–85; A.M.S. Transl. **13** (1960) 265–346.
7. A. Brown and P. R. Halmos, "Algebraic properties of Toeplitz operators," to appear.
8. J. W. Calkin, "Two-sided ideals and congruences in the ring of bounded operators in Hilbert space," *Ann. Math.* **42** (1941) 839–873.
9. W. F. Donoghue, "On the numerical range of a bounded operator," *Mich. Math. J.* **4** (1957) 261–263.
10. W. F. Donoghue, "The lattice of invariant subspaces of a completely continuous quasi-nilpotent transformation," *Pac. J. Math.* **7** (1957) 1031–1035.
11. P. L. Duren, "Extension of a result of Beurling on invariant subspaces," *Trans. Amer. Math. Soc.* **99** (1961) 320–324.
12. P. R. Halmos, "Normal dilations and extensions of operators," *Summa Brasil.* **2** (1950) 125–134.
13. P. R. Halmos, "Spectra and spectral manifolds," *Ann. Soc. Pol. Math.* **24** (1952) 43–49.
14. P. R. Halmos, "Commutators of operators," *Amer. J. Math.* **74** (1952) 237–240.
15. P. R. Halmos, "Commutators of operators, II," *Amer. J. Math.* **76** (1954) 191–198.
16. P. R. Halmos, "Shifts on Hilbert spaces," *J. Reine Angew Math.* **208** (1961) 102–112.
17. P. R. Halmos, "Numerical ranges and normal dilations," to appear.
18. P. Hartman and A. Wintner, "On the spectra of Toeplitz's matrices," *Amer. J. Math.* **73** (1950) 359–366.
19. P. Hartman and A. Wintner, "The spectra of Toeplitz's matrices," *Amer. J. Math.* **76** (1954) 867–882.
20. F. Hausdorff, "Der Wertvorrat einer Bilinearform," *Math. Zeit.* **3** (1919) 314–316.
21. E. D. Hellinger, "Spectra of quadratic forms in infinitely many variables," *Math. Monographs,* Evanston, Ill. (1941) Northwestern University, 133–172.
22. H. Helson and D. Lowdenslager, "Prediction theory and Fourier series in several variables," *Acta Math.* **99** (1958) 165–202.
23. G. K. Kalisch, "On similarity, reducing manifolds, and unitary equivalence of certain Volterra operators," *Ann. Math.* **66** (1957) 481–494.

24. V. L. Klee, "Convex bodies and periodic homeomorphisms in Hilbert space," *Trans. Amer. Math. Soc.* **74** (1953) 10–43.
25. D. C. Kleinecke, "On operator commutators," *Proc. Amer. Math. Soc.* **8** (1957) 535–536.
26. P. D. Lax, "Translation invariant spaces," *Acta. Math.* **101** (1959) 163–178.
27. P. Masani and N. Wiener, "The prediction theory of multivariate stochastic processes," *Acta Math.* **98** (1957) 111–150, and **99** (1958) 93–137.
28. B. Sz.-Nagy, "Prolongements des transformations de l'espace de Hilbert qui sortent de cet espace," Appendix to *Leçons d'analyse fonctionelle*, by F. Riesz and B. Sz.-Nagy, Budapest (1955) Akadémiai Kiadó.
29. B. Sz.-Nagy, "Sur les contractions de l'espace de Hilbert, II," *Acta Univ. Szeged.* **18** (1957) 1–14.
30. J. von Neumann and P. Jordan, "On inner products in linear, metric spaces," *Ann. Math.* **36** (1935) 719–723.
31. C. R. Putnam, "On Toeplitz matrices, absolute continuity, and unitary equivalence," *Pac. J. Math.* **9** (1959) 837–846.
32. F. Riesz and M. Riesz, *Ueber die Randwerte einer analytischen Funktion*, Quatrième Congrès de Mathématiciens Scandinaves, Stockholm (1916) 27–44.
33. M. Rosenblum, "The absolute continuity of Toeplitz's matrices," *Pac. J. Math.* **10** (1960) 987–996.
34. G. C. Rota, "On models for linear operators", *Comm. Pure Appl. Math.* **13** (1960) 469–472.
35. L. A. Sakhnovich. "On the reduction of Volterra operators to the simplest form and on inverse problems," *Izv. Akad. Nauk SSR. Ser. Mat.* **21** (1957) 235–262.
36. D. E. Sarason, *The theorem of F. and M. Riesz on the absolute continuity of analytic measures*, University of Michigan thesis (1963).
37. J. J. Schäffer, "On unitary dilations of contractions," *Proc. Amer. Math. Soc.* **6** (1955) 322.
38. R. Schatten, *Norm ideals of completely continuous operators*, Berlin (1960) Springer.
39. M. Schreiber, "Unitary dilations of operators," *Duke Math. J.* **23** (1956) 579–594.
40. J. G. Stampfli, *On operators related to normal operators*, University of Michigan thesis (1959).
41. M. H. Stone, *Linear transformations in Hilbert space*, New York (1932) American Mathematical Society.
42. O. Toeplitz, "Das algebraische Analogon zu einem Satze von Fejér," *Math. Zeit.* **2** (1918) 187–197.
43. H. Widom, "Inversion of Toeplitz Matrices, II," *Ill. J. Math.* **4** (1960) 88–99.
44. H. Wielandt, "Ueber die Unbeschränktheit der Operatoren der Quantenmechanik," *Math. Ann.* **121** (1949) 21.
45. A. Wintner, "Zur Theorie der beschränkten Bilinearformen," *Math. Zeit.* **30** (1929) 228–282.
46. A. Wintner, "The unboundedness of quantum-mechanical matrices," *Phys. Rev.* **71** (1947) 738–739.

Reprinted from the
AMERICAN MATHEMATICAL MONTHLY
Vol. 77, No. 5, pp. 457–464, May 1970

# FINITE-DIMENSIONAL HILBERT SPACES

P. R. HALMOS, Indiana University

**Prologue.** I used to like linear algebra because it gave me a motivation for the study of operators on Hilbert space and because it gave me insight into the algebraic skeleton of operator theory, which made that study easier. Now I like what I learned about Hilbert space because it keeps shedding light on more and more new aspects of linear algebra and because it succeeds in keeping that classical subject alive and exciting. The purpose of this report is to illustrate the latter point by describing three non-trivial parts of finite-dimensional linear algebra, the original impetus for which came from operator theory on infinite-dimensional Hilbert space. The subjects are (1) an algebraic characterization of pairs of subspaces of a finite-dimensional Hilbert space, (2) a geometric characterization of linear transformations in terms of rotations and projections (dilation theory), and (3) a statement of some fragmentary results and challenging open problems about lattices of invariant subspaces.

A finite-dimensional Hilbert space is, by definition, a finite-dimensional unitary space (complex inner product space). The only prerequisite for an intelligent reading of this paper is acquaintance with the language, notation, and principal facts of finite-dimensional unitary geometry; see, for instance, [6]. As for the insistence on complex numbers: the geometric language of Hilbert space is motivated by the real case, but the algebraic hurdles are most easily overcome if complex numbers are allowed. The customary way out (followed here) is to use complex coefficients, and, at the same time, continue to use real language; this does not seem to lead to serious or permanent confusion.

**Two subspaces.** What are all the different ways in which a subspace can be placed in a finite-dimensional Hilbert space? The question is vague, but it has a reasonably definite answer. If $H = C^n$ (where $C$ is the complex number field), then one way to get an $r$-dimensional subspace of $H$ ($0 \leq r \leq n$) is to form the set of all those vectors whose last $n-r$ coordinates vanish. More to the point is that to within isomorphism this is the only way; every $r$-dimensional subspace of $C^n$ can be obtained from this one by a suitable rotation.

What are all the different ways in which two subspaces, or three, or any number can be placed in a finite-dimensional Hilbert space? The difficulty of the answer seems to increase with the number. The preceding paragraph shows that there is no difficulty about putting *one* subspace $M$ into a space $H$; just put it down, anywhere, let $M^\perp$ fall where it may, and there is nothing left to ask. Before the position of *two* subspaces $M$ and $N$ in $H$ can be said to be known, many questions must be asked and answered. Is $M$ included in $N$? Is $M$ orthogonal to $N$? Does $M^\perp$ have a nontrivial intersection with $N^\perp$? If the relation between $M$ and $N$ is not describable in the simple terms of inclusion and orthogonality, does it make sense to ask for the "angle" between them? Such questions were first raised by Dixmier [3]; the point of view described below is somewhat different and more recent [11]. As for three subspaces, or more, the mind boggles. There is, in fact, reason to believe that the problem of three subspaces will be out of human reach for a long time to come. A comment of Chandler Davis [2] indicates that if we knew all about three subspaces, then we could learn more about unitary equivalence than, apparently, we are meant to know.

In the study of pairs of subspaces there are four thoroughly uninteresting cases, the ones in which both $M$ and $N$ are either 0 or $H$. In the most general case the entire space is the direct sum of five subspaces:

$$M \cap N,\ M \cap N^{\perp},\ M^{\perp} \cap N,\ M^{\perp} \cap N^{\perp},$$

and the rest. The parts of $M$ and $N$ in the first four are "thoroughly uninteresting". In "the rest", the orthogonal complement of the span of the first four, $M$ and $N$ are in *generic position*, in the sense that all four of the special intersections listed above are equal to 0.

The simplest example of two subspaces in generic position consists of two distinct non-orthogonal lines in a plane, and there is no loss of generality in taking one of them to be the first coordinate axis. To get a useful generalization, suppose that $T$ is a non-singular linear transformation on a finite-dimensional Hilbert space $K$, write $H = K \oplus K$, let $M$ be the "horizontal axis" consisting of all vectors of the form $\langle f, 0 \rangle$ in $H$, and let $N$ be the graph of $T$, i.e., the set of all vectors of the form $\langle f, Tf \rangle$ in $H$. The assertion that $M$ and $N$ are in generic position needs a little proof. The first step is to show that $M \cap N = 0$. Indeed, how can an $\langle f, 0 \rangle$ be equal to a $\langle g, Tg \rangle$? Answer: only if $Tg = 0$, whence $g = 0$ (because $T$ is non-singular), and therefore $f = 0$. For the rest of the proof it is necessary to know $M^{\perp}$ (trivial: all $\langle 0, f \rangle$) and $N^{\perp}$ (easy and standard computation: all $\langle -T^*f, f \rangle$). From this it is easy to deduce that $M^{\perp} \cap N^{\perp} = 0$: since $T^*$ is just as non-singular as $T$, the proof just given applies again. The equations $M \cap N^{\perp} = 0$ and $M^{\perp} \cap N = 0$ are trivial.

The basic result in the theory of two subspaces is that this way of constructing pairs of subspaces in generic position is, to within unitary equivalence, the only way. More precisely: *if $M$ and $N$ are subspaces in generic position in a finite-dimensional Hilbert space $H$, then there exists a finite-dimensional Hilbert space $K$, and there exists a non-singular linear transformation $T$ on $K$, such that the pair $\langle M, N \rangle$ is unitarily equivalent to the pair $\langle K \oplus 0$, graph $T \rangle$.*

What follows is an outline of the proof; with suitable analytic caution the proof is generalizable to the infinite-dimensional case. Let $P$ be the projection with range $M$. Assertion: the restriction of $P$ to $N$ is a non-singular linear transformation from $N$ onto $M$. Suppose, indeed, that $Pg = 0$ for some $g$ in $N$. It follows that $g \in M^{\perp} \cap N$, and hence (generic position) that $g = 0$; this proves that the kernel of $P$ in $N$ is 0. To prove that the image $PN$ is equal to $M$, suppose that $f \in M$ and $f \perp PN$. This means that if $g \in N$, then $0 = (f, Pg) = (Pf, g) = (f, g)$, so that $f \in M \cap N^{\perp}$. It follows (generic position) that $f = 0$; the proof of the assertion is complete.

The existence of a non-singular linear transformation from $N$ onto $M$ implies that $M$ and $N$ have the same dimension. (This could have been proved more quickly, but the slower approach is needed for the rest of the proof anyway.) Since $M$ and $M^\perp$ on the one hand and $N$ and $N^\perp$ on the other hand enter the hypotheses with perfect symmetry, it follows that all four of these subspaces have the same dimension. Since this applies to $M$ and $M^\perp$ in particular, there exists an isometric linear mapping from $M$ onto $M^\perp$; the idea from now on is to identify each element of $M^\perp$ with the element of $M$ that it corresponds to.

Now put $K = M$. To define $T$ at an element $f$ of $K$, recall first that $f = Pg$ for a uniquely determined vector $g$ in $N$, project $g$ into $M^\perp$, and let $Tf$ be the element of $M$ that is identified with the element of $M$ so obtained. (A simple 2-dimensional picture should make that long-winded sentence crystal clear.) The verification that $K$ and $T$ do what is expected of them is straightforward.

Given a line in the plane, distinct from both the horizontal and the vertical axes, rotate the plane through the negative of half the angle of inclination. The given line and the horizontal axis become, after the rotation, a line and its reflection through the (new) horizontal axis. This half-angle rotation can be generalized to yield a different and useful representation for a pair of subspaces in generic position; the result is that any such pair is unitarily equivalent to a pair of the form $\langle \text{graph } T_0, \text{graph } (-T_0)\rangle$, for a suitable linear transformation $T_0$. From this representation, in turn, it is easy to recapture Dixmier's main theorem on pairs of subspaces in generic position: the result is that a *single* Hermitian transformation, namely the sum of the two projections whose ranges are the given subspaces, constitutes a complete set of unitary invariants for the pair.

**Unitary dilations.** Suppose that $H$ is a subspace of a finite-dimensional Hilbert space $K$, and let $P$ be the projection from $K$ onto $H$. Each linear transformation $B$ on $K$ induces in a natural way a linear transformation $A$ on $H$ defined for each $f$ in $H$ by

$$Af = PBf.$$

Under these conditions the transformation $A$ is called the *compression* of $B$ to $H$, and $B$ is called a *dilation* of $A$ to $K$. This geometric definition of compression and dilation is to be contrasted with the customary concepts of restriction and extension: if it happens that $H$ is invariant under $B$, then it is not necessary to project $Bf$ back into $H$ (it is already there), and, in that case, $A$ is the restriction of $B$ to $H$, and $B$ is an extension of $A$ to $K$. Restriction-extension is a special case of compression-dilation, the special case in which the linear transformation on the larger space leaves the smaller space invariant.

Compressions and dilations can be usefully described in terms of matrices. If $K$ is decomposed into $H$ and $H^\perp$, and, correspondingly, transformations on $K$ are written in terms of matrices (whose entries are transformations on $H$, and on $H^\perp$, and between the two), then a necessary and sufficient condition that $B$ be a dilation of $A$ is that the matrix of $B$ have the form

$$\begin{pmatrix} A & X \\ Y & Z \end{pmatrix}.$$

The purpose of dilation theory is to get information about difficult transformations by finding their easy dilations. The program is spectacularly successful. Unitary transformations (rotations) are among the easiest to deal with, and it turns out that, except for an easily adjusted normalization, every transformation has a unitary dilation. Some normalization is clearly necessary: if $B$ is unitary, then $\|Bf\| = \|f\|$ for every vector $f$, and it follows that $\|Af\| \leqq \|f\|$ for every vector $f$; in other words, if $A$ has a unitary dilation, then $A$ must not increase the norm of any vector. (In the appropriate geometrical technical term, $A$ must be a *contraction*.) That much normalization is sufficient: *every contraction has a unitary dilation.*

As a heuristic guide to the proof, consider the very special case in which the given Hilbert space is 1-dimensional real Euclidean space and the dilation space $K$ is the plane. In that case the given contraction is a scalar $\alpha$ (with $|\alpha| \leqq 1$), and, in geometric terms, the assertion is that multiplication by $\alpha$ (on the line) can be achieved by a suitable rotation (in the plane), followed by projection (back to the line). A picture makes all this crystal clear again.

The proof in the general case can be obtained by first transcribing the synthetic proof just outlined to analytic form, and then imitating the analytic geometry with matrices in the place of numbers. The conceptual problems that the program encounters are familiar ones, and so are their solutions; see [5] for the details.

The least unitary looking contraction is 0, but, of course, even it has a unitary dilation; one such is

$$\begin{pmatrix} 0 & 1 \\ 1 & 0 \end{pmatrix}.$$

This dilation does not have many useful algebraic properties. It is not necessarily true, for instance, that the square of a dilation is a dilation of the square; indeed, the square of the dilation of 0 exhibited above is

$$\begin{pmatrix} 1 & 0 \\ 0 & 1 \end{pmatrix},$$

which is not a dilation of the square of 0. Is there a unitary dilation of 0 that is fair to squares? The answer is yes:

$$\begin{pmatrix} 0 & 0 & 1 \\ 1 & 0 & 0 \\ 0 & 1 & 0 \end{pmatrix}$$

is an example. The square of this dilation is

$$\begin{pmatrix} 0 & 1 & 0 \\ 0 & 0 & 1 \\ 1 & 0 & 0 \end{pmatrix},$$

which is a dilation of the square of 0. Unfortunately, however, this dilation is not perfect either; its cube is

$$\begin{pmatrix} 1 & 0 & 0 \\ 0 & 1 & 0 \\ 0 & 0 & 1 \end{pmatrix},$$

which is not a dilation of the cube of 0. That is (in self-explanatory language): the $3\times 3$ matrix is a 2-dilation of 0, but not a 3-dilation.

Question: does every contraction on an $n$-dimensional Hilbert space have a unitary $k$-dilation for every $k$? Answer: yes, on a Hilbert space of dimension $n(k+1)$; an elementary proof was given by Egerváry [4]. More is true: Nagy proved [15] that every contraction $A$ has a unitary dilation $B$ such that $B^k$ is a dilation of $A^k$ simultaneously for every positive integer $k$. (Nagy's paper came between [5] and [4].) The result is true for infinite-dimensional Hilbert spaces too, but its impressive generality has a price: even if the given space $H$ is finite-dimensional, the dilation space $K$ may have to be infinite-dimensional.

To show how dilation theory can be used, consider the finite-dimensional special case of a beautiful and powerful analytic theorem of von Neumann [18]. The assertion is that if $A$ is a contraction and if $q$ is a polynomial such that $|q(z)| \leqq 1$ whenever $|z| = 1$, then $q(A)$ is a contraction. The proof via dilation theory goes as follows: If the degree of $q$ is $k$, find a unitary $k$-dilation $B$ of $A$. It follows then that $q(B)$ is a dilation of $q(A)$, and hence it is sufficient to prove that $q(B)$ is a contraction. (In other words, dilation theory reduces the problem to the consideration of unitary transformations only.) But that is trivial: a unitary $B$ has a diagonal matrix, whose diagonal entries are complex numbers of modulus 1; since the corresponding matrix of $q(B)$ has diagonal entries whose moduli are not greater than 1, the desired conclusion becomes obvious.

**Reflexive lattices.** The set of all subspaces of a finite-dimensional Hilbert space $H$ is a lattice (with respect to the operations of intersection and span) with zero element 0 and unit element $H$. Certain of its sublattices, the ones called *reflexive*, are of interest in linear algebra. The definition of reflexivity requires of a lattice $\mathfrak{L}$ that a two-step process performed on $\mathfrak{L}$, which always yields a lattice of subspaces at least as large as $\mathfrak{L}$, should, in fact, yield exactly $\mathfrak{L}$, and nothing more. The two steps are these: (1) form all linear transformations that leave invariant each subspace of $\mathfrak{L}$, and then (2) form all subspaces of $H$ that are invariant under all those linear transformations.

Here is an example. Suppose that $H$ is 2-dimensional, and let $\mathcal{L}$ consist of 0, $H$, and two distinct lines (i.e., 1-dimensional subspaces of $H$). To say of a linear transformation that it leaves invariant each subspace in $\mathcal{L}$ is to say just that it has two prescribed eigenvectors, and hence that its matrix with respect to the basis they form is diagonal. Since the only subspaces simultaneously invariant under all such diagonal transformations are the ones in $\mathcal{L}$, the lattice $\mathcal{L}$ is reflexive indeed.

Here is a non-example. Suppose again that $H$ is 2-dimensional, and let $\mathcal{L}$ consist of 0, $H$, and three distinct lines. It is very hard for a linear transformation on $H$ to have three distinct eigenvectors; the only linear transformations that can do it are the scalar multiples of the identity. Such scalar multiples, on the other hand, leave invariant every subspace of $H$. The two-step process applied to this $\mathcal{L}$ drastically enlarges $\mathcal{L}$; instead of the 5-element lattice $\mathcal{L}$, the enlargement is the infinite lattice of all subspaces of $H$.

The non-example of the preceding paragraph fails to be reflexive the worst way anything can; the enlargement it effects is maximal. Another way of saying the same thing is that a linear transformation that leaves invariant every subspace of the lattice is necessarily a scalar. A lattice that is non-reflexive in this extreme way is called *transitive.*

A basic open problem about operator theory on infinite-dimensional Hilbert spaces is to characterize all reflexive lattices and all transitive (extremally non-reflexive) lattices. A little progress has been made, but not very much. The finite-dimensional specialization of what is known amounts to two statements: (1) every chain (totally ordered set) of subspaces is reflexive, and (2) every Boolean algebra of subspaces is reflexive. The infinite-dimensional case of (1) is Ringrose's generalization [**17**] of a result of Kadison and Singer [**14**]. For a statement of the appropriate infinite-dimensional formulation of (2) see [**10**]; the proof has not been published yet. In the finite-dimensional case both results become almost trivial.

As for transitive lattices, even less is known. The example above (the one with three lines) is in a certain sense degenerate. From the point of view of projective geometry, which is quite appropriate here, the space of that example has dimension 1, not 2, and the example does not help to answer the question whether higher-dimensional examples exist at all.

An interesting unpublished observation of J. E. McLaughlin shows that they do exist; one such, in $C^n$, consists of all those subspaces that are invariant under the formation of complex conjugates. More explicitly: call a subspace $M$ of $C^n$ *symmetric* in case $M$ contains, along with each of its vectors, the vector whose coordinates are obtained from the given one by complex conjugation. Assertion: the set of all symmetric subspaces is a transitive lattice. The proof requires a moment's thought, but there is nothing profound about it.

The example of the preceding paragraph yields many examples. Given a finite-dimensional Hilbert space, coordinatize it (i.e., establish an isomorphism between it and $C^n$); the lattice of symmetric subspaces with respect to that coordinatization is a transitive lattice. Question: are all transitive lattices obtainable in this way?

The answer is no for two reasons, both trivial, and the heart of the question is still unanswered. *First*, a topological distinction arises: are the lattices under discussion closed or not? (There is only one reasonable topology for the space of subspaces; the question makes unambiguous sense.) If a lattice is dense in a transitive lattice, then it itself is transitive; the question loses no vigor at all if attention is restricted to closed lattices only. As long as restrictions are in order, here is one spot where the complex field makes life more complicated, not less; the question retains all its interest if attention is restricted to real spaces only. *Second*, for spaces of even dimension $2n$ there is a construction that yields a transitive lattice whose non-trivial elements are all of dimension $n$. Such a lattice imitates an already observed misbehavior; it is isomorphic to a sublattice of the lattice of the projective line.

The result of the indicated specializations is the following question: is every closed transitive lattice of subspaces of an odd-dimensional Euclidean space (real inner product space) equal to the lattice of all subspaces? The answer does not seem to be known.

*Note added in proof*. K. J Harrison (Monash University) has recently discovered a new transitive lattice of 18 elements that shows that the answer to the question as it stands is no. A modification, however, restores the question: just add the hypothesis that the atoms of the lattice span the whole space. Harrison's discovery makes the problem of determining all transitive lattices even more challenging than before.

**Epilogue.** Three topics were discussed above to illustrate the thesis that operator theory on Hilbert space yields non-trivial questions and answers about finite matrices. Choosing the examples was not an easy task; many more are available than can be included in one lecture, or one paper, of reasonable length.

I could have chosen the theorem about "near" projections (if the projections onto two subspaces are near enough in norm, then the subspaces have the same dimension) [**8**, Problem 43]. I could have discussed the "power inequality" ($w(A^n) \leqq (w(A))^n$) for the numerical range [**8**, Problem 176], which started in the theory of partial differential equations and ended as a problem about matrices, a problem that refused to become trivial even in the $2\times 2$ case. The topological properties of sets of reducible and irreducible operators were of recent research interest [**9**, **16**], and so also were partial isometries [**7**, **12**]; in both these cases new and interesting facts about the finite-dimensional case emerged. The theory of matrices whose entries are matrices still has some life in it; thus, for instance, the "ultra-invariant" subspaces of binormal operators ($2\times 2$ matrices whose entries are commutative normal operators; see [**1**]) are still being studied (by R. G. Douglas and C. M. Pearcy), and there are small, amusing, and until recently unnoticed questions even about determinants. (Sample: if all four ways of forming the formal determinant of a $2\times 2$ matrix whose entries are matrices yield an invertible matrix, does the matrix itself have to be invertible? See [**13**].)

The subjects just given honorable mention, as well as the three actually discussed in detail, have been receiving serious research attention in the course of the last twenty years (and many still are), and they have all led to questions that could and should have been asked in the finite-dimensional case long before, but as a matter of historical fact they were not. The reason perhaps is that the powerful tools of finite-dimensional linear algebra are too good; they sometimes conceal the elegant and intricate structure that the difficulties of the infinite-dimensional theory bring out. Most of the problems of operator theory can be formulated in the finite-dimensional case, and there are two reasons why it is good to do so. The old reason is that the finite can suggest what should and should not be tried with the infinite; the new reason is the joy of seeing the infinite inspire and guide the finite and contribute to a new flowering of an old subject.

### References


1. A. Brown, The unitary equivalence of binormal operators, Amer. J. Math., 76 (1954) 414–434.
2. C. Davis, Generators of the ring of bounded operators, Proc. Amer. Math. Soc., 6 (1955) 970–972.
3. J. Dixmier, Position relative de deux variétés linéaires fermées dans un espace de Hilbert, Rev. Sci., 86 (1948) 387–399.
4. E. Egerváry, On the contractive linear transformations of $n$-dimensional vector space, Acta Sci. Math. (Szeged) 15 (1953) 178–182.
5. P. R. Halmos, Normal dilations and extensions of operators, Summa Brasil. Math., 2 (1950) 125–134.
6. ———, Finite-dimensional Vector Spaces, Van Nostrand, Princeton, 1958.
7. ——— and J. E. McLaughlin, Partial isometries, Pacific J. Math., 13 (1963) 585–596.
8. ———, A Hilbert Space Problem Book, Van Nostrand, Princeton, 1967.
9. ———, Irreducible operators, Michigan Math. J., 15 (1968) 215–223.
10. ———, Invariant subspaces, Abstract spaces and approximation, Proc. MRI Oberwolbach (1968).
11. ———, Two subspaces, Trans. Amer. Math. Soc., 144(1969) 381–389.
12. ——— and L. J. Wallen, Powers of partial isometries, J. Math. Mech., 19 (1970) 657–663.
13. ———, Advanced problem 5727, this MONTHLY, 77 (1970) 409.
14. R. V. Kadison and I. M. Singer, Triangular operator algebras, Amer. J. Math., 82 (1960) 227–259.
15. B. Sz.-Nagy, Sur les contractions de l'espace de Hilbert, Acta Sci. Math. (Szeged) 15 (1953) 87–92.
16. H. Radjavi and P. Rosenthal, The set of irreducible operators is dense, Proc. Amer. Math. Soc., 21 (1969) 256.
17. J. R. Ringrose, On some algebras of operators, Proc. London Math. Soc., 15 (1965) 61–83.
18. J. von Neumann, Eine Spektraltheorie für allgemeine Operatoren eines unitären Raumes. Math. Nachr., 4 (1951) 258–281.

# CHAPTER II

Reprinted from the
AMERICAN MATHEMATICAL MONTHLY
Vol. LI, No. 9, pp. 493–510, Nov. 1944

# THE FOUNDATIONS OF PROBABILITY

P. R. HALMOS, Syracuse University

**1. Introduction.** Probability is a branch of mathematics. It is not a branch of experimental science nor of armchair philosophy, it is neither physics nor logic. This is not to say that the experimenter and the philosopher should not discuss probability from their points of view. They should, and they do. The situation is analogous to that in geometry. No one denies that the physicist and the philosopher have made valuable contributions to our understanding of the space concept, nor, in spite of this, that geometry is a rigorous part of modern mathematics.

Like Euclidean geometry, and for that matter like most mathematical theories, probability has four aspects: axiomatization, development, coordinatization, and application. We proceed to explain our use of these words.

"Axiomatization" is clear. We all know that the study of geometry begins with a list of undefined terms and a list of postulates. It is important in this connection to remember two facts. First: the selection of the list of terms and postulates is not entirely arbitrary, but is derived only after a thorough examination of our intuitive notions of the subject. Second: the selection of terms and postulates is not uniquely determined. When several different axiomatizations of the same subject exist then only extra mathematical considerations, such as practical convenience or personal prejudice, can lead us to prefer one among the many. The greater part of this paper is devoted to a prepostulational examination of probability. The axiomatic system to which this examination leads is not the only possible approach to probability, but it is the approach which has been adopted by the majority of workers in this field.

By "development" we mean simply the main part of the theory, the definitions and theorems which chiefly occupy the professional mathematician. "Coordinatization" is a general process the most familiar instance of which is the proof of the equivalence of the synthetic and analytic aspects of Euclidean geometry. The isomorphism of a finite group to a group of permutations and the representation of an algebra by matrices are further examples of this process. Properly speaking coordinatization is just one of the theorems belonging to development, but a theorem of such fundamental implications that it effects basic changes in the appearance, methods, and results of the entire theory.

The hardest philosophical problem in geometry as well as in probability is the problem of "application." Do the theorems derived from the postulates reflect any light on the physical world which suggested them, and if so, how and why?

The purpose of this paper is exposition, exposition intended to convince the professional mathematician that probability is mathematics. To this end we shall discuss the four features just enumerated. The paper contains almost no proofs, very few precise definitions and theorems, and many heuristic derivations. Despite however the small number of rigorous statements, they form the foundation on which the remainder is built. For the convenience of the reader they are italicized. If these italicized statements are lifted from their context and read consecutively, they will furnish at least a partial answer to the question "what is probability?"

**2. Boolean algebra.** The principal undefined term in probability theory is "event." Intuitively speaking an event is one of the possible outcomes of some physical experiment.

To take a rather popular example consider the experiment of rolling an ordinary six-sided die and observing the number $v(=1, 2, 3, 4, 5, \text{ or } 6)$ showing on the top face of the die. "The number $v$ is even"—"it is less than 4"—"it is equal to 6"—each such statement corresponds to a possible outcome of the experiment. From this point of view there are as many events associated with the experiment as there are combinations of the first six positive integers taken any number at a time. If for the sake of aesthetic completeness and later convenience we consider also the impossible event, "the number $v$ is not equal to any of the first six positive integers," then there are altogether $2^6$ admissible events associated with the experiment of the rolling die. For the purpose of studying this example in more detail let us introduce some notation. We write $\{246\}$ for the event "$v$ is even," $\{123\}$ for "$v$ is less than 4," and so on. The impossible event and the certain event $(=\{123456\})$ deserve special names: we reserve for them the symbols $o$ and $e$ respectively.

Everyday language concerning events uses such phrases as these: "two events $a$ and $b$ are incompatible or mutually exclusive," "the event $a$ is the opposite of the event $b$ or complementary to $b$," "the event $a$ consists of the simultaneous occurrence of $b$ and $c$," "the event $a$ consists of the occurrence of at least one of the two events $b$ and $c$." Such phrases suggest that there are relations between events and ways of making new events out of old that should certainly be a part of their mathematical theory.

The notion of complementary event is probably closest to the surface. If $a$ is an event we denote the complementary event by $a'$: an experiment one of whose outcomes is $a$ will be said to result in $a'$ if and only if it does not result in $a$. Thus if $a=\{246\}$ then $a'=\{135\}$. We may also introduce combinations of events suggested by the logical concepts of "and" and "or." With any two events $a$ and $b$ we associate their "join" $a\cup b$ (also called union or sum and often denoted by $a+b$), and their "meet" $a\cap b$ (or intersection or product, often denoted by $ab$). Here $a\cup b$ occurs if and only if at least one of the two events $a$ or $b$ occurs, while $a\cap b$ occurs if and only if both $a$ and $b$ occur. Thus if $a=\{246\}$ and $b=\{123\}$ then $a\cup b=\{12346\}$ and $a\cap b=\{2\}$.

The operations $a'$, $a\cup b$, and $a\cap b$ satisfy some simple algebraic laws. It is clear for example that both the expressions $a\cup b$ and $a\cap b$ are independent of the order of the terms (commutative law), and that neither of the expressions $a\cup b\cup c$ and $a\cap b\cap c$ depends on the order in which the two indicated operations are performed (associative law). These facts are intuitively obvious from the verbal definition of the operations and are easily verified in any finite case such as the rolling die. There are many other similar identities satisfied by these methods of combining events: the following is a list of the most important ones.

$$o' = e \qquad (a')' = a \qquad e' = o$$

$$(a \cap b)' = a' \cup b' \qquad (a \cup b)' = a' \cap b'$$

$$a \cap a' = o \qquad a \cup a' = e$$

$$o \cap a = o \qquad o \cup a = a$$

$$e \cap a = a \qquad e \cup a = e$$

$$a \cap b = b \cap a \qquad a \cup b = b \cup a$$

$$(a \cap b) \cap c = a \cap (b \cap c) \qquad (a \cup b) \cup c = a \cup (b \cup c)$$

$$a \cap (b \cup c) = (a \cap b) \cup (a \cap c) \qquad a \cup (b \cap c) = (a \cup b) \cap (a \cup c)$$

A system $B$ of elements $o, a, b, \cdots, e$ in which operations $a'$, $a \cup b$, and $a \cap b$ are defined in such a way that each of the above list of identities is satisfied is called a "Boolean algebra." For the traditional theory of probability, concerned with simple gambling games such as the rolling die, in which the total number of possible events is finite, the above heuristic reduction of events to elements of a Boolean algebra is adequate. For situations arising in modern theory and practice, and even for the more complicated gambling games, it is necessary to make an additional assumption. This assumption, in descriptive terms, is that the operations $\cup$ and $\cap$, assumed defined for two elements and immediately extended by mathematical induction to any finite number, should make sense also for an infinite sequence. In other words it is desirable to have an interpretation for symbols such as $a_1 \cup a_2 \cup \cdots$ and $a_1 \cap a_2 \cap \cdots$. In order to phrase precisely this assumption of infinite operations it is necessary to use a few simple facts from the theory of Boolean algebras.

If $a$ and $b$ are any two elements of the Boolean algebra $B$ which satisfy the relation $a \cup b = b$ (or the equivalent relation $a \cap b = a$) we shall write $a \subset b$ and say that "$a$ is smaller than $b$" or "$a$ is contained in $b$" or "$a$ implies $b$." The intuitive interpretation of this relation is as follows: the event $a$ implies the event $b$, or is contained in the event $b$, if the occurrence of $a$ is a sub-case of the occurrence of $b$. Thus in the example of the die $\{123\} \subset \{1234\}$ and "$v = 2$"$\subset$"$v$ is even." The technical significance of the relation $\subset$ is that the operations $\cup$ and $\cap$ may be defined in terms of it. For example $a \cup b$ is the smallest of all elements which contain both $a$ and $b$. In more detail: given $a$ and $b$, consider all $c$'s for which both $a \subset c$ and $b \subset c$. The assertion concerning $a \cup b$ is two fold: first, $a \cup b$ is an admissible $c$, and second, for any admissible $c$ we have $a \cup b \subset c$. As an example consider $a = \{12\}$ and $b = \{24\}$. The elements $\{1234\}$, $\{1246\}$, $\{12456\}, \cdots$ all have the property of containing both $a$ and $b$. However the element $\{124\}$, which also has that property, is smaller than any other such element, and it is in fact true that $\{12\} \cup \{24\} = \{124\}$.

Motivated by the relation between $\cup$ and $\subset$ we now proceed as follows. Let $B$ be a Boolean algebra. If for every infinite sequence $a_1, a_2, \cdots$ of elements of

$B$ there exists among the elements containing all the $a_n$ a smallest one, say $a$, we say that $B$ is a $\sigma$-algebra and we write $a = a_1 \cup a_2 \cup a_3 \cup \cdots$. Not every Boolean algebra is a $\sigma$-algebra; the assumption that $B$ is one (the hypothesis of countable additivity) is an essential restriction.

Perhaps an example, though a somewhat artificial one, might illustrate the need for the added assumption. Suppose that a player determines to roll a die repeatedly until the first time that the number showing on top is 6. Let $a_n$ be the event that the first 6 appears only on the $n$th roll. The event $a = a_1 \cup a_2 \cup a_3 \cup \cdots$ occurs if and only if the game ends in a finite number of rolls. The occurrence of the opposite event $a'$ is at least logically (even if not practically) conceivable and it seems reasonable to want to include a discussion of it in a general theory of probability. Numerous examples of this kind together with some rather deep lying technical reasons justify therefore the following statement.

*The mathematical theory of probability consists of the study of Boolean $\sigma$-algebras.*

This is not to say that all Boolean $\sigma$-algebras are within the domain of probability theory. In general statements concerning such algebras and the relations between their elements are merely qualitative: probability theory differs from the general theory in that it studies also the quantitative aspects of Boolean algebras. In the next section we shall describe and motivate the introduction of numerical probabilities.

**3. Measure algebra.** When we ask "what is the probability of a certain event?" we expect the answer to be a number, a number associated with the event. In other words probability is a numerically valued function $P$ of events $a$, that is of elements of a Boolean $\sigma$-algebra $B$, $P = P(a)$. On intuitive and practical grounds we demand that the number $P(a)$ should give information about the occurrence habits of the event $a$. If in a large number of repetitions of the experiment which may result in the event $a$ we observe that $a$ actually occurs only a quarter of the time (the remaining three quarters of the experiments resulting therefore in $a'$) we may attempt to summarize this fact by saying that $P(a) = 1/4$. Even this very rough first approximation to what is desired yields some suggestive clues concerning the nature of the function $P$.

If, to begin with, $P(a)$ is to represent the proportion of times that $a$ is expected to occur, then $P(a)$ must be a positive real number, in fact a number in the unit interval $0 \leqq P(a) \leqq 1$. The extreme value 0 has a special significance. Since the impossible event $o$ will never occur, it is clear that we must write $P(o) = 0$. Conversely however if an event $a$ refuses ever to occur, we are tempted to declare its occurrence impossible and thus from the relation $P(a) = 0$ to deduce $a = o$. The other extreme value of $P(a)$ has of course a similar interpretation: $P(a) = 1$ if and only if $a = e$.

The relation between proportion and probability has further consequences. Suppose that $a$ and $b$ are mutually exclusive events—say $a = \{1\}$ and $b = \{246\}$ in the example of the die. (In the algebraic theory mutually exclusive events correspond to "disjoint" elements of the Boolean algebra $B$, that is to elements

$a$ and $b$ for which $a \cap b = o$.) In this case the proportion of times that the join $a \cup b (= \{1246\}$ for the example) occurs is clearly the sum of the proportions associated with $a$ and $b$ separately. If an ace shows up one-sixth of the time and an even number half the time, then the proportion of times in which the top face is either an ace or an even number is $\frac{1}{6} + \frac{1}{2}$. It follows therefore that the function $P$ cannot be completely arbitrary—it is necessary to subject it to the condition of additivity, that is to require that if $a \cap b = o$ then $P(a \cup b)$ should be equal to $P(a) + P(b)$.

We are now separated from the final definition of probability theory only by a seemingly petty (but in fact very important) technicality. If $P(a)$ is an additive function of the sort just described on a Boolean $\sigma$-algebra $B$, and if $a_1, a_2, \cdots, a_n$ is any finite set of pairwise disjoint elements of $B$ (this means that for $i \neq j$, $a_i \cap a_j = o$) then it's easy to prove by mathematical induction that $P(a_1 \cup a_2 \cup \cdots \cup a_n) = P(a_1) + P(a_2) + \cdots + P(a_n)$. If however $a_1, a_2, a_3, \cdots$ is an infinite sequence of pairwise disjoint elements then it may or may not be true that $P(a_1 \cup a_2 \cup a_3 \cup \cdots) = P(a_1) + P(a_2) + P(a_3) + \cdots$. The general condition of countable (that is, finite or enumerably infinite) additivity is a further restriction on the probability measure $P$—a restriction without which modern probability theory could not function. It is a tenable point of view that our intuition demands infinite additivity just as much as finite additivity. At least however infinite additivity does not contradict any of our intuitive ideas and the theory built on it is sufficiently far developed to assert that the assumption is justified by its success. We shall therefore adopt this assumption as our final postulate.

*Numerical probability is a measure function, that is a finite, nonnegative, and countably additive function $P$ of elements in a Boolean $\sigma$-algebra $B$, such that if the null and unit elements of $B$ are $o$ and $e$ respectively then $P(a) = 0$ is equivalent to $a = o$ and $P(a) = 1$ is equivalent to $a = e$.*

In the next section we shall discuss a general method of constructing examples of probability measures.

**4. Measure space.** Let $\omega_j (j = 1, \cdots, 6)$ be the point on the real axis whose directed distance from the origin is $j$, and let $\Omega$ be the set whose elements are these six points. Consider the system $B^*$ of all subsets of $\Omega$. (The empty set $o$ and the full set $e = \Omega$ are counted as belonging to $B^*$.) With any element $a$ of $B^*$ (that is, with any subset of $\Omega$) we may associate the complementary element (set) consisting of exactly those points $\omega_j$ which do not belong to $a$. Similarly with any two subsets $a$ and $b$ of $\Omega$ we may associate their union (the set of points belonging to either $a$ or $b$ or both), and their intersection (the set of points belonging simultaneously to $a$ and $b$). It is easy to verify that under the operations of complementation ($a'$), formation of unions ($a \cup b$), and formation of intersections ($a \cap b$), the system $B^*$ forms a Boolean algebra, in fact, though somewhat vacuously, a $\sigma$-algebra. Suppose moreover that for each $j = 1, \cdots, 6$, $p_j$ is a positive number such that $p_1 + \cdots + p_6 = 1$. Then we may define $P(a)$ for any subset $a$ of $\Omega$, to be the sum of those $p_j$ whose $\omega_j$ belongs to $a$. Thus if

$a = \{135\}$ then $P(a) = p_1 + p_3 + p_5$; if $a = o$ then $P(a) = 0$. The function $P$ and the algebra $B^*$ satisfy all the assumptions of probability theory and the reader has doubtless recognized that this $B^*$ and $P$ were implicit in our earlier discussion of the rolling die. It is often customary on philosophical and practical grounds to discuss only the case $p_1 = \cdots = p_6 = \frac{1}{6}$. We shall say a word about this special case later; for the moment it is sufficient to point out that any other choice of the $p_j$ furnishes an equally acceptable probability structure and does in fact constitute the mathematical theory of some carefully loaded die.

The above example of a Boolean algebra can be generalized: we attempt next to obtain a similar but more geometrical example. For this purpose we again choose a set $\Omega$, but, instead of a finite set, we choose a set with infinitely many points, in fact all the points of a continuum. To be specific let us choose for $\Omega$ the points $\omega$ of a square of unit area in the Cartesian plane. In analogy with the preceding example we consider the system $B^*$ of all subsets of $\Omega$ and define complement, union, and intersection as before. Once more $B^*$ is a Boolean $\sigma$-algebra; it is not however the one on which we shall base our probability theory. (It can be shown that it is not possible to define a probability measure $P$ with the desired properties on $B^*$.) We shall instead consider a certain subsystem (sub-algebra) of $B^*$, constructed as follows:

We begin with the system $R$ of all rectangles contained in $\Omega$ (where for the sake of definiteness we consider closed rectangles, that is sets consisting of the interior plus the perimeter of a rectangle). The system $R$ is not closed under the Boolean operations: in general not even a finite (let alone a countably infinite) union or intersection of rectangles is itself a rectangle, and similarly the complement of a rectangle isn't one. We have therefore to enlarge the system $R$ to a system $R'$ including all complements and countable unions and intersections of elements of $R$. It turns out that even this is not enough: $R'$ is still not a Boolean algebra, and the extension process has to be continued. If however the extension process is continued sufficiently (and this happens to mean transfinitely) often, we reach eventually a Boolean $\sigma$-algebra $B$ of subsets of $\Omega$. (The algebra $B$ is important in analysis: sets of $B$ are called the Borel sets of the square.)

We face next the task of defining $P$. For those familiar with the theory of Lebesgue measure it will suffice to say that we define $P(a)$, for each $a$ in $B$, to be the Lebesgue measure of the set $a$. It is not difficult to get an intuitive idea of how $P$ is defined. If $a$ is a rectangle (that is an element of $R$) we define $P(a)$ to be the area of $a$. If $a$ is an element of $R'$ we proceed to determine $P(a)$ in accordance with the requirement of countable additivity. Thus for example if $b$ is the complement of a rectangle $a$, we write $P(b) = 1 - P(a)$, and if $b$ is the union of a finite or infinite sequence of disjoint rectangles $a_1, a_2, \cdots$ we define $P(b) = P(a_1) + P(a_2) + \cdots$. By repeating this extension process ad transfinitum we succeed eventually in defining $P(a)$ for every $a$ in $B$.

There is an objection to the construction just described. If the set $a$ consists of a single point then it is intuitively obvious (and follows easily from the rigorous definition of $P$) that $P(a)$ (= the area of $a$) is zero. More generally if $a$

consists of any finite or enumerably infinite set of points we still have $P(a)=0$, and it is even possible (if for example $a$ is a line segment) to have $P(a)=0$ for sets $a$ containing uncountably many points. This definitely contradicts our explicitly formulated axiom that $P(a)=0$ should happen if and only if $a=o$. The customary way to get around this difficulty is by redefining the notion of equality that occurs in the equation $a=o$. It is proposed that we agree to consider as identical two subsets of $\Omega$ whose difference has probability zero. (In technical language, we consider, instead of the sets $a$, equivalence classes of sets modulo the class of sets of probability zero.) Through this agreement we are committed in particular to identifying any set of probability zero with the empty set $o$, and it follows therefore that in the reduced algebra $B$ (that is, the algebra obtained from $B$ by making the suggested identifications) all the axioms of probability are valid.

The long and tortuous process just described is very general. If $\Omega$ is any space (such as an interval or a cube) on a certain $\sigma$-algebra $B$ of subsets of which a countably additive measure $P$ is defined (such as length or volume), subject only to the restriction that the measure of all $\Omega$ is equal to 1, we obtain from $B$ and $P$ a system satisfying all the axioms of probability theory by the process of identification according to sets of measure zero. Thus there are as many probability systems as there are examples of "measure spaces."

The reason for the introduction of measure spaces into a discussion of probability theory is not merely to give examples. It can in fact be shown that the two theories (measure and probability) are coextensive. More precisely:

*If $B$ is any Boolean $\sigma$-algebra and $P$ a probability measure on $B$, then there exists a measure space $\Omega$ such that the system $B$ is abstractly identical with an algebra of subsets of $\Omega$ reduced by identification according to sets of measure zero, and the value of $P$ for any event $a$ is identical with the values of the measure for the corresponding subsets of $\Omega$.*

Hence measure is probability and probability is measure and, in virtue of the theorem just stated, the entire classical theory of measure and integration may be and has been carried over and used to give rigorous proofs of probability theorems.

**5. Measure vs. probability.** Having discussed the extent to which probability and measure are the same, we now dedicate a few words to describing the extent to which they are different. One feature that differentiates the two theories is that in the general theory of measure it is usual to admit the possibility that the measure of the entire space is infinite. This possibility is not admissable in probability theory. As long, however, as the measure of the whole space is finite it is always possible to introduce a scale factor which makes it equal to 1, and hence it is always possible to think of it (even if somewhat artificially) as a "probability space." Thus for example the language and notation of probability may be and have been used in such seemingly widely separated parts of mathematics as ergodic theory, topological groups, and integral geometry.

Even however if the infinite case is ruled out, it is a conspicuous fact that most theorems in which the word measure is used (rather than the word probability) have a very different appearance from the theorems of probability theory. The best way to explain the difference between measure and probability is to liken it to the difference between analytic and synthetic geometry. It isn't stretching a point too far to say that the representation of a probability algebra by a measure space is similar to the introduction of coordinates into geometry. Synthetic and analytic geometry are of course abstractly identical in the sense that any theorem in the one domain may be stated and proved in the language and machinery of the other—may be, but isn't. The theorems in the two fields differ in their intuitive content. It is natural to discuss linear transformations in analytic geometry and the nine point circle in synthetic geometry—and even though the interchange is possible, it isn't desired. The abstract identity of the two fields is however an extremely useful fact, exploited mostly by the synthetic side which often finds it convenient to lean on the analytic crutch. Similarly, probability is measure, and research in the field would be very greatly hampered if we were not permitted to use this analytic crutch—but the notions suggested by probability, the notions which are important and intuitive and natural inside the field, appear sometimes extremely special and artificial in the frame work of general measure theory.

In this section and the preceding ones we have treated axiomatization and coordinatization. We proceed now to development. In the following sections we shall define the basic concepts of probability theory, and discuss in particular those which serve in the sense described above to give to probability its distinguishing flavor.

**6. Independent events.** In order to motivate the definitions of the concepts to be studied in the sequel we return to the example of the die. For simplicity we make the classical assumption that any two faces are equally likely to turn up and that consequently the probability of any particular face showing is $\frac{1}{6}$. Consider the events $a=\{246\}$ and $b=\{12\}$. The first notion we want to introduce, the notion of conditional probability, can be used to answer such questions as these: "what is the probability of $a$ when $b$ is known to have occurred?" In the case of the example: if we know that $v$ is less than 3, what can we say about the probability that $v$ is even? The adjective "conditional" is clearly called for in the answer to a question of this type: we are evaluating probabilities subject to certain preassigned conditions.

To get a clue to the answer consider first the event $c=\{2\}$ and ask for the conditional probability of $a$, given that $c$ has already occurred. The intuitive answer is perfectly clear here, and is independent as it happens of any such numerical assumptions as the equal likelihood of the faces. If $v$ is known to be 2 then $v$ is certainly even, and the probability must be 1. What made the answer easy was the fact that $c$ implied $a$. The general question of conditional probability asks us to evaluate the extent (measured by a numerical probability or proportion) to which the given event $b$ implies the unknown event $a$. Phrased in this way the question almost suggests its own answer: the extent to which $b$ is contained in $a$ can be measured by the extent to which $a$ and $b$ are likely to occur simultaneously, that is by $P(a\cap b)$. Almost—not quite. The trouble is that $P(a\cap b)$ may be very small for two reasons: one is that not much of $b$ is contained in $a$, and the other is that there isn't very much of $b$ altogether. In other words it isn't merely the absolute size of $a\cap b$ that matters: it's the relation or proportion of this size to the size of $b$ that's relevant.

We are led therefore to define the conditional probability of $a$, given that $b$ has occurred, in symbols $P_b(a)$, as the ratio $P(a\cap b)/P(b)$. For $a=\{246\}$ and $c=\{2\}$ this gives the answer we derived earlier, $P_c(a)=1$; for $a=\{246\}$ and $b=\{12\}$ we get the rather reasonable figure $P_b(a)=\frac{1}{2}$. In other words if it's known that $v$ is either 1 or 2 then $v$ is even or odd (that is equal to 1 or equal to 2) each with probability $\frac{1}{2}$.

Consider now the following two questions: "$b$ happened, what is the chance of $a$?" and simply "what is the chance of $a$?." The answers of course are $P_b(a)$ and $P(a)$ respectively. It might happen, and does in the example given above, that the two answers are the same, that in other words knowledge of $b$ contributes nothing to our knowledge of the probability of $a$. It seems natural in this situation to use the word "independent": the probability distribution of $a$ is independent of the knowledge of $b$. This motivates the precise definition: two events $a$ and $b$ are independent if $P_b(a)=P(a)$. The definition is transformed into its more usual form and at the same time gains in symmetry if we recall the definition of $P_b(a)$. In symmetric form: $a$ and $b$ are independent in the sense of probability (statistically or stochastically independent) if and only if $P(a\cap b) = P(a)P(b)$.

**7. Repeated trials.** Suppose next that we wish to make two independent trials of the same experiment—say, for example, to roll an honest die twice in succession. We shall presently exploit the precise definition of independence to clarify the notion of independent trials; first however it's worth while to remark on the intuitive content of the concept. Suppose that in a crude attempt to even things up we resolve on the following procedure: if the first die shows an even number we choose for the second experiment a die on which all the numbers are odd, and vice versa. The two experiments are not independent of each other in this case: whereas the a priori probability of getting an even number with the second die is $\frac{1}{2}$, the conditional probability of getting an even number with the second die, given that the first one showed an odd number, is one. We say that the two experiments are performed independently of each other only if the conditions under which the second experiment is to be performed are unaffected by the outcome of the first experiment.

If an experiment consists of two rolls of a die we don't expect the reported outcome of the experiment to be a number $v$, but rather a pair of numbers $(v_1, v_2)$. The measure space $\Omega$ associated with the two-fold experiment consists not of 6 but of 36 points. (It is convenient to imagine these points laid out along the regular pattern of a $6\times 6$ square.) The problem is to determine how the probability is distributed among these points. For a clue to the answer consider the events $a=$"$v_1<3$" and $b=$"$v_2<4$." We have $P(a)=\frac{1}{3}$ and $P(b)=\frac{1}{2}$; hence if we interpret the independence of the trials to mean the independence of any two events such as $a$ and $b$ we should have $P(a\cap b)=\frac{1}{6}$. If in the suggested diagram for the measure space associated with this discussion we encircle the points belonging to $a\cap b$ we get the following figure.

| $v_2$ \ $v_1$ | 1 | 2 | 3 | 4 | 5 | 6 |
|---|---|---|---|---|---|---|
| 1 | ⊙ | ⊙ | · | · | · | · |
| 2 | ⊙ | ⊙ | · | · | · | · |
| 3 | ⊙ | ⊙ | · | · | · | · |
| 4 | · | · | · | · | · | · |
| 5 | · | · | · | · | · | · |
| 6 | · | · | · | · | · | · |

We see therefore that the formula $P(a\cap b)=P(a)P(b)$ appears analogous to the fact that the area of a rectangle is the product of the lengths of its sides.

We say therefore, if the analytic description of an experiment is given by a measure space $\Omega$ with a Boolean $\sigma$-algebra $B$ of subsets on which a probability measure $P$ is defined, that the analytic description of the experiment consisting of two independent trials of the given experiment is as follows. The space of points $\omega$ is replaced by the space of pairs of points $(\omega_1, \omega_2)$ (the so called product space $\Omega\times\Omega$), $B$ is replaced by the Boolean $\sigma$-algebra generated by the "rectangular" sets of the form $\{\omega_1$ is in $a_1$, $\omega_2$ is in $a_2\}$ where $a_1$ and $a_2$ belong to $B$, and the probability measure on this space of pairs is determined by the requirement that its value for rectangular sets of the kind described should be given by the product $P(a_1)P(a_2)$. The ideas involved in this procedure are not essentially original nor characteristic of probability theory: they are the same as the ideas involved in defining the area of plane sets in terms of the length of linear sets. There is of course a theorem hidden in this definition—a theorem which asserts that a probability measure satisfying the stated product requirement indeed exists and is in fact uniquely determined by this requirement.

What we can do once, we can do again. Just as two repetitions of an experiment gave rise to ordered pairs $(\omega_1, \omega_2)$, similarly any finite number of repetitions (say $n$) give rise to the space of ordered $n$-tuples $(\omega_1, \omega_2, \cdots, \omega_n)$, with a multiplicatively determined probability measure. The procedure can be extended also to infinity: the analytic model of an infinite sequence of independent repetitions of an experiment is a measure space $\Omega$ whose points $\omega$ are infinite sequences $\{\omega_1, \omega_2, \omega_3, \cdots\}$. Even if an actually infinite sequence of repetitions of an experiment is practically unthinkable, there is a point in considering the infinite dimensional space $\Omega$. The point is that many probability statements are assertions concerning what happens in the long run—assertions which can be made precise only by carefully formulated theorems concerning limits. Hence even if practice yields only approximations to infinity, it is the infinite sequence space $\Omega$ that is the touchstone whereby the mathematical theory of probability can be tested against our intuitive ideas. The first and most important such long run statement is described in the following paragraphs.

Suppose that an experiment is capable of producing an event $a$ with probability $p$, and suppose that an infinite sequence of independent trials of this experiment is performed. We consider therefore the space of all sequences $\omega = \{\omega_1, \omega_2, \omega_3, \cdots\}$ where for each $n$, $\omega_n$ may or may not belong to $a$. Once the experiments have been performed so that we are given a particular point $\omega$ we may start asking numerical questions. We may ask for example: out of the first $n$ trials of the basic experiment how many resulted in $a$? This means: out of the first $n$ coordinates $\omega_1, \omega_2, \cdots, \omega_n$ of $\omega$ how many belong to $a$? The answer to this question depends obviously on $n$ and just as essentially on the particular sequence $\omega$—let us denote it by $m_n(\omega)$.

Now what does out intuition say? The usual statement (one which we have already exploited in our heuristic derivation of the notion of probability) is that the ratio of the number of successes to the total number of trials should be approximately equal to the probability of the event being tested. In our notation this seems to mean that for large $n$ the ratio $m_n(\omega)/n$ should be close to the constant $p = P(a)$. The question arises: for which $\omega$'s should this be true? Not surely for all of them. For the sequence space $\Omega$ contains sequences none of whose coordinates belong to $a$, and for such a sequence $\omega$, $m_n(\omega)$ is zero for all $n$. The best that we have a right to demand is that the $\omega$'s for which our statement is not true should be equivalent to the empty set of $\omega$'s in the sense of probability—that is that their totality should have probability zero. And this is true.

To sum up: we have just derived the statement (not the proof) of the most important special case of the so called strong law of large numbers. In mathematical language the assertion of this law is that as $n \rightarrow \infty$, $\lim m_n(\omega)/n$ exists and is equal to $p (= P(a))$ except for a set of $\omega$'s of measure zero. In more classical terms: it is almost certain that the "success ratios" converge to the probability of the event being tested.

**8. Random variables.** In order to gain a more thorough understanding of the law of large numbers and at the same time to introduce the language in which most of the theorems of probability theory are stated, we proceed to discuss the notion of a random variable.

"A random variable is a quantity whose values are determined by chance." What does that mean? The word "quantity" is meant to suggest magnitude—numerical magnitude. Ever since rigor has come to be demanded in mathematical definitions it has been recognized that the word "variable," particularly a variable whose values are "determined" somehow or other, means in precise language a function. Accordingly a random variable is a function: a function whose numerical values are determined by chance. This means in other words that a random variable is a function attached to an experiment—once the experiment has been performed the value of the function is known. The spatial model of probability is extremely well adapted to making this notion still more precise. If the analytic correspondent of an experiment is a measure space $\Omega$ then any possible outcome of the experiment is by definition represented by a point $\omega$ in this space. Hence a function of outcomes is a function of $\omega$'s: a random variable is a real valued function defined on a probability space $\Omega$.

The preceding sentence does not yet constitute our final definition of a random variable. For suppose that $x=x(\omega)$ is a function on the space $\Omega$. We shall call $x$ a random variable only if probability questions concerning the values of $x$ can be answered. An example of such a question is: what is the probability that $x$ is between $\alpha$ and $\beta$? In measure theoretic language: what is the measure of the set of those $\omega$'s for which the inequality $\alpha \leqq x(\omega) \leqq \beta$ is satisfied? In order for such questions to be answerable it is necessary and sufficient that the sets that occur in them belong to the basic $\sigma$-algebra $B$ of $\Omega$. A function $x(\omega)$ for which this is true for every interval $(\alpha, \beta)$ is called "measurable." Accordingly we make the following definition:

*A random variable is a measurable function defined on a measure space with total measure* 1.

Instances of random variables can be found even in that part of our discussion which preceded their definition. The quantity $v$ associated with the rolling die is an example, as are also the quantities $v_1$ and $v_2$ associated with the two fold repetition of this experiment. To obtain some further examples, consider any fixed event $a$ which may result from an experiment and let the random variable $x$ be the number of times that $a$ actually occurs. If the experiment is performed only once then $x$ has only two possible values: 1 if $a$ occurs and 0 otherwise. More generally if the experiment is repeated $n$ times the random variable $x$ becomes the function $m_n(\omega)$ introduced in the discussion of the law of large numbers.

**9. Expectation, variance, and distribution.** Let us consider in detail the random variable $v$ associated with an honest die. The possible values of $v$ are the first six positive integers. The arithmetic mean of these values, that is the number $(1+ \cdots +6)/6$, is of considerable interest in probability theory. It is called the average, or mean value, or expectation of the random variable $v$ and it is denoted by $E(v)$. If the die is loaded so that the probability $p_j$ associated with $j$ is not necessarily $\frac{1}{6}$ then the arithmetic mean is replaced by a weighted average: in this case $E(v)=1 \cdot p_1+ \cdots +6 \cdot p_6$. It is well known that the analogs of such weighted sums in cases where the number of values of the function (random variable) need not be finite are given by integrals. The kind of integral that enters into probability theory is similar in every detail to the Lebesgue integral and we shall not reproduce its definition here.

*If the measurable function $x(\omega)$ is integrable then its expectation $E(x)$ is by definition the value of its integral extended over the entire domain $\Omega$.*

As a useful though extremely special case we mention that if $x$ is a counting variable of the sort mentioned in the preceding paragraph ($x=1$ if a certain even $a$ occurs and $x=0$ otherwise) then $E(x)=P(a)$.

It is obviously of interest to ask not only what is the expected value of a random variable $x$ but also how closely the values of $x$ are clustered about its expected value. The customary measure of clustering of a random variable $x$ is one inspired by the method of least squares and called the "variance" or "dispersion" of $x$.

*The variance of $x$ is the expression $\sigma^2(x)=E(x-\alpha)^2$, where $\alpha=E(x)$.*

(The square root of the variance is called the "standard deviation.") In words: take the square of the deviation of $x$ from its expected value $\alpha$, and use the sum (weighted sum, integral) of these squared deviations as a measure of clustering. Since a sum of squares vanishes only if each term does, the vanishing of the variance indicates that $x$ is identically equal to its expected value (except perhaps for a set of probability zero). In general, the smaller the variance the closer the values of $x$ lie to $E(x)$.

Such numbers as $E(x)$ and $\sigma^2(x)$ yield partial information about the distribution of the values of $x$. Complete information would mean an answer to every question of the form "what is the probability that $x$ lies in the interval $(\alpha, \beta)$?" In order to deal with such questions we introduce the notion of distribution function.

*The distribution function $F_x(\lambda)$ of a random variable $x$ is a function of a real variable $\lambda$ defined for each $\lambda$ to be the probability that $x<\lambda$.*

These functions can be used to answer every probability question concerning random variables; for example the expression $F_x(\beta)-F_x(\alpha)$ represents the probability that $x$ belong to the (half open) interval $\alpha \leqq x<\beta$, and the Stieltjes integrals $\int_{-\infty}^{\infty}\lambda dF_x(\lambda)$ and $\int_{-\infty}^{\infty}\{\lambda-E(x)\}^2 dF_x(\lambda)$ represent the expectation and variance of $x$ respectively. Distribution functions are useful because being comparatively simple real functions of real variables they are amenable to treatment by the methods of classical analysis. It is the whole purpose of a large part of probability theory to find the distribution functions of certain random variables.

**10. Independent variables.** Let us consider next two random variables $x$ and $y$ which are comparable in the sense that they are both represented by measurable functions on the same measure space $\Omega$, so that $x=x(\omega)$ and $y=y(\omega)$. It is easy to see that the function $E(x)$, being defined by an integral, is homogeneous of degree 1 and additive, that is $E(\lambda x)=\lambda E(x)$ for every real constant $\lambda$ and $E(x+y)=E(x)+E(y)$. Similarly the variance $\sigma^2(x)$ is homogeneous of degree 2, that is $\sigma^2(\lambda x)=\lambda^2\sigma^2(x)$. One way to prove this latter fact is to make use of the following identity connecting $\sigma^2$ and $E$:

$$\sigma^2(x) = E(x)^2 - E^2(x), \tag{1}$$

(where for later convenience we write $E(x)^2$ for $E(x^2)$ and $E^2(x)$ for $\{E(x)\}^2$). This identity in turn follows from the definition of $\sigma^2$. Since $\sigma^2(x)=E(x-\alpha)^2$

where $\alpha = E(x)$, we have $\sigma^2(x) = E(x^2 - 2\alpha x + \alpha^2) = E(x^2) - 2\alpha E(x) + \alpha^2$. (We used here the fact that the expected value of a constant is equal to that constant.) The identity (1) follows by substituting for $\alpha$ its value $E(x)$. Letting the formalism guide us we may inquire whether $\sigma^2$ is additive, that is whether or not the identity

$$\sigma^2(x + y) = \sigma^2(x) + \sigma^2(y) \tag{2}$$

is valid. The answer in general is no. In order to investigate conditions under which (2) is true we proceed to a brief discussion of some possible relations between pairs of random variables.

Let $a$ and $b$ be two independent events and let $x$ and $y$ be the associated counting random variables (so that $x$ for example is 1 if and only if $a$ occurs and $x = 0$ otherwise). The product random variable $xy$ in this case can be equal to 1 if and only if both $a$ and $b$ occur, so that $xy$ is the counting variable of $a \cap b$. Since $E(x) = P(a)$, $E(y) = P(b)$, and similarly $E(xy) = P(a \cap b)$, we have in this special case

$$E(xy) = E(x)E(y). \tag{3}$$

The validity of this formula is sufficiently important in the applications of probability to bear a name of its own: two random variables, not necessarily the counting variables of a pair of independent events, satisfying it are called "uncorrelated." The reason for the terminology is that the coefficient of correlation $r = r(x, y)$ of two random variables $x$ and $y$ is defined by $r = \{E(xy) - E(x)E(y)\}/\sigma^2(x)\sigma^2(y)$; this coefficient vanishes if and only if (3) holds.

It is now easy to state the facts concerning the formula (2): it is valid if and only if (3) is. In other words the variance is additive for a pair of random variables if and only if the expectation is multiplicative, that is if and only if they are uncorrelated. For the proof we merely expand the left member of (2), thus:

$$\begin{aligned}\sigma^2(x + y) &= E(x + y)^2 - E^2(x + y) \\ &= \{E(x)^2 - 2E(xy) + E(y)^2\} - \{E^2(x) - 2E(x)E(y) + E^2(y)\} \\ &= \sigma^2(x) + \sigma^2(y) - 2\{E(xy) - E(x)E(y)\}.\end{aligned}$$

Let us now return to the pair of counting variables $x$ and $y$ associated with two independent events $a$ and $b$. Because of the independence of $a$ and $b$, any probability statement concerning $y$ is unaffected by our knowledge of ignorance of the value of $x$. More precisely, any two events defined by $x$ and $y$, for example the events "$x = 0$" and "$y = 1$," are independent. If in general any two events by two random variables $x$ and $y$ respectively, that is any two events defined by inequalities of the form $\alpha \leqq x \leqq \beta$ and $\gamma \leqq y \leqq \delta$, are independent events, no matter what $\alpha$, $\beta$, $\gamma$ and $\delta$ are, we say that $x$ and $y$ are independent random variables. It is not too difficult to generalize what we proved about the special case of counting variables: for independent random variables the expectation, if it exists, is multiplicative and consequently the variance is additive. In still other words: independence implies absence of correlation—a proposition which certainly sounds natural enough.

One word of caution before we leave this brief introduction to the notion of independence for random variables. What we defined was the independence of two random variables. It would be natural to try to define the independence of a finite or infinite sequence of random variables $x_1, x_2, \cdots$, by the requirement that any pair be independent. Natural, but as it happens, not very useful. The correct definition replaces two-term products by many-term products in the following way.

*The random variables $x_1, x_2, \cdots$, are independent if the probability of the simultaneous occurrence of any finite number of the events defined by $\alpha_n \leqq x_n \leqq \beta_n$ is the product of the separate probabilities, no matter what real constants the $\alpha$'s and $\beta$'s are.*

It is easy to construct examples to show that this notion is indeed different from the notion of pairwise independence.

**11. Law of large numbers.** We are now in a position to reformulate and generalize the strong law of large numbers in terms of random variables. Let the sequence space of points $\omega = \{\omega_1, \omega_2, \cdots\}$ be the analytic model of the infinite repetition of an experiment one of whose possible outcomes is the event $a$. Let $a_n$ be the event "$\omega_n$ belongs to $a$" or equivalently the event "the $n$th experiment results in $a$," and let $x_n = x_n(\omega)$ be the counting variable associated with $a_n$. In this context that means that $x_n(\omega)$ has the value 1 for all those sequences $\omega\{\omega_1, \omega_2, \cdots\}$ for which the $n$th coordinate $\omega_n$ belongs to $a$, and $x_n(\omega)$ has the value 0 otherwise. What significance has the sum $x_1 + \cdots + x_n$? Since a particular term $x_j$ contributes one unit to this sum if and only if the $j$th experiment results in $a$, it is clear that the value of the sum, for any sequence $\omega$, is the number of those coordinates among the first $n$ coordinates of $\omega$ which do belong to $a$. But this is exactly the function we denoted above by $m_n(\omega)$. Hence our version of the law of large numbers is equivalent to the assertion that the averages $(x_1 + \cdots + x_n)/n$ converge (except possibly for a set of $\omega$'s of probability zero) to the constant $p = P(a)$. For the generalization of this result that we are about to formulate it is worth while to observe that $p = E(x_n)$ is also equal to the common value of the expectations of the $x$'s.

The sequence of random variables $x_1, x_2, \cdots$ has two important properties which are sufficient to ensure the validity of the law of large numbers. One of these properties is independence. It follows very easily from the fact that the experiments yielding the values of the various $x$'s are independently performed, that the variables $x_1, x_2, \cdots$ are indeed independent. The other essential property of the sequence is usually expressed by the statement that the random variables $x_n$ all have the same distribution. The definition of this concept is as follows.

*Two random variables $x$ and $y$ have the same distribution if for every interval $(\alpha, \beta)$ the probabilities of the two events $\alpha \leqq x \leqq \beta$ and $\alpha \leqq y \leqq \beta$ are equal, or equivalently if the distribution functions $F_x(\lambda)$ and $F_y(\lambda)$ are identical.*

In our particular case it is the fact that the probability that $\omega_n$ belong to $a$ is the same for all $n$ (namely $P(a)$) that implies that the $x_n$ all have the same distribution. That independence and equidistribution are indeed the crucial hypotheses for the law of large numbers is shown by the following general formulation of that law.

*If $x_1, x_2, \cdots$ is a sequence of independent random variables with the same distribution, and if the expectations $E(x_n)$ exist and have the value $\alpha$ (necessarily the same for all $n$) then the averages $x_1+\cdots+x_n/n$ converge as $n\to\infty$ (except perhaps on a set of probability zero) to the constant $\alpha$.*

**12. Central limit theorem.** Sums (such as $x_1+\cdots+x_n$) of independent random variables with the same distribution occur very often in probability theory. It is of considerable practical importance to investigate the precise distribution of such sums and if possible the limiting behavior of these distributions. We assume concerning the $x$'s that their expectations and variances both exist and write $E(x_j)=\alpha$, $\sigma^2(x_j)=\beta$. It follows from the independence and equidistribution of the $x$'s that $E(x_1+\cdots+x_n)=n\alpha$ and $\sigma^2(x_1+\cdots+x_n)=n\beta$. At first sight this seems like a discouraging phenomenon: if both the expectation and the variance become infinite, how can we expect a reasonable asymptotic behavior from the much more delicate distribution function? But the way out of the difficulty is easy: by a translation and a change of scale (different to be sure for each $n$) it is possible to normalize the sum $x_1+\cdots+x_n$ so that its expectation is 0 and its variance 1 for every positive integer $n$. To get the expectation to be 0 we merely subtract its actual value, $n\alpha$, from the sum—the additivity of the expectation ensures the desired result. To get the variance to be 1 we divide by a constant factor. It is important to recall that the variance is homogeneous of degree 2, so that the constant factor will be not $n\beta$ but $\sqrt{n\beta}$. We arrive thus at the normalized sums

$$\frac{x_1+\cdots+x_n-n\alpha}{\sqrt{n\beta}}$$

and inquire again after the distribution function of this random variable and the limit of such distribution functions. The answer here is known and is embodied in the so called central limit theorem (or Laplace-Liapounoff theorem) stated as follows.

*If $x_1, x_2, \cdots$ is a sequence of independent random variables with the same distribution, expectation $\alpha$, and variance $\beta$, then the distribution functions of the modified sums $(x_1+\cdots+x_n-n\alpha)/\sqrt{n\beta}$ converge as $n\to\infty$ to a fixed distribution function, the same no matter what the original distribution of the $x$'s is. In more detail, the limit as $n\to\infty$ of the probability of the event defined by the inequality*

$$\frac{x_1+\cdots+x_n-n\alpha}{\sqrt{n\beta}}<\lambda$$

*exists and is equal to*

$$G(\lambda)=\frac{1}{\sqrt{2\pi}}\int_{-\infty}^{\lambda}e^{-u^2/2}du.$$

*The distribution function $G(\lambda)$ is called the Gaussian or normal distribution.*

With this statement we end our discussion of the development of probability theory and turn to a few remarks connected with the problem of application.

**13. Determination of initial probabilities.** When the mathematician announces that the probability of an event is a certain number, he is immediately faced with two questions. First the practical man asks what is the practical meaning of a probability statement? How should one act on it? If the mathematician succeeds in answering this question then the philosopher wants to know the reason for the answer. What establishes the connection between mathematical theory and practice? Our remarks in what follows will bear on these very old and very difficult questions only incidentally—they are dedicated mainly to a smaller problem of the theory, but one which frequently worries the layman.

The problem is how the probability of concretely given events is really defined. It is all very well to talk about Boolean algebras and measure theory, but what is the probability that a coin will fall heads up? What the layman realizes and what we now wish to emphasize is that the mathematician has not answered any such questions. He cannot. He can no more say that the probability of obtaining two heads in succession with a coin is $\frac{1}{4}$ than he can say that the volume of a cube is 8. The volume of a cube is given by a formula. If the hypotheses under which the formula applies are verified and if the variables entering into the formula are given specific values then the volume of a cube can be calculated. In exactly the same sense the mathematical theory of probability is a collection of formulae which enable us to calculate certain probabilities assuming that certain other ones are given. If we know that the probability of obtaining heads with a certain coin is $\frac{1}{2}$ and if we know that two successive tosses of the coin were performed independently then we can assert that the probability of getting two heads is $\frac{1}{4}$.

Despite the fact that probability theory shares with all other mathematical theories its inability to state a conclusion without hypotheses, the above answer to the layman's question will probably seem unsatisfactory to many readers. There must be some reason why most people believe that the probability of heads is $\frac{1}{2}$. It is often even proved. The usual proof is based on symmetry arguments, or equivalently on the principle of sufficient reason. (Why should heads have any greater likelihood of appearing than tails?) Do these proofs have any mathematical validity?

The answer is definitely yes. In some cases it is more pleasing to the intuition or more convenient for practice to formulate our hypotheses purely qualitatively. In almost all such cases the hypotheses take the form of invariance—the probabilities entering into the problem are required to be invariant under a certain group of transformations. It often turns out then that an existence and uniqueness theorem is true, that is it can be proved that there exists one and only one probability measure satisfying the stated hypotheses. Theorems of this type are certainly a part, an increasingly important part, of the theory of probability, and as long as their hypotheses are clearly formulated and recognized as hypotheses, the professional mathematician is the last person to sneer at them. Their advantage at the level of elementary pedagogy seems to lie in the fact that the statement "heads and tails are equally likely" is easier to grasp intuitively than the statement "the probability of heads is $\frac{1}{2}$."

We see thus that a mathematical statement on probability has to have certain either explicitly or implicitly given probabilities to begin with. In practice the physicist (or actuary, or anyone else interested in applying the theory) obtains these initial numbers experimentally. If he wants to know what is the probability of a coin falling heads up, he tosses the coin a large number of times and then uses the law of large numbers to assure himself that he may use the obtained frequency ratio as an approximation to the correct value of the probability. Or he may observe that the values of a random variable are obtained as the sum of a large number of independent variables each with a negligible variance and thus be led to introduce the normal distribution. Such approximative procedures are of course common to all parts of applied mathematics.

**14. Conclusion.** Our exposition is finished. If the reader has been patient enough to read this far he may be curious enough to read farther. Our scanty bibliography will furnish a basis for such reading. For certainly not all probability theory is contained in this paper, nor as yet in any collection of books or papers. There is still much room in the field for the exercise of the analytic ingenuity and abstract generality of both classical and modern mathematics. If this paper will be instrumental in persuading mathematicians that probability is mathematics, and in causing some to look into the subject more deeply than they had previously thought worth while, it will have more than accomplished its purpose.

## BIBLIOGRAPHY

G. Birkhoff, Lattice Theory, New York, 1940.

H. Cramér, Random Variables and Probability Distributions, Cambridge, 1937.

A. Khintchine, Asymptotische Gesetze der Wahrscheinlichkeitsrechnung, Berlin, 1933.

A. Kolmogoroff, Grundbegriffe der Wahrscheinlichkeitsrechnung, Berlin, 1933.

S. Saks, Theory of the Integral, Warsaw, 1937.

J. V. Uspensky, Introduction to Mathematical Probability, New York, 1937

Reprinted from the AMERICAN MATHEMATICAL MONTHLY
Vol. 83, No. 7, August-September 1976
pp. 503–516

# AMERICAN MATHEMATICS FROM 1940 TO THE DAY BEFORE YESTERDAY*

J. H. EWING, W. H. GUSTAFSON, P. R. HALMOS,
S. H. MOOLGAVKAR, W. H. WHEELER, AND W. P. ZIEMER

**Preface.** What is the best way to present the small fragment of history described by the title above? Should this report occupy itself mainly with the statistics of the growth of Mathematical Reviews?, with the lives of mathematicians?, with lists of books and papers?, or with retracing the influences and implications that led from the bridges of Königsberg first to *analysis situs* and then to homological algebra? We decided to do none of these things, but, instead, to tell as much as possible about mathematics, the live mathematics of today. To do so within prescribed boundaries of time and space, we present the subject in the traditional "battles and kings" style of history. We try to describe some major victories of American mathematics since 1940, and mention the names of the winners, with, we hope, enough explanation (but just) to show who the enemy was. The descriptions usually get as far as statements only. We omit all proofs, but we sometimes give a brief sketch of how a proof might go. A sketch can be one sentence, or two or three paragraphs; its purpose is more to illuminate than to convince.

Progress in mathematics means the discovery of new concepts, new examples, new methods, or new facts. Schwartz's concept of distribution, Milnor's example of an exotic sphere, Cohen's method of forcing, and the Feit–Thompson theorem about simple groups are surely major by any standards. It was no trouble to find such victories to include in our list; the difficulty was to decide what to exclude. We formulated some rough rules (e.g., theorems, not theories); since at least some aspects of applied mathematics were covered by other presentations, we restricted our attention to pure mathematics; we excluded work that had neither root, nor branch, nor flower in the U.S.; and, in deciding which of two candidates to keep, we leaned toward the one of greater general interest. ("Of general interest" is not quite the same as "famous", but it's close.)

We ended up with ten "battles and kings", and we think that they draw a fair picture of what's been happening. We do not say that our ten are greater than any others, nor that they are necessarily maximal in the mathematical sense of not being lesser than any others. We do say that they would all appear, and would be discussed with respect, in any responsible history of our place and time. The total number of such "non-omittable" victories is certainly greater than ten; it may be twenty or even forty. The choice of our ten was influenced by the limits of our competence and by our personal preferences; that could not be helped. Anyone else would very likely have selected a different set of

---

* An expurgated version of this paper was delivered (by P. R. H.) as an invited address at the San Antonio meeting of the Mathematical Association of America on January 24, 1976.The authors are grateful to W. Ambrose, G. Bennett, J. L. Doob, L. K. Durst, I. Kaplansky, R. Narasimhan, I. Reiner, and F. Treves for help, including advice, references, and, especially, encouragement.

The preparation of this paper was supported in part by a grant from the National Science Foundation.

ten. We hope and think, however, that everyone's list would have a large overlap with ours, and that the local differences would not essentially alter the global picture.

In history, every moment influences its successors; to restrict attention to a time interval may be often necessary, and sometimes possible, but it is rarely natural. In the same way, every place influences all others. Since the topology of the surface of our globe is much more intricate than that of the time line, to restrict attention to one country is almost impossible. The history of mathematics is no exception: trying to describe what happened *here*, we frequently yield to the pressure of distant influences and discuss what happened *there*. We were able to stay reasonably close to our original charge just the same; if fractional credits are assigned, something like 8.25 of the ten accomplishments described below can be called American. It might be of interest to observe also that over half of the original papers we refer to appeared in the *Annals of Mathematics.*

The order of the presentations might have been based on any of several principles (e.g., what actually happened first?, what is a prerequisite for what?). We decided to arrange them in order of complexity of the underlying category, or, in other words, very roughly speaking, in order of distance from the foundations of mathematics. At the end of each section there is a small list of pertinent references. The list is intentionally incomplete. All it contains is one (or, if necessary, two or three) of the earliest papers in which the discovery appears, and a more recent exposition of the discovery whenever we could find one.

**Continuum Hypothesis.** All mathematics is derived from set theory (or, in any event, many of us believe it is) and the manipulation of sets is a simple, natural exercise (or, in any event, students have very little trouble catching on to it). Everything that any working mathematician ever needs to know about sets (and a few extra things that he never thought he needed to know) could be summarized on one printed page (or three or four printed pages, if motivation is wanted along with the formalism). Such a page would state the basic ways of making new sets out of old (e.g., the formation of sets consisting of specified elements, the formation of unions of sets of sets, and the formation of the power set, i.e., the set of all subsets, of a set); it would describe the basic properties of sets (e.g., that two sets are equal if and only if each is a subset of the other, and that no set has elements that are themselves sets that have elements continued on downwards *ad infinitum*); and it would state (as an assumption or as a conclusion, but in either case as a description of the universe that sets live in) that infinite sets exist. These basic set-theoretic statements might be regarded either as obvious factual observations or as an axiomatic description of the ZF (Zermelo–Fraenkel) structure. In either case it would be a simple matter to code them in the language of a suitable (not very complicated) computer. Such a machine could easily be taught all the rules of inference that mathematicians ever use. If, in addition, its basic data were increased by two more statements, it could, in principle, easily print out all known mathematics (and a lot that is not yet known).

The two statements that history has subjected to extra scrutiny are AC (the axiom of choice) and GCH (the generalized continuum hypothesis). AC says that, for each set $X$, there is a function $f$ from the power set of $X$ into $X$ itself such that $f(A) \in A$ for each non-empty subset $A$ of $X$; GCH says that each subset of the power set of an infinite set $X$ is in one-to-one correspondence either with some subset of $X$ or with the entire power set — there is nothing in between.

Is AC true? The question has often been likened to a similar one about Euclid's parallel postulate. In both cases there is a more or less pleasant axiom system and a less pleasant, more complicated, non-obvious additional axiom. If the extra axiom is a consequence of the basic ones, it is true, and all is well; if its negation is a consequence of the basic ones, it is false, and, for better or for worse, the question is definitively answered. The same question can, of course, be asked about GCH. It has long been known that GCH implies AC; in view of this there is an obvious connection between the two answers.

The answers are subtle and profound intellectual achievements. Gödel proved (1940) that AC and GCH are not false (i.e., that they are consistent with the axioms of ZF), and Paul Cohen proved (1964) that they are not true (i.e., that they are independent of ZF).

Gödel argued by the construction of a suitable model. If, he said, ZF is consistent, so that there is a universe $V$ of sets satisfying the basic axioms of ZF, then, he proved, there is a "sub-universe" that also satisfies them, and in which, moreover, both AC and GCH are true. The sub-universe Gödel constructed was the class $L$ of "constructible" sets. (The word is given a very liberal but completely precise meaning; roughly speaking, the constructible sets are the ones that can be obtained from the empty set by a transfinite sequence of elementary set-theoretic constructions.) The class $L$ is a substructure of $V$ in the familiar mathematical sense of that word: the objects of $L$ are some of the objects of $V$, and the relation $\in$ among them is the restriction of the set-theoretic $\in$ in $V$ to the objects of $L$. The existence of a model such as $L$ (constructed out of a hypothetically consistent model $V$) proves the consistency of AC and GCH the same way as the existence of the Euclidean plane proves the consistency of the parallel postulate.

Cohen's argument was similar but harder. It is reminiscent of Felix Klein's construction of a Lobachevskian plane by endowing a Euclidean disk with a new metric. Cohen started with a suitable model of ZF and adjoined new objects to it. The new objects are "classes" (but not sets) in the old model. The adjunctions proceed by a new method called "forcing", which, once it was discovered, was found to be applicable in many parts of set theory. Cohen's proof constructs an infinite sequence of better and better finite approximations to the new objects. Roughly speaking, each property of the new model is "forced" by properties of the old model and one of the approximations. Depending on how the details are adjusted, the end result can be a model of ZF in which AC is false, or a model of ZF in which AC is true but even the classical un-generalized continuum hypothesis CH is false. (CH is GCH for a countably infinite set.) Conclusion: AC and CH are independent of ZF.

REFERENCES. [1] P. J. Cohen, The independence of the continuum hypothesis, Proc. N. A. S., 50(1963) 1143–1148 and 51(1964) 105–110. [2] P. J. Cohen, Set theory and the continuum hypothesis, Benjamin, New York, 1966(MR 38 # 999). [3] J. B. Rosser, Simplified independence proofs, Academic Press, New York, 1969(MR 40 # 2536). [4] T. J. Jech, Lectures in set theory, with particular emphasis on the method of forcing, Springer, Berlin, 1971(MR 48 # 105).

**Diophantine Equations.** The continuum hypothesis was the subject of Hilbert's first problem (in the famous list of 23 problems that he proposed in 1900); Hilbert's tenth problem concerned the solvability of Diophantine equations. The problem was to design an algorithm, a computational procedure, for determining whether an arbitrarily prescribed polynomial equation with integer coefficients has integer solutions. It is in some respects more natural and sometimes technically easier to discuss the *positive* integer solutions (solutions in $\mathbb{Z}_+$) of polynomial equations with *positive* integer coefficients. Caution: that does not mean equations such as $p(x) = 0$ only. The problem includes the search for $x$'s such that $p(x) = q(x)$; more generally, it includes the search for $n$-tuples $(x_1, \ldots, x_n)$ such that $p(x_1, \ldots, x_n) = q(x_1, \ldots, x_n)$; and, in complete generality, it means the search for $n$-tuples $(x_1, \ldots, x_n)$ for which there exist $m$-tuples $(y_1, \ldots, y_m)$ such that

$$p(x_1, \ldots, x_n, y_1, \ldots, y_m) = q(x_1, \ldots, x_n, y_1, \ldots, y_m).$$

For each $p$ and $q$ (in $n + m$ variables) the solution set, in the latter sense, is called a "Diophantine set" in $\mathbb{Z}_+^n$.

What does it mean to say that there is an algorithm for deciding solvability? A reasonable way to answer the question is to offer a definition of computability for sets and functions, and then to define an algorithm in terms of computability.

When does a function from $\mathbb{Z}_+$ to $\mathbb{Z}_+$, or, more generally, a function from $\mathbb{Z}_+^n$ to $\mathbb{Z}_+$ deserve to be called "computable"? There is general agreement on the definition nowadays: computable functions (also called "recursive" functions) are the ones obtained from certain easy functions (constant, successor, coordinate) by three procedures (composition, minimalization, primitive recursion). The details do not matter here (they won't be used anyway); it might be comforting to know, however, that they are not at all difficult. A set (in $\mathbb{Z}_+$, or, more generally, in $\mathbb{Z}_+^n$) will be called computable in case its characteristic function is computable. Consequence: a set (in $\mathbb{Z}_+^n$) is computable if and only if its complement is computable.

Consider now all polynomial equations (in the sense described above), and let $\{E_1, E_2, E_3, \ldots\}$ be an enumeration of them. (In order for what follows to be in accord with the intuitive concept of an algorithm, the enumeration should be "effective" in some sense. That can be done, and it is relatively easy.) The indices $k$ for which $E_k$ has a solution (in the sense described above) form a subset $S$ of $\mathbb{Z}_+$. The Hilbert problem (is there an algorithm?) can be expressed as follows: is $S$ a computable set? The answer is no. The answer was a long time coming; it is the result of the cumulative efforts of J. Robinson (1952), M. Davis (1953), H. Putnam (1961), and Y. Matijasevič (1970).

The central concept in the proof is that of a Diophantine set, and the major step proves that every computable set is Diophantine. The techniques make ingenious use of elementary number theory (e.g., the Chinese remainder theorem, and a part of the theory of Fibonacci numbers, or, alternatively, of Pell's equation). The proof exhibits some interesting Diophantine sets whose Diophantine character is not at all obvious (e.g., the powers of 2, the factorials, and the primes).

One way to prove that $S$ (the index set of the solvable equations) is not computable is by contradiction. If $S$ were computable, then it would follow (by a slight bit of additional argument) that each particular Diophantine set (i.e., the solution set of each particular equation) is computable, and hence (by the "major step" of the preceding paragraph) that the complement of every Diophantine set is Diophantine. The contradiction is derived by exhibiting a Diophantine set whose complement is not Diophantine.

This last step uses a version of the familiar Cantor diagonal argument. The idea is "effectively" to enumerate all Diophantine subsets of $\mathbb{Z}_+$, as $\{D_1, D_2, D_3, \ldots\}$, say, prove that the set $D^* = \{n \colon n \in D_n\}$ is Diophantine (that takes some argument), and, finally, to prove that the complement $\mathbb{Z}_+ - D^* = \{n \colon n \notin D_n\}$ is not Diophantine (that's where Cantor comes in).

REFERENCES. [1] J. Robinson, Existential definability in arithmetic, Trans. A. M. S., 72(1952) 437–449 (MR14–4). [2] M. Davis, Arithmetical problems and recursively enumerable predicates, J. Symb. Logic, 18(1953) 33–41 (MR14–1052). [3] M. Davis, H. Putnam, and J. Robinson, The decision problem for exponential Diophantine equations, Ann. Math., 74(1961) 425–436 (MR24 # A3061). [4] Y. Matijasevič, The Diophantineness of enumerable sets (Russian), Dokl. Akad. Nauk S. S. S. R., 191(1970) 279–282; improved English translation, Soviet Math. Doklady, 11(1970) 354–358 (MR41 # 3390). [5] M. Davis, Hilbert's tenth problem is unsolvable, this MONTHLY, 80 (1973) 233–269 (MR47 # 6465).

**Simple Groups.** So much for the foundations. The next subject up the ladder is algebra; in the present instance, group theory.

Every group $G$ has two obvious normal subgroups, namely $G$ itself and, at the other extreme, the subgroup 1. A group is called "simple" if these are all the normal subgroups it has.

Simple groups are like prime numbers in two ways: they have no proper parts, and every finite group can be constructed out of them. (By general agreement the trivial positive integer 1 is not called a prime, but the trivial group 1 is called simple. Too bad, but that's how it is.)

Suppose, indeed, that $G$ is finite, and let $G_1$ be a maximal normal subgroup of $G$. (To say that $G_1$ is maximal means that $G_1$ is a proper normal subgroup of $G$ that is not included in any other proper normal subgroup of $G$.) If $G$ is simple, then $G_1 = 1$; in any event, the maximality of $G_1$ implies that the quotient group $G/G_1$ is simple. The relation between $G$, $G_1$, and $G/G_1$ (group, normal subgroup, quotient group) is sometimes expressed by saying that $G$ is an extension of $G/G_1$ by $G_1$. In this language, every finite group (except the trivial group 1) is an extension of a simple group by a group of strictly smaller order. The statement is a group-theoretic analogue of the number-theoretic one that says that every positive integer (except 1) is the product of a prime by a strictly smaller positive integer.

If $G_1$ is not trivial, the preceding paragraph can be applied to it; the result is a maximal normal subgroup $G_2$ in $G_1$, such that $G_1$ is an extension of the simple group $G_1/G_2$ by $G_2$. The procedure can be repeated so long as it produces non-trivial subgroups; the end-product is a chain

$$G = G_0 \supset G_1 \supset G_2 \supset \cdots \supset G_n = 1$$

(a "composition series") with the property that each $G_i/G_{i+1}$ is simple ($i = 0, \ldots, n-1$). A great part of the problem of getting to know all finite groups reduces in this way to the determination of all finite simple groups. (The celebrated Jordan-Hölder-Schreier theorem is the comforting reassurance that, to within isomorphism, the composition factors $G_i/G_{i+1}$ are uniquely determined by $G$, except for the order in which they occur.)

The abelian ones among the finite simple groups are easy to determine: they are just the cyclic groups of prime order. That's easy. What's hard is to find all non-abelian ones. Some examples of simple groups are easy to come by; among permutation groups, for instance, the most famous ones are the alternating groups of degree 5 or more. The known simple groups did not exhibit any pattern, and even the simplest questions about them resisted attack. Burnside conjectured, for instance, that every non-abelian simple group has even order, but that conjecture stood as an open problem for more than 50 years.

In a spectacular display of group-theoretic power, Feit and Thompson (1963) settled Burnside's conjecture (it is true). The proof occupies an entire issue (over 250 pages) of the *Pacific Journal.* It is technical group theory and character theory. Some reductions in it have been made since it appeared, but no short or easy proof has been discovered. The result has many consequences, and the methods also have been used to attack many other problems in the theory of finite groups; a subject that was once pronounced dead by many has shown itself capable of a vigorous new life.

REFERENCE. W. Feit and J. G. Thompson, Solvability of groups of odd order, Pac. J. Math., 13(1963) 775–1029 (MR29 # 3538).

**Resolution of Singularities.** Algebra becomes richer, and harder, when it is mixed with and applied to geometry; one of the richest mixtures is the old but very vigorous subject known as algebraic geometry. This section reports the solution of an old and famous problem in that subject.

Let $k$ be an algebraically closed field, and let $k^n$ be, as usual, the $n$-dimensional coordinate space over $k$. (The heart of the matter in what follows will be visible to those who insist on sticking to the field of complex numbers in the role of $k$.) An "affine algebraic variety" $V$ in $k^n$ is the locus of common zeros of a collection of polynomials in $n$ variables with coefficients in $k$. Since only the zeros matter, the collection itself is not important; it can be replaced by any other collection that yields the same locus. Thus, if $R$ is the ring of *all* polynomials in $n$ variables with coefficients in $k$, and if $I$ is the ideal in $R$ generated by the prescribed collection, then $I$ will define the same variety; there is, therefore, no loss of generality in assuming that the collection was an ideal to begin with.

The objects of interest on varieties are their "singular points". Intuitively, these are points where the "tangent vectors" are not as they should be. Consider, for example, the curves defined by

$$y^2 = x^3 + x^2 \quad \text{and} \quad y^2 = x^3.$$

(Since the ground field was restricted to be algebraically closed, the *real* planar curves with these equations are not the right things to look at, but they are more lookable at than the complex curves, which lie in the complex plane. Warning: the complex plane has four real dimensions. To the algebraic geometer, the familiar "complex plane" of analysis is the complex *line.*) The first of these comes in to the origin from the first quadrant with slope 2, has a loop in the left half plane, and goes out from the origin to the fourth quadrant with slope $-2$; it has the origin as a double point. The other one comes in to the origin from the first quadrant with slope 0, and goes out the same way to the fourth quadrant; it has the origin as a cusp.

The effective way to deal with singular points begins by giving a purely algebraic description of them. Consider, for this purpose, the ring $R_V$ of polynomial functions on $V$ (i.e., the restrictions of the polynomials in $R$ to $V$). If $N_V$ is the ideal of $R$ consisting of the polynomials that vanish on $V$, then, clearly, $R_V = R/N_V$. Each point $\alpha = (\alpha_1, \ldots, \alpha_n)$ of $V$ induces a maximal ideal $N_\alpha$ in $R$ (consisting of the set of polynomials that vanish at $\alpha$); clearly $N_V \subset N_\alpha$.

The next step (in the program of defining singular points algebraically) is to form a new ring that studies the local behavior of functions near $\alpha$. The idea is (very roughly) this. (i) Consider pairs $(U, f)$, where $U$ is a "neighborhood" of $\alpha$ and $f$ is a rational function with no poles in $U$. (ii) Define an equivalence relation for pairs by writing $(U, f) \sim (U', f')$ exactly when there is a neighborhood $U''$ of $\alpha$, included in $U \cap U'$, such that $f = f'$ on $U''$. (iii) The equivalence classes ("germs") form a ring (with, for example,

$$[(U, f)] + [(U', f')] = [(U \cap U', f + f')]),$$

called the "local ring" of $V$ at $\alpha$.

From the algebraic point of view, the preceding topological considerations are just heuristic; they will now be replaced by an algebraic construction. The process is, appropriately, called "localization". (i) Consider pairs $(f, g)$, where $f$ and $g$ are in $R$ and $g \notin N_\alpha$. (ii) Define an equivalence relation for pairs by writing $(f, g) \sim (f', g')$ exactly when there is an $h$ not in $N_\alpha$ such that $h \cdot (fg' - gf') = 0$. (iii) Write $f/g$ for the equivalence class of $(f, g)$. The equivalence classes form a ring $R_\alpha$ (with the usual rules of operations for fractions). The ring $R_\alpha$ is indeed a "local ring" in the customary algebraic sense: it has a *unique* maximal ideal, namely the one formed by the elements of $R_\alpha$ that vanish at $\alpha$.

To motivate the next step, pretend, again, that the subject is not algebraic geometry, but analytic geometry. In that case $R_\alpha$ would consist of Taylor series at $\alpha$ convergent near $\alpha$, and the ideal $N_\alpha$ of germs vanishing at $\alpha$ would consist of the Taylor series at $\alpha$ with vanishing constant term. The linear terms of a Taylor series are, in some sense, first order differentials. One way to capture just those terms is to "ignore" higher order terms. More precisely: consider the ideal $N_\alpha^2$, which, in the analytic case, consists of the Taylor series with vanishing constant term and vanishing linear term, and form $N_\alpha/N_\alpha^2$.

The definition is now easy to formulate. The "dimension" $d$ of $V$ is, by definition, the minimum of the dimensions (over the field $k$, of course) of all the quotient spaces $N_\alpha/N_\alpha^2$; a point $\alpha$ is "singular" when $\dim(N_\alpha/N_\alpha^2) > d$. It is not difficult to see that for the two curves mentioned as examples above, the origin is indeed a singular point in the sense of this definition.

One of the main problems of algebraic geometry is to "get rid of" singular points. For this purpose the discussion is restricted to "irreducible" varieties, i.e., to the ones for which $R_V$ is an integral domain, or, equivalently, $N_V$ is a prime ideal. In that case, form the field of fractions $F_V$ of $R_V$. Two varieties $V$ and $W$ are "birationally equivalent" if $F_V$ and $F_W$ are isomorphic. This means roughly that $V$ and $W$ parametrize one another by rational mappings at all but finitely many places. The problem of "resolution of singularities" is that of finding a non-singular variety birationally equivalent to $V$.

The subject has a long history. Curves were handled by Max Noether in the 19th century. Surfaces were the subject of much geometric discussion by the Italian school; a rigorous proof was found by R. J. Walker (1935). For varieties of arbitrary dimension, over fields of characteristic 0, the final victory was inspired by Zariski's work; it was won by Hironaka (1964).

REFERENCE. H. Hironaka, Resolution of singularities of an algebraic variety over a field of characteristic zero, Ann. Math., 79(1964) 109–326 (MR 33 # 7333).

**Weil Conjectures.** The mathematician's work is often most difficult (and most rewarding) when he reasons by analogy, when he guesses that *this* situation ought to be just like *that* one. In 1949 A. Weil, reasoning in this way, proposed three conjectures that have profoundly influenced the development of algebraic geometry over the past 25 years.

The conjectures appeared in a paper entitled "Numbers of solutions of equations in finite fields", which was ostensibly a survey of previous work. Counting the number of solutions of a polynomial equation in several variables over a finite field was a classical problem, investigated by Gauss, Jacobi, Legendre, and others, but Weil took a new point of view. To understand his approach, consider the special case of the homogeneous equation

$$(\dagger) \qquad a_0x_0^n + a_1x_1^n + \cdots + a_rx_r^n = 0,$$

where the coefficients $a_i$ are in the prime field $F$ of $p$ elements. The basic problem is to count the number of solutions in $F$, but, to number theorists, it is just as important to count the number of solutions in any finite extension field of $F$. Recall that, for every positive integer $k$, there is a unique extension field $F_k$ of $F$ with $p^k$ elements. What Weil did was to count the number of solutions of (†) in each field $F_k$, and then code that information in a generating function.

To do this economically, examine the solution set of an equation such as (†). There is, of course, always the trivial solution, where all the $x_i$ are zero; that one is justly regarded as trivial. If $(x_0, x_1, \ldots, x_r)$ is a non-trivial solution, and if $0 \neq c \in F_k$, then $(cx_0, cx_1, \ldots, cx_r)$ is also a non-trivial solution. Each non-trivial solution generates in this way $p^k - 1$ others, and there is no virtue in counting them all separately. It is natural, therefore, to consider the $r$-dimensional "projective space" $P^r(F_k)$, i.e., the set of non-trivial ordered $(r+1)$-tuples of elements of $F_k$, where two are identified if one is a scalar multiple of the other. (This is exactly analogous to the familiar real and complex projective spaces.) The problem in these terms is to count the number of "points" in $P^r(F_k)$ that are "solutions" of (†).

That is precisely what Weil did. He let $N_k$ be the number of solutions of (†) in $P^r(F_k)$, considered the generating function $G$,

$$G(u) = \sum_{k=1}^{\infty} N_k u^{k-1},$$

and proved a remarkable statement: $G$ is the logarithmic derivative of a *rational* function. That is: there exists a rational function $Z$ such that

$$\sum_{k=1}^{\infty} N_k u^{k-1} = \frac{d}{du} \log Z(u),$$

or, in other words, if

$$Z(u) = \exp\left( \sum_{k=1}^{\infty} \frac{N_k}{k} u^k \right),$$

then $Z$ is rational. The function $Z$ satisfies a functional equation analogous to the one satisfied by the Riemann zeta function, and it is appropriate to refer to $Z$ as the zeta function associated with the equation (†). Motivated by classical problems that the Riemann zeta function gave rise to, Weil studied and was able to determine many properties of the zeros and the poles of $Z$.

Here is where Weil's paper reaches its climax. Weil wanted to extend the results about (†) to algebraic varieties in $P^r(F_k)$, i.e., to the solution sets of *systems* of homogeneous equations in $r$ variables. The notion of a zeta function, originally defined by Riemann, was extended by Dedekind to algebraic number fields, by Artin to function fields, and now, by Weil, to algebraic varieties. (The varieties to be considered should be non-singular. It doesn't matter here what the general definition of that condition is; for most fields it can be defined as usual by requiring that the Jacobian of the system of equations have maximal rank at every point.) Given a system of equations, with coefficients in $F$, let $N_k$ be, as before, the number of solutions in $P^r(F_k)$. Weil advanced the following conjectures. One: the function $Z$, defined as before by

$$Z(u) = \exp\left( \sum_{k=1}^{\infty} \frac{N_k}{k} u^k \right),$$

is rational. Two: $Z$ satisfies a particular functional equation, which, as before, bears a striking resemblance to the one satisfied by the Riemann zeta function. Three: the reciprocals of the zeros and the poles of $Z$ are algebraic integers and their absolute values are powers of $\sqrt{p}$. (This is called the generalized Riemann hypothesis.)

All this might seem far removed from what is normally thought of as geometry, and, although several examples were known, it might seem that Weil made his conjectures on strikingly little evidence. What was really behind the conjectures? The answer is contained in the last paragraph of Weil's paper, where he suggests that there is an analogy between the behavior of these varieties (for fields of characteristic $p$) and that of the classical varieties (for the field of complex numbers).

In 1960 Dwork established the rationality conjecture (without the condition of non-singularity). The final triumph came in 1974: using twenty years' of results of the Grothendieck school, Deligne established all the Weil conjectures, and, perhaps more importantly, proved that there is a beautiful connection between the theory of varieties over fields of characteristic $p$ and classical algebraic geometry. "God ever geometrizes", said Plato, and "God ever arithmetizes", said Jacobi; the Weil conjectures show, better than anything else, how He can do both at once.

REFERENCES. [1] A. Weil, Numbers of solutions of equations in finite fields, Bull. A. M. S., 55(1949) 497–508 (MR 10–592). [2] P. Deligne, La conjecture de Weil I, Inst. Haute Études Sci. Publ. Math., No. 43(1974) 273–307 (MR 49 # 5013). [3] J. A. Dieudonné, The Weil conjectures, The Mathematical Intellingencer, No. 10(September 1975) 7–21.

**Lie Groups.** So much for algebra, with or without geometry. The next subject points toward some of the later analytic ones by mixing algebra with topology. The result, like a few other outstanding results of mathematics, seems to get something for nothing, or, at the very least, to get quite a lot for an astonishingly low price: One of the most famous results of this kind occurs in the early part of courses on complex function theory: it asserts that a differentiable function on an open subset of the complex plane is necessarily analytic.

Hilbert's fifth problem asked for such a something-for-nothing result. The context is the theory of topological groups. A topological group is a set that is both a Hausdorff space and a group, in such a way that the group operations

$$(x, y) \mapsto xy \quad \text{and} \quad x \mapsto x^{-1}$$

are continuous. A typical example is the set of all $2 \times 2$ real matrices of the form $\begin{pmatrix} x & y \\ 0 & 1 \end{pmatrix}$ with $x > 0$; the topological structure is that of the right half plane (all $(x, y)$ with $x > 0$), and the multiplicative structure is the usual one associated with matrices. Equivalently: define multiplication in the right half plane by

$$(x, y) \cdot (x', y') = (xx', xy' + y);$$

since

$$(x, y)^{-1} = \left(\frac{1}{x}, \frac{-y}{x}\right),$$

it is clear that both multiplication and inversion are continuous.

This example has an important special property: it is "locally Euclidean" in the sense that every point has a neighborhood that is homeomorphic to an open ball in (2-dimensional) Euclidean space. (Equivalently: every point has a "local coordinate system".) An even more important special property of the example is that the group operations, regarded as functions on the appropriate Euclidean space, are not only continuous but even analytic. If a group is locally Euclidean, i.e., if it can be "coordinatized" at all, then there are many ways of coordinatizing it; if at least one of them is such that the group operations are analytic, the group is called a "Lie group". Hilbert's fifth problem was this: is every locally Euclidean group a Lie group?

The analogy of this problem with the one in complex function theory is quite close. It is relatively elementary that a twice-differentiable function is analytic; it has been known for a long time that if a topological group has sufficiently differentiable coordinates, then it has analytic ones.

Immediately after the discovery of Haar measure, von Neumann (1933) applied it to prove that the answer to Hilbert's question is yes for compact groups. A little later Pontrjagin (1939) solved the abelian case, and Chevalley (1941) solved the solvable case. (Sorry about that, but "solvable" is a technical word here and its use is unavoidable.)

The general case was solved in 1952 by Gleason and, jointly, by Montgomery and Zippin; the answer to Hilbert's question is yes. What Gleason did was to characterize Lie groups. (Definition: a topological group "has no small subgroups" if it has a neighborhood of the identity that includes no subgroups of order greater than 1. Characterization: a finite-dimensional locally compact group with no small subgroups is a Lie group.) Montgomery and Zippin used geometric-topological tools (and Gleason's theorem) to reach the desired conclusion.

Warning: the subject cannot be considered closed. The question can be generalized in ways that are both theoretically and practically valuable. Groups can be replaced by "local groups", and abstract groups can be replaced by groups of transformations acting on manifolds. The best kind of victory is the kind that indicates where to look for new worlds to conquer, and the one over Hilbert's fifth problem was that kind.

REFERENCES. [1] A. Gleason, Groups without small subgroups, Ann. Math., 56(1952) 193–212 (MR 14–135). [2] D. Montgomery and L. Zippin, Small subgroups of finite-dimensional groups, Ann. Math., 56(1952) 213–241 (MR 14–135). [3] D. Montgomery and L. Zippin, Topological transformation groups, Interscience, New York, 1955 (MR 17–383).

**Poincaré Conjecture.** A "manifold" is a topological space (a separable Hausdorff space to be exact) that is locally Euclidean. Manifolds have been the central subject of topology for many years, and still are. Hilbert's fifth problem was about group manifolds; the Poincaré conjecture is about the connectedness properties of smooth manifolds. A "differential manifold" is a manifold endowed with local coordinate systems such that the change of coordinates from one coordinate neighborhood to an overlapping one is smooth. "Smooth" in this context is a generally accepted abbreviation for $C^\infty$, i.e., for infinitely differentiable.

The axioms of Euclidean plane geometry characterize the plane. This kind of activity (find the central core of a subject, abstract it, and use the result as an axiomatic characterization) is frequent and useful in mathematics. Since spheres are the principal concept of a large part of topology, it is natural to try to subject them too to the axiomatic approach. The attempt has been made, and, to a large extent, it was successful.

The 1-sphere, for instance (i.e., the circle), is a compact, connected 1-manifold (i.e., a manifold of dimension 1), and that's all it is: to within a homeomorphism every compact connected 1-manifold is a 1-sphere.

For the 2-sphere, the facts are more complicated: both the 2-sphere $S^2$ and the torus $T^2$ $(= S^1 \times S^1)$ are compact connected 2-manifolds, and they are not homeomorphic to each other. To distinguish $S^2$ from $T^2$ and, more generally, from a sphere with many handles, it is necessary to observe that, although both $S^2$ and $T^2$ are connected, $S^2$ is more connected. In the appropriate technical language, $S^2$ is "simply connected" and $T^2$ is not. The relevant definitions go as follows. Suppose that $X$ and $Y$ are topological spaces and that $f$ and $g$ are continuous functions from $X$ to $Y$; write $I$ for the unit interval $[0, 1]$. The functions $f$ and $g$ are "homotopic" if there exists a continuous function $h$ from $X \times I$ to $Y$ such that $h(x, 0) = f(x)$ and $h(x, 1) = g(x)$ for all $x$. (Intuitively: $f$ can be continuously deformed to $g$.) The space $Y$ is simply connected if every continuous function from $S^1$ to $Y$ is homotopic to a constant. (Intuitively: every closed curve can be shrunk to a point.) Once this concept is at hand, the characterization of the 2-sphere becomes easy to state: to within a homeomorphism, every compact, connected, simply connected 2-manifold is a 2-sphere.

The discussion of dimensions 1 and 2 does not yet provide a firm basis for guessing the general case, but it does at least make the following concept plausible. There is a way of defining "$k$-connected" that generalizes "connected" ($k = 0$) and "simply connected" ($k = 1$): just replace $S^1$ in the definition of simple connectivity by $S^j$, $j = 0, 1, \ldots, k$. Thus: a space $Y$ is $k$-connected if, for each $j$ between 0 and $k$ inclusive, every continuous function from $S^j$ to $Y$ is homotopic to a constant.

The general Poincaré conjecture is that if a smooth compact $n$-manifold is $(n-1)$-connected, then it is homeomorphic to $S^n$. For $n = 1$ and $n = 2$ the result has been known for a long time; the big recent step was the proof of the assertion for all $n \geqq 5$. The proof was obtained by Smale (1960). Shortly thereafter, having heard of Smale's success, Stallings gave another proof for $n \geqq 7$ (1960) and Zeeman extended it to $n = 5$ and $n = 6$ (1961). For $n = 3$ (the original Poincaré conjecture) and for $n = 4$ the facts are not yet known.

Actually Smale proved a much stronger result. He showed how certain manifolds could be obtained by gluing disks together. His results provide a starting point for a classification of simply connected manifolds.

REFERENCES. **[1]** S. Smale, The generalized Poincaré conjecture in higher dimensions, Bull. A.M.S., 66(1960) 373–375 (MR 23 # A2220). **[2]** J. R. Stallings, Polyhedral homotopy-spheres, Bull. A.M.S., 66(1960) 485–488 (MR 23 # A2214). **[3]** E. C. Zeeman, The generalized Poincaré conjecture, Bull. A.M.S., 67(1961) 270 (MR 23 # A2215). **[4]** S. Smale, Generalized Poincaré's conjecture in dimensions greater than four, Ann. Math., 74(1961) 391–406 (MR 25 # 580).

**Exotic Spheres.** A "diffeomorphism" between two differential manifolds is a homeomorphism such that both it and its inverse are smooth. Homeomorphism is an equivalence relation between manifolds; the equivalence classes (homeomorphism classes) consist of manifolds with the same topological properties. Similarly, diffeomorphism is an equivalence relation between differential manifolds, and the equivalence classes (diffeomorphism classes) consist of manifolds with the same differential properties. Are these concepts really different? Is diffeomorphism really more stringent than homeomorphism? The answer is yes, even for topologically very well-behaved manifolds, but that is far from obvious. An example constructed by Milnor in 1956 came as a surprise, and, according to Hassler Whitney, that single, isolated example led to the modern flowering of differential topology.

Milnor's example is the 7-sphere. For every positive integer $n$, the $n$-sphere $S^n$ is embedded in Euclidean $(n+1)$-space in a natural way, and thus has a natural differential structure. Milnor showed that there exists a differential manifold that is homeomorphic but not diffeomorphic to $S^7$; such a manifold has come to be called an "exotic" 7-sphere.

To prove the assertion, there are three problems to solve: (1) find a candidate, (2) prove that it is homeomorphic to $S^7$, and (3) prove that it is not diffeomorphic to $S^7$. The first problem was easy (with hindsight); the candidate was a space (a 3-sphere bundle over the 4-sphere) that had been familiar to topologists for a number of years. Milnor solved the second problem using Morse theory. A Morse function on a differential manifold is a real-valued smooth function with only non-degenerate critical points. The $n$-sphere has a Morse function with exactly two critical points (project onto the last coordinate and consider two poles). A theorem of G. Reeb's goes in the other direction: if a differential manifold has a Morse function with exactly two critical points, then it is homeomorphic to a sphere. Milnor showed that his candidate had such a Morse function. The third problem was the hardest. Here Milnor used two facts: first, that $S^7$ is the boundary of the unit ball in $\mathbb{R}^8$, and, second, that his candidate was presented as the boundary of an 8-dimensional manifold $W$. If the candidate were diffeomorphic to $S^7$, then, using the diffeomorphism, one could glue the unit ball onto $W$ and obtain an 8-dimensional manifold that (as Milnor showed) cannot exist.

Once it was known that exotic 7-spheres could exist, it was natural to ask how many there were, i.e., how many diffeomorphism classes there were. Milnor and Kervaire showed that there are 28.

What about the other spheres? Again Milnor and Kervaire showed that the set of differential $n$-spheres (modulo diffeomorphism) could be made into a finite abelian group, with the "natural" sphere as the zero element; the group operation is the "connected sum", which is the natural gluing together of manifolds. The group is trivial for $n < 7$; it has order 28 for $n = 7$, order 2 for $n = 8$, order 8 for $n = 9$, order 6 for $n = 10$, and order 992 for $n = 11$. For $n = 31$, there are over sixteen million (diffeomorphism classes of) exotic spheres.

There are two systematic ways of constructing exotic spheres. The first is Milnor's "plumbing" construction (joining holes by tubes), which presents exotic spheres as boundaries of manifolds assembled by cutting and pasting. The other method (due to Brieskorn, Pham, and others) gives preassembled examples. For each finite sequence $(a_1,\ldots,a_n)$ of positive integers, let $\Sigma(a_1,\ldots,a_n)$ be the set of those zeros of the polynomial $z_1^{a_1} + \cdots + z_n^{a_n}$ that lie on the unit sphere in complex $n$-space. Milnor gave precise criteria on the $n$-tuple that ensure that this manifold is homeomorphic to a sphere of the appropriate dimension (which is $2n-3$, by the way). For example, as $k$ runs from 1 to 28, the manifolds $\Sigma(3, 6k-1, 2, 2, 2)$ provide representatives for the 28 different diffeomorphism classes of 7-spheres.

REFERENCES. [1] J. W. Milnor, On manifolds homeomorphic to the 7-sphere, Ann. Math., 64(1956) 399–405 (MR 18–498). [2] J. W. Milnor, Differential topology, Lectures on Modern Mathematics vol. II, pp. 165–183, Wiley, New York, 1964 (MR 31 # 2731).

**Differential Equations.** Differential concepts play an important role everywhere, including pure algebra and, as above, topology. Differential equations are what make the world go around, and anyone who wants to predict and perhaps partly to change how the world goes around must know about differential equations and their solutions.

Differential equations are classified in a curiously primitive manner according to the number of independent variables that are involved in differentiation, and the way in which the unknown functions enter. The classification is "one" and "many" in the one case, and "good" and "not-so-good" in the other, or, in terms of the corresponding adjectives that apply to the equations, "ordinary" and "partial" in the one case, and "linear" and "non-linear" in the other. This report is concerned with linear equations only, and partial ones at that; ordinary ones make just a brief appearance at the beginning, to set the stage.

The beginnings of the theory of ordinary linear differential equations are simple and satisfactory; they can be found in elementary textbooks. If $p$ is a polynomial

$$p(\xi) = \sum_{j=0}^{k} a_j \xi^j,$$

and if $D = d/dx$, then $P = p(D)$ is a differential operator, and $Pu = g$ (for given $g$ and unknown $u$) is the typical linear O.D.E. with constant coefficients. If $g$ is continuous (a reasonable, useful, but much too special assumption), then the equation always has a solution. The conclusion remains true even for variable coefficients (i.e., in case the $a_j$'s themselves are functions of $x$), provided they are subjected to some mild restrictions. It is, for instance, sufficient that the $a_j$'s be continuous and that the "principal" coefficient $a_k$ have no zeros.

For partial differential equations even the beginnings are non-trivial and new, and, for instance, even the theory for constant coefficients belongs to the most recent period of research. The formulation of the problem is easy enough: consider a polynomial in several variables $\xi_1,\ldots,\xi_n$, and obtain a differential operator $P$ by replacing $\xi_j$ by $\partial/\partial x_j$; the problem is to solve $Pu = g$ for $u$.

To avoid some not especially enlightening and not especially useful epsilontic hairsplitting, it has become customary to take $g$ (and to seek $u$) in either the most or the least restrictive class of objects in sight. The most restrictive class consists of the smooth (infinitely differentiable) functions on whatever domain is under consideration ($\mathbb{R}^n$, an open set in $\mathbb{R}^n$, a manifold); the other extreme is represented by

Laurent Schwartz's distributions. (The motivation of distribution theory is that functions $f$ induce linear functionals $\phi \mapsto \int \phi(x)f(x)dx$ on $C^\infty$. A "distribution" is a suitably continuous linear functional, not necessarily one induced by a function. The analogy between the generalization and its source suggests an appropriate definition of differentiation for distributions, and with that definition the theory of partial differential equations is off and running.)

Partial differential equations is an old subject and a widely applied one, and it is astonishing that the basic theorem is as recent as it is; it seems only the day before yesterday that Ehrenpreis (1954) and Malgrange (1955) proved that every linear P.D.E. with constant coefficients is solvable. If the right hand side is smooth, there is a smooth solution; even if the right hand side is allowed to be an arbitrary distribution, there is a distribution solution. The subject is exhaustively treated in Ehrenpreis's book (1962) and can be regarded as closed.

So far so good; the proofs are harder than for O.D.E.'s, but the facts are pleasant. The theory for variable (i.e., function) coefficients is much harder, much less known, and nowhere near finished. Two exciting contributions to it in the late 1950's showed that old guesses and old methods were woefully inadequate.

As for old guesses: Hans Lewy produced (1957) an inspired and amazingly simple example of a P.D.E. with variable (but *very* smooth) coefficients that has no solutions at all. Lewy's polynomial is of degree 1,

$$p(x, \xi) = a_1\xi_1 + a_2\xi_2 + a_3\xi_3,$$

where the coefficients $a_1$, $a_2$, $a_3$ are functions of three variables $x_1$, $x_2$, $x_3$, and, in fact, the first two are constants:

$$a_1 = -i, \quad a_2 = 1, \quad a_3 = -2(x_1 + ix_2).$$

The corresponding differential operator is, of course,

$$P = -i\frac{\partial}{\partial x_1} + \frac{\partial}{\partial x_2} - 2(x_1 + ix_2)\frac{\partial}{\partial x_3}.$$

What Lewy proved is that for almost every $g$ in $C^\infty$ (in the sense of Baire category) the equation $Pu = g$ is satisfied by no distribution whatever.

At about the same time (1958) Calderón studied the uniqueness of the solution of certain important partial differential equations (under suitable initial conditions). He showed, in effect, that if $Pu = 0$, with $u = 0$ for $t \leqq 0$ (intuitively, "$t$" here is time), then, locally $u$ remains 0 for some positive time. Calderón's methods were transplanted from harmonic analysis; they introduced singular integrals into the subject, whence, a little later, came pseudo-differential operators and Fourier integral operators. These ideas have dominated the subject ever since.

Hörmander analyzed and generalized Lewy's example (1960). What makes it work, he pointed out, was that the coefficients are complex; what is fundamental is the behavior of the commutator of $P$ and $\bar{P}$. The operator $\bar{P}$ here is obtained simply by replacing each coefficient by its complex conjugate. (In operator language: $\bar{P}u = \overline{(P\bar{u})}$.) More precisely: consider, for each polynomial in $(\xi_1, \ldots, \xi_n)$ its "principal part", i.e., the part that involves the terms of highest degree only. (For Lewy's example there is no other part.) If $p(x, \xi)$ is the principal part, write $b(x, \xi)$ for the "Poisson bracket",

$$b(x, \xi) = \sum_j \left( \frac{\partial p}{\partial \xi_j} \frac{\partial \bar{p}}{\partial x_j} - \frac{\partial p}{\partial x_j} \frac{\partial \bar{p}}{\partial \xi_j} \right).$$

Assertion: if, for some $(x^0, \xi^0)$, the principal part $p(x^0, \xi^0)$ vanishes but the Poisson bracket $b(x^0, \xi^0)$ does not, then $p$ is, in the sense of Lewy, not solvable in any open set containing $x^0$. It is easy to see that the Lewy example is covered by the Hörmander umbrella. Indeed: since

$$p = -i\xi_1 + \xi_2 - 2(x_1 + ix_2)\xi_3,$$

$$\bar{p} = i\xi_1 + \xi_2 - 2(x_1 - ix_2)\xi_3,$$

elementary computation yields

$$b = 8i\xi_3,$$

and it becomes clear that for every $x = (x_1, x_2, x_3)$ there is a $\xi = (\xi_1, \xi_2, \xi_3)$ such that $p(x, \xi) = 0$ and $b(x, \xi) \neq 0$.

REFERENCES. [1] L. Ehrenpreis, Solution of some problems of division, Amer. J. Math., 76 (1954) 883–903 (MR 16–834). [2] B. Malgrange, Existence et approximation des solutions des équations aux dérivées partielles et des équations de convolution, Ann. Inst. Fourier, Grenoble, 6 (1955) 271–355 (MR 19–280). [3] H. Lewy, An example of a smooth linear partial differential equation without solution, Ann. Math., 66(1957) 155–158 (MR 19–551). [4] A. P. Calderón, Uniqueness in the Cauchy problem for partial differential equations, Amer. J. Math., 80(1958) 16–36 (MR 21 # 3675). [5] L. Hörmander, Differential operators of principal type, Math. Ann., 140(1960) 124–146 (MR 24 # A434). [6] L. Hörmander, Linear partial differential equations, Springer, New York, 1969 (MR 40 # 1687). [7] L. Ehrenpreis, Fourier analysis in several complex variables, Wiley, New York, 1970 (MR 44 # 3066).

**Index Theorem.** The Atiyah–Singer index theorem (1963) spans two areas of mathematics, topology and analysis, and that's not an accident of technique but in the nature of the subject: the span is what it's all about. Theorems with such a broad perspective are usually the ones that are the most useful and the most elegant, and the index theorem is no exception. The very breadth of the theorem requires, however, that an expository sketch of it proceed obliquely. In what follows we describe, first and mainly, a historical and conceptual precursor, the Riemann–Roch theorem, and then indicate, briefly, how the Atiyah–Singer theorem generalizes it.

The classical Riemann–Roch theorem deals with the dual nature (topological and analytic) of a Riemann surface. Every compact Riemann surface is homeomorphic to a (two-dimensional) sphere with handles. The number of handles, the "genus", completely determines the topological character of the surface; that part is easy. The analytic structure is more complicated. It consists of a covering by a finite number of open sets and of explicit homeomorphisms from the complex plane $\mathbb{C}$ to each open set, which define holomorphic functions on the overlaps. (It is convenient and harmless to use the homeomorphisms to identify each open set in the covering with an open set in $\mathbb{C}$; that is tacitly done below.) If, for example, the surface is the sphere (with no handles), think of $\mathbb{C}$ as slicing through the equator, and use stereographic projections (toward the north and south poles) as the homeomorphisms. There are two open sets here, the complement of the north pole and the complement of the south pole; the holomorphic function on the overlap is given by $w(z) = 1/z$.

A smooth function on a Riemann surface can be viewed as a set of functions on, say, the open unit disk in $\mathbb{C}$ (one for each of the open sets of the covering) that are smooth ($C^\infty$) and transform into one another under the changes of variables induced by the overlaps. (If, that is, $f$ and $g$ are two of these functions, and $w$ is the transformation on the disk induced by going via the appropriate homeomorphism to the open set corresponding to $f$ and coming back from the overlap with the open set corresponding to $g$, then $f(z) = g(w(z))$.) The function on the Riemann surface is called holomorphic (or meromorphic) if each of these functions on the disk is holomorphic (or meromorphic). Another necessary concept for the analytic study of a Riemann surface is that of a smooth differential: that is an expression of the form $p(x, y)dx + q(x, y)dy$, where $p$ and $q$ are complex-valued smooth functions that, on the overlaps, satisfy the chain rule for change of variables. A holomorphic differential is one of the form $f(z)dz$, where $f$ is holomorphic and $dz = dx + idy$. (In the notation used above, the overlap relation for these differentials becomes $f(z)dz = g(w)dw = g(w(z))w'(z)dz$; the functions $f$ and $g$ no longer merely transform into one another, but are altered by the contribution of the differentials as well.)

The analytic properties of a Riemann surface are the properties of the holomorphic (and meromorphic) functions and differentials that it possesses. A well-known result is that the only holomorphic functions on a compact Riemann surface are constants: that is essentially what Liouville's theorem says. The Riemann–Roch theorem says much more. In its simplest form it deals with a compact Riemann surface $S$ of genus $g$, and $n$ points $z_1, \ldots, z_n$ on $S$. Let $F$ be the vector space of meromorphic functions on $S$ with poles of order not greater than 1 at each $z_i$ (and nowhere else); let $D$ be the vector space of holomorphic differentials with zeros of order not less than 1 at each $z_i$ (and possibly elsewhere). Conclusion:

$$\dim F - \dim D = 1 + n - g.$$

(In the special case of the classical Liouville theorem, $g = 0$, $n = 0$, and $\dim D = 0$.) The important aspect of the conclusion is that a quantity described completely in *analytic* terms can be computed from nothing but *topological* data.

In the special case $n = 0$, $F$ is the vector space of holomorphic functions on $S$ (so that $\dim F = 1$) and $D$ is the space of all holomorphic differentials. There is a linear map, conventionally denoted by $\bar{\partial}$, from the vector space of all smooth functions on $S$ to the vector space of all smooth differentials: write

$$\bar{\partial} f = \frac{\partial f}{\partial \bar{z}} d\bar{z}$$

in each of the open sets of the prescribed covering. The map $\bar{\partial}$ is an example of a differential operator. The kernel of $\bar{\partial}$ consists precisely of the functions satisfying the Cauchy–Riemann equations; in other words

$$\ker \bar{\partial} = F.$$

The cokernel of $\bar{\partial}$ (the quotient space of the space of all smooth differentials modulo the image of $\bar{\partial}$) is similarly identifiable with $D$. The conclusion of the Riemann–Roch theorem takes, in this case, the form

$$\dim \ker \bar{\partial} - \dim \operatorname{coker} \bar{\partial} = 1 - g.$$

The Atiyah–Singer theorem is a generalization of the Riemann–Roch theorem in that it too states that a certain analytically defined number (the "analytic index") can be computed in terms of topological data. Which aspects are generalized? All. To begin with, the Riemann surface is replaced by an arbitrary compact smooth manifold $M$ of arbitrary dimension. The vector spaces of smooth functions and smooth differentials are replaced by vector spaces of smooth sections of complex vector bundles over $M$ (in fact, complexes of vector bundles). The map $\bar{\partial}$, finally, is replaced by a differential operator $\Delta$, which satisfies a certain invertibility condition (called ellipticity). It follows that both $\ker \Delta$ and $\operatorname{coker} \Delta$ are finite-dimensional; the difference of the two dimensions is the analytic index. The conclusion is that the analytic index can be computed in terms of topological invariants (the "topological index"), which are very sophisticated generalizations of the genus.

Even in its relatively short life the Atiyah–Singer index theorem has had important and interesting consequences, and has been proved in at least three enlighteningly different ways. A recent one depends on the study of the heat equation on a manifold.

REFERENCES. [1] M. F. Atiyah and I. M. Singer, The index of elliptic operators on compact manifolds, Bull. A. M. S., 69(1963) 422–433 (MR 28 # 626). [2] R. S. Palais, Seminar on the Atiyah-Singer index theorem, Ann. Math. Studies, No. 57, Princeton University Press, Princeton, 1965 (MR 33 # 6649).

**Epilogue.** Concepts, examples, methods, and facts continue to be discovered; problems get reformulated, placed in new contexts, better understood, and solved every day. We hope that the ten examples above have communicated at least a part of the breadth, depth, excitement, and power of the mathematics of our time. Mathematics is alive, and it's here to stay.

DEPARTMENT OF MATHEMATICS, INDIANA UNIVERSITY, BLOOMINGTON, IN 47401.

Reprinted from the AMERICAN MATHEMATICAL MONTHLY
Vol. 84, No. 9, November 1977
pp. 714–716

# BERNOULLI SHIFTS

P. R. HALMOS

Many of the important problems of mathematics are of this kind: when are two objects the "same"? Example: when are two matrices similar?, when are two groups isomorphic?, when are two topological spaces homeomorphic?, when are two continuous mappings homotopic?

Ergodic theory (or at least a large part of it) is the study of measure-preserving transformations, and one of the important problems of ergodic theory is to determine when two measure-preserving transformations are "conjugate". If, for instance, $S$ is translation by one unit to the right on the real line ($Sx = x + 1$) and $T$ is its inverse ($Tx = x - 1$), then $S$ and $T$ are conjugate elements of the group of all measure-preserving transformations of the line; the reflection $Q(Qx = -x)$ transforms $S$ onto $T(Q^{-1}SQ = T)$. If, for another example, $S$ is rotation by 1 radian on the perimeter of the unit circle and $T$ is rotation by 2 radians, then it is perhaps intuitively plausible that $S$ and $T$ are *not* conjugate, and indeed they are not, but the proof needs slightly more sophisticated methods. For a final example, consider the same $S$ (rotation by 1 radian), but, this time, let $T$ be translation modulo $2\pi$ by one unit to the right on the interval $[0, 2\pi)$. In this last example $S$ and $T$ act on different spaces, but, plainly, the difference is merely notational; if $Q$ is the mapping $x \to e^{ix}$ from $[0, 2\pi)$ to the unit circle, then $Q^{-1}SQ = T$. The point of this example is to emphasize that conjugacy is a relation between measure-preserving transformations acting on possibly different spaces.

A special case of the conjugacy problem that was unsolved for a long time concerned "Bernoulli shifts". The prototypical Bernoulli shift is the measure-theoretic model of coin tossing. The space is the set of all two-way infinite sequences of 0's and 1's (heads and tails, from the beginning of time to eternity). Measure is defined by the familiar requirements that the probabilities of heads and tails be 1/2 and 1/2, and that the tosses on different days be independent of one another. The transformation is the index-shift one unit to the right (so that, for instance, the history of the coin experiment that results in heads today but results in tails on all past and future days is mapped onto the history that registers heads tomorrow and tails all other days). A more general Bernoulli shift allows $n$ outcomes ($n \geqq 2$), with possibly unequal probabilities $p_1, \ldots, p_n$, but is otherwise formally the same. (The most general Bernoulli shift allows certain infinite experiments too.) The conjugacy problem for Bernoulli shifts is to decide when the shift built on $(p_1, \ldots, p_n)$ is conjugate to the one built on $(q_1, \ldots, q_m)$.

The first step toward the solution was taken by Kolmogorov (1958) and Sinai (1959). It depends on an ingenious conjugacy invariant, called *entropy*, suggested by statistical mechanics and information theory.

The definition of entropy goes like this. Suppose that $X$ is a space with measure $m$. Define the entropy of a finite partition $\{A_1, \ldots, A_m\}$ of $X$ to be the number $-\sum_{i=1}^{m} m(A_i)\log m(A_i)$. If $T$ is a measure-preserving transformation on $X$, consider the partition obtained from the $A_i$'s by adjoining to them the $T^{-1}A_i$'s. (Example: if $X$ is the heads-tails sequence space, and $T$ is the shift, and if the partition $\{A_1, A_2\}$ is the two-way classification "today it's heads" and "today it's tails", then the finer partition is "today it's heads and tomorrow it's heads", "today it's heads and tomorrow it's tails", etc.) Calculate the entropy of the refined partition, and divide by 2; apply $T$ again, calculate the entropy, and divide by 3; etc. The limit exists (that's not hard to prove); it is called the entropy of $T$ relative to the originally chosen partition. The entropy of $T$ is the supremum of all these relative entropies (for all possible partitions). The entropy of a partition corresponds to the amount of information that can be gained by performing an experiment; the limiting entropy relative to a partition corresponds to the average information per day that can be gained by repeated performances of the experiment; the entropy of a transformation corresponds to the maximum amount of information per day that can be gained, from the given model, by performing a finite experiment.

It is plausible and provable that the entropy of the Bernoulli shift built on $(p_1, \ldots, p_n)$ is the same as the entropy of the easiest partition associated with the shift (that is, $-\sum_{i=1}^{m} p_i \log p_i$). This is the Kolmogorov–Sinai solution of one of the unsolved problems of conjugacy: it implies that some very different looking Bernoulli shifts (e.g., the one for a coin and the one for a die) are indeed different. In other words: some Bernoulli shifts are *not* conjugate.

Is entropy a complete conjugacy invariant? If two transformations have the same entropy, are they necessarily conjugate? The answer is no. Next question: does the answer change to yes if the transformations to be considered are suitably and usefully restricted? For instance: is it true that if two Bernoulli shifts have the same entropy, then they are conjugate? The answer is yes. This is the first and greatest step in a sequence of results obtained by Ornstein (1970). It answered an old question, and, apparently, encouraged many scholars to attack the subject with new vigor.

**Reference**

D. Ornstein, Bernoulli shifts with the same entropy are isomorphic, Advances in Math., 4 (1970) 337–352 (MR 41 # 1973).

DEPARTMENT OF MATHEMATICS, UNIVERSITY OF CALIFORNIA, SANTA BARBARA, CA 93106.

Reprinted from the AMERICAN MATHEMATICAL MONTHLY
Vol. 85, No. 1, January 1978
pp. 33–34

## FOURIER SERIES

P. R. HALMOS

It is a historical misfortune (which was responsible for almost 200 years of barking up the wrong tree) that Fourier series were discovered before convergence. Fourier series are a vital part of much of both classical and modern analysis; they are important for both abstract theory and concrete applications. They arise in topological groups and in operator theory; they have their origins in problems about vibrating strings and about heat conduction.

In their most classical manifestation, Fourier series have to do with numerical-valued functions (it is best to let them be complex-valued) on the line $(-\infty, +\infty)$ that are periodic of period $2\pi$ and integrable on the interval $[0, 2\pi]$. The Fourier series of such a function $f$ is the series

$$\sum_{n=-\infty}^{+\infty} \alpha_n e^{inx},$$

where

$$\alpha_n = \frac{1}{2\pi}\int_0^{2\pi} f(x)e^{-inx}dx.$$

(Since $e^{inx} = \cos nx + i \sin nx$, there is an alternative way of writing Fourier series; it uses sines and cosines and has the index running in one direction only. This alternative, real, version is geometrically more intuitive, but the complex exponential version given above is algebraically simpler to manipulate.)

Trigonometric polynomials (in either real or complex form) are familiar objects and are computationally accessible; surely nothing but good could come out of representing more difficult functions as limits of such polynomials. It seemed natural, therefore, to hope that the "sum" of the Fourier series associated with a function $f$ would be "equal" to $f$, and, in any event, to ask for which functions that would occur. The answer, it was hoped, was that good functions have good series, and the history of the subject has been strongly influenced by that hope.

When limits began to be understood, "sum" and "equal" were interpreted in the sense of pointwise convergence; the more fruitful and usable concepts of weak convergence and convergence with respect to a norm came along only after the mathematical community was irretrievably committed to research in the pointwise direction.

How good does a good function have to be? Differentiability is good enough, but, it turns out, continuity is not; there are continuous functions whose Fourier series diverges at a point, and, in fact, at many points. If convergence is replaced by summability in the sense of Cesàro averages, then Fejér's theorem saves the day; in *that* sense the Fourier series of every continuous function $f$ converges to $f$ at every point. Assertions such as these are considered relatively easy nowadays; they occur in most textbooks on the subject.

How bad can an integrable function be? Answer: very bad. Kolmogorov showed that if all that is assumed is that $f \in L^1[0, 2\pi]$ (i.e., that $f$ is integrable on $[0, 2\pi]$), then it could happen that the Fourier series of $f$ diverges almost everywhere (1923), or even everywhere (1926).

The biggest question along these lines was asked by Lusin, and it remained unanswered for 50 years: if $f \in L^2[0, 2\pi]$ (note the exponent 2 in place of 1), does it follow that the Fourier series of $f$ converges to $f$ almost everywhere? Repeated failure to prove the affirmative answer led to the official state religion among the cognoscenti in the 1950's and 1960's: the answer must be no.

The answer is yes. The first proof is due to Carleson (1966). A remarkable feature of Carleson's achievement is that it uses no unknown techniques; it just uses the known ones better. It depends on an ingenious push-me-pull-you way of selecting subintervals. It is as if Carleson had power enough to replace everyone else's $\varepsilon$ by $\varepsilon^2$, and that did the trick.

**References**

1. L. Carleson, On convergence and growth of partial sums of Fourier series, Acta. Math., 116 (1966) 135–157 (MR 33 # 7774).

2. R. A. Hunt, Almost everywhere convergence of Walsh-Fourier series of $L^2$ functions, Actes, Congrès Intern. Math., Nice (1970) Tome 2, 655–661.

Reprinted from the AMERICAN MATHEMATICAL MONTHLY
Vol. 85, No. 2, February 1978
pp. 95–96

# ARITHMETIC PROGRESSIONS

P. R. HALMOS AND C. RYAVEC

A set of positive integers with some algebraic structure, however rudimentary, is better to work with than one without any — it is likely to give more number-theoretic information. From this point of view arithmetic progressions are good sets, and a set that contains many arithmetic progressions is better than one that does not. It is reasonable to conjecture that "large" sets contain many arithmetic progressions.

A famous theorem of van der Waerden asserts that if the set of positive integers is partitioned into two subsets, then at least one of them must be large, in the sense that it contains arbitrarily long arithmetic progressions. (Caution: a set that contains arbitrarily long arithmetic progressions may fail to contain an infinitely long one. Example: the sequence

$$11, 101, 102, 1001, 1002, 1003, 10001, 10002, 10003, 10004, \ldots,$$

made up of the blocks $10^i + j$, with $i = 1, 2, 3, \ldots$ and $j = 1, \ldots, i$.)

Van der Waerden's theorem is implied by the following assertion: to each positive integer $k$ there corresponds a positive integer $n(= n(k))$ such that if the set $\{1, \ldots, n\}$ is partitioned into two subsets, then at least one of them contains an arithmetic progression of $k$ terms. (The first non-trivial example is easily obtained just by listing all possibilities: if $k = 3$, then $n = 9$.)

Many results in number theory concern the connection between primes and arithmetic progressions. It is good to know the extent to which primes are regularly distributed in arithmetic progressions, and it would be good to know whether the set of primes contains arbitrarily long arithmetic progressions — but the latter is a long-standing unsolved problem.

Erdös and Turán (1936) had once hoped to attack this problem (and others) by showing that if a sequence is sufficiently dense, then it must contain arbitrarily long arithmetic progressions. To make this precise, fix $k$ and $n$ and ask: how many numbers between 1 and $n$ (inclusive) are needed to guarantee that they will contain an arithmetic progression with $k$ terms? (If $k = 3$ and $n = 9$, the answer is 5.) Equivalently: what is the largest number $r(=r_k(n))$ such that some set of $r$ numbers between 1 and $n$ does *not* contain an arithmetic progression with $k$ terms? (The value of $r_3(9)$ is 4.)

The problem of long arithmetic progressions of primes would be solved by a proof that $r_k(n)<\pi(n)$, where $\pi(n)$ is the number of primes less than or equal to $n$. It would be good enough, of course, to show that for each $k$ the inequality holds for all sufficiently large $n$.

Erdös and Turán observed that

$$r_k(m+n) \leqq r_k(m)+r_k(n);$$

they inferred that, for each $k$, the sequence $\{(1/n)r_k(n)\}$ has a limit $c_k$ ; and, finally, they conjectured that $c_k=0$ for each $k$, i.e., that

$$\lim_n \frac{1}{n} r_k(n)=0.$$

The conjecture is simple and elegant and would have pleasant consequences, but it turned out to be extraordinarily difficult to settle. Sample consequence: every set of positive density contains arbitrarily long arithmetic progressions. In detail: if $E$ is a set of positive integers, if $a_n$ is the number of elements of $E$ between 1 and $n$, and if $\lim_n(1/n)a_n>0$, then $E$ contains arbitrarily long arithmetic progressions. Reason: the truth of the conjecture implies that, for each $k$, the inequality $r_k(n)<a_n$ holds for all sufficiently large $n$.

For $k=3$ the Erdös–Turán conjecture was proved in 1954 by K. F. Roth; for $k=4$, using van der Waerden's theorem, E. Szemerédi proved it in 1967. The proof was so intricate (some called it the apotheosis of the elementary method in number theory and combinatorics) that there was reluctance (to put it mildly) to attempt the case $k=5$ till substantial simplifications could be made in the case $k=4$. The next step was taken by Roth (1970), who put the proof on an analytic basis and, incidentally, removed the use of van der Waerden's theorem, but none of that seemed to help.

In 1972 Szemerédi announced that he had solved the problem in full generality. Erdös had offered a prize of $1000.00 for a solution, but to collect it Szemerédi needed to solve another formidable problem, namely the problem of writing the proof down so as to make it comprehensible to others. A preliminary exposition of Szemerédi's proof was written by A. Hajnal (who became sufficiently convinced that he told Erdös that he was willing to buy the proof from Szemerédi for $500.00). An abstract of Szemerédi's result appeared in 1974 and the full version a year later.

The leading idea of Szemerédi's proof can probably not be said better in a short paragraph than the way Szemerédi himself put it. "The basic objects with which the proof of the main theorem deals are not just arithmetic progressions themselves but rather generalizations of arithmetic progressions called *m-configurations.* Roughly speaking, a 1-configuration is just an arithmetic progression; an $m$-configuration is an "arithmetic progression" of $(m-1)$-configurations. In a nutshell, one can show that for any given set of integers $R$ of positive upper density, a very long $m$-configuration which intersects $R$ in a moderately regular way must always contain a shorter (but still quite long) $(m-1)$-configuration which intersects $R$ in an even more regular way. In this way we eventually conclude that $R$ must contain arbitrarily long 1-configurations, i.e., arithmetic progressions, and we are done."

**Reference**

E. Szemerédi, On sets of integers containing no $k$ elements in arithmetic progression, Proc. Int. Congress Math., Vancouver, 2 (1974) 503–505; Acta Arithmetica, 27 (1975) 199–245 (MR 51 # 5547).

Reprinted from the AMERICAN MATHEMATICAL MONTHLY
Vol. 85, No. 3, March 1978
p. 182

## INVARIANT SUBSPACES

P. R. Halmos

The main reason for the success of finite-dimensional complex linear algebra is the existence of eigenvalues and eigenvectors. An eigenvector of a transformation spans a 1-dimensional subspace invariant under the transformation; it makes possible a study of the transformation on the whole space via a study of its behavior on smaller and therefore more manageable subspaces. The existence of eigenvalues is a deep fact, derived by techniques far from the spirit of linear algebra. What it comes down to is that an eigenvalue, a geometric concept, is the same as a zero of the characteristic polynomial, an algebraic concept, and the existence of such zeroes is guaranteed by the fundamental theorem of algebra, an analytic tool.

Neither the methods nor the results hinted by the preceding paragraph extend to Hilbert space, which is the simplest, most natural, and most useful infinite-dimensional generalization of finite-dimensional vector spaces. There is no simple generalization of chracteristic polynomials, what there is has no useful relation to eigenvalues, there are transformations that have no eigenvectors at all, and, for all anyone knows, non-trivial invariant subspaces may fail to exist.

That, in fact, is the invariant subspace problem. If $A$ is an operator (bounded linear transformation) on a separable infinite-dimensional Hilbert space $H$, does there exist a subspace $M$ of $H$ (closed linear manifold) different from both $O$ and $H$ and invariant under $A$ ($AM \subset M$)? If the topological restrictions (boundedness and closure) are removed, the answer (yes) becomes easy. If the space is too small (finite-dimensional), the answer (yes) is classical; if the space is too large (non-separable), the answer (yes again) is trivial.

The most important class of operators for which the answer has been known to be yes for a long time is the class of Hermitian (and, more generally, normal) operators; this fact is a corollary of the infinite-dimensional version of the process of diagonalizing a Hermitian matrix (the spectral theorem).

The next most important class is connected with the concept of compactness (complete continuity). An operator on a Hilbert space is *compact* if it maps the unit ball onto a compact set. The first major invariant subspace theorem was obtained by Aronszajn and Smith (1954): using an idea of von Neumann, they proved that every compact operator has non-trivial invariant subspaces. The method depends on approximation by operators of finite rank.

The Aronszajn–Smith technique seemed to be so sharply focused on its particular purpose that for a dozen years it resisted even mild generalization; it was, for instance, not known whether the conclusion remained true for operators whose square is compact. Now that is known; the extension to polynomially compact operators was obtained by Bernstein and Robinson (1966). They presented their result in the metamathematical language called non-standard analysis, but, as it was realized very soon, that was a matter of personal preference, not necessity.

One of the next problems in this connection that received much attention can be expressed this way: if two compact operators commute, must they have a non-trivial invariant subspace in common? The Aronszajn–Smith–Bernstein–Robinson technique could be stretched to do more than it did at first, but it did not seem to be elastic enough to stretch this far. The final advance (which may close the subject of invariant subspaces of compact operators) was made by Lomonosov (1973); he proved that for every non-zero compact operator there is a non-trivial subspace that is invariant not only under that operator but under every operator that commutes with it. The main tool in the proof is the Schauder fixed point theorem for continuous mappings of a compact convex set into itself.

The invariant subspace problem for operators that have nothing to do with compact ones is still open.

**References**

1. V. I. Lomonosov, Invariant subspaces of the family of operators that commute with a completely continuous operator, Funkcional. Anal. i Priložen., 7(1973) 55–56.

2. H. Radjavi and P. Rosenthal, Invariant subspaces, Springer-Verlag, New York–Heidelberg, 1973.

Reprinted from the
AMERICAN MATHEMATICAL MONTHLY
Vol. 85, No. 4, pp. 256–257, April 1978

# SCHAUDER BASES

P. R. HALMOS

Euclidean spaces have three basic properties: they are (1) vector spaces, with (2) a concept of length, and (3) a concept of angle. Banach spaces, whose intensive study began in the 1930's, constitute a generalization that retains requirements (1) and (2) but does not insist on (3). (Precisely: a Banach space is a real or complex vector space endowed with a norm, with respect to which, as a metric space, it is complete.)

Generalizations can turn out to be too general (as wise hindsight sometimes shows); they can lead to complicated inbred questions (do properties $A$, $C$, $E$, and $G$ imply properties $B$, $D$, $F$, or $H$?), and to pathological counterexamples, that interest only the most dedicated taxonomist. Banach spaces are sometimes accused of such excessive generality. It is, however, undeniable that Banach space theory has illuminated the study of integration (via $L^p$ spaces) and continuity (via spaces of continuous functions). Another sign that Banach spaces may not be too wildly general is that even the finite-dimensional ones are of interest, and, in fact, some experts feel that if we knew all about all finite-dimensional Banach spaces, we would know all about all of them. Many of the recent spectacular theorems in the field are based on deep finite-dimensional lemmas.

One of the earliest questions about Banach spaces was the basis problem, raised by Banach himself in his book (1932). A sequence $\{x_1, x_2, x_3, \ldots\}$ of elements in a Banach space $X$ is a (Schauder) *basis* for $X$ in case every $x$ in $X$ has a unique (!) expression of the form $\sum_{n=1}^{\infty} \alpha_n x_n$. (That is, to each $x$ there corresponds a unique sequence $\{\alpha_1, \alpha_2, \alpha_3, \ldots\}$ of scalars such that $\lim_n \| x - \sum_{j=1}^{n} \alpha_j x_j \| = 0$.) It is easy to see that if a Banach space has a basis, then it is separable. The basis problem, which was outstanding for 40 years, was the converse: does every separable Banach space have a basis? Each space that ever came up in analysis had one, and yet a proof that that had to be so remained elusive.

A classically important concept in the study of Banach spaces is that of a compact (completely continuous) operator, i.e., a linear transformation $T: Y \to X$ between Banach spaces, with the property that the closure of the image (under $T$) of the unit ball (in $Y$) is compact (in $X$). The easy compact operators are the ones of finite rank (i.e., the ones for which the range of $T$ is finite-dimensional); the next easiest ones are the (uniform) limits of operators of finite rank. If $X$ is a "good" Banach space, then every compact operator into $X$ is such a limit (in technical language: "$X$ has the approximation property"), and, in particular, if a Banach space has a basis, then it has the approximation property.

The basis problem was solved by Enflo (1973). The solution is negative: there exists a separable Banach space that does not have the approximation property. The technique is constructive; it is a combinatorial way of constructing and putting together infinitely many finite-dimensional Banach spaces.

## Reference

Per Enflo, A counterexample to the approximation problem in Banach spaces, Acta Math., 130 (1973) 309–317.

THE AMERICAN MATHEMATICAL MONTHLY
Vol. 85, No. 5, May 1978

# THE SERRE CONJECTURE

W. H. Gustafson, P. R. Halmos, J. M. Zelmanowitz

What does the first row of an invertible matrix look like? That depends: where are the entries? If the entries are real numbers (or, for that matter, elements of an arbitrary field), the first row of an invertible matrix can be anything except $(0,0,\ldots,0)$. (If the given row of length $n$ has a non-zero entry, assume, with no loss of generality, that that entry is in the first position, and construct the square matrix of size $n$ with that first row, the rest of the first column all 0's, and the lower right corner equal to the identity matrix of size $n-1$.)

If the matrices about which the question is being asked are restricted to have integer entries, the answer is less obvious. Thus, for instance, $(2,4,6)$ cannot be the first row of an invertible integer matrix, because the determinant of any such matrix must be $\pm 1$, whereas the determinant of an integer matrix whose first row is $(2,4,6)$ is necessarily even. There is a clue here: if the entries of a prescribed row have a non-trivial common divisor, then that row cannot occur in an invertible matrix. It turns out that the necessary condition thus discovered (that the greatest common divisor of the prescribed row be 1) is sufficient as well: any row satisfying it *can* occur in an invertible matrix.

The last assertion is true for the ring of integers, or, for that matter, for any principal ideal domain: the condition that a row $(a_1,a_2,\ldots,a_n)$ must satisfy is the existence of a related column $(t_1,t_2,\ldots,t_n)$ such that $a_1t_1+a_2t_2+\cdots+a_nt_n=1$. Such rows are called unimodular rows. (The $t$'s exist if and only if the greatest common divisor of the $a$'s is 1. Equivalently: the $t$'s exist if and only if the ideal generated by the $a$'s is the entire domain. Note that in case the domain is a field the conditions are equivalent to $(a_1,a_2,\ldots,a_n)\neq(0,0,\ldots,0)$.)

There are various approaches to the proof of sufficiency; one that works smoothly for Euclidean domains goes as follows. Regard $(a_1,a_2,\ldots,a_n)$ as a matrix with one row and $n$ columns, and use the existence of the $t$'s to infer the possibility of performing a sequence of elementary column operations on it so as to convert it to $(1,0,\ldots,0)$. The performance of an elementary column operation has the same effect as multiplying the given matrix on the right by certain elementary matrices. If the product of all the multipliers is $U$, so that

$$(a_1,a_2,\ldots,a_n)\cdot U=(1,0,\ldots,0),$$

then

$$(a_1,a_2,\ldots,a_n)=(1,0,\ldots,0)\cdot U^{-1},$$

so that $U^{-1}$ is an invertible matrix with first row $(a_1,a_2,\ldots,a_n)$.

It is now tempting to jump to a general algebraic conclusion: if $R$ is a commutative ring with unit, and if $(a_1,a_2,\ldots,a_n)$ is a unimodular row over $R$, then $(a_1,a_2,\ldots,a_n)$ is fit to be the first row of an invertible matrix over $R$. For $n=1$ the conclusion is trivial, and for $n=2$ it is almost equally trivial: given $(a_1,a_2)$ and $(t_1,t_2)$ with $a_1t_1+a_2t_2=1$, the matrix

$$\begin{pmatrix} a_1 & a_2 \\ -t_2 & t_1 \end{pmatrix}$$

does the trick.

For $n \geqslant 3$, however, the conclusion is false; the standard counter-example makes contact with a well-known part of elementary topology. Let $R$ be the ring of all real-valued continuous functions defined on the 2-sphere, i.e., on the locus of the equation $x^2+y^2+z^2=1$ in $\mathbf{R}^3$. If $a_1(x,y,z)=x$, $a_2(x,y,z)=y$, and $a_3(x,y,z)=z$, then $(a_1,a_2,a_3)$ is a unimodular row (because $x^2+y^2+z^2=1$). The row $(a_1,a_2,a_3)$ cannot, however, occur in an invertible matrix over $R$. Reason: the second row $(b_1,b_2,b_3)$ of such a matrix would be, at each point $(x,y,z)$, linearly independent of the first, and, therefore, would have a non-zero projection in the plane tangent to the sphere at $(x,y,z)$. This is impossible: there is no non-singular continuous tangent vector field on the sphere (or you can't comb a porcupine).

Serre (1955) made the conjecture that for certain important special rings the general conclusion is true; the rings are the polynomial rings $R=k[x_1,\ldots,x_m]$ in $m$ variables, with coefficients in a field $k$. The original formulation of Serre's conjecture had to do with modules over these $R$'s. Recall that a module over $R$ is "a vector space with respect to scalars from $R$". A module over $R$ is of special importance if it has a finite basis (equivalently, if it is isomorphic to $R^m$ for some $m$). If a module over $R$ has a basis, and is the direct sum of two submodules, do they too necessarily have bases? A module with a basis is called *free*, and a direct summand of a free module is called *projective*; the original formulation of Serre's conjecture was that every projective module over $k[x_1,\ldots,x_m]$ is free.

The subject makes surprising contact with topology. A *vector bundle* over a base space $X$ is a generalization of the concept of the Cartesian product of $X$ with a vector space. Intuitively, a bundle is a "twisted" Cartesian product, in the sense in which a Möbius strip is a twisted cylinder. The Cartesian product itself is the easiest vector bundle (called the *trivial* bundle); the generalized ones retain the projection map onto $X$, just as if they were Cartesian products, but are like Cartesian products only locally.

A *section* of a vector bundle is a continuous function $s$ from $X$ to the bundle, such that $s$ followed by the projection onto $X$ is the identity mapping on $X$. The vector structure of the bundle makes possible a natural definition of addition of sections. If, moreover, $R$ is the ring of scalar-valued continuous functions on $X$, then the vector structure of the bundle makes possible a natural definition of multiplication of a section by an element of $R$. In other words, the set of all sections is a module over $R$. If the base space is at all decent, it turns out that this module is always projective; the module is free if and only if the bundle is trivial. There is, thus, a correspondence between projective modules and vector bundles, and, in that correspondence, the free modules correspond to trivial bundles.

What is known in the topological context is that a bundle over $m$-dimensional Euclidean space $\mathbf{R}^m$ is in a natural sense always (isomorphic to) the trivial bundle. How much of that conclusion continues to make sense and remains true in an algebraic context? If, in other words, $k$ is an arbitrary field, does it make sense to speak of vector bundles over $k^m$? The answer is yes; a topology that gives the phrase its sense (the Zariski topology) is definable in purely algebraic terms. (A closed set in $k^m$ is the locus of common zeros of a set of polynomials in $m$ variables.) A vector bundle over $k^m$, with "fiber" given by the vector space $k^n$, is a "twisted" version of $k^m \times k^n$. The role of the ring of scalar-valued continuous functions on the base space $k^m$ is played by the ring $R=k[x_1,\ldots,x_m]$ of polynomials. The sections form a projective module over $R$; Serre's conjecture is that such a module is necessarily free, i.e., that such a bundle is necessarily trivial.

The conjecture remained open for more than 20 years. The first non-trivial step was Seshadri's (1958); he proved the conjecture for $m=2$. A later and important step was taken by Horrocks (1964) who proved an analogous result for local rings (rings with only one maximal ideal). The final step was taken simultaneously and independently by Quillen and Suslin (1976).

Quillen in effect reduced the general case to the one treated by Horrocks. The essence of Quillen's method is induction on the number $m$ of variables. The step from $m-1$ to $m$ is similar to the procedure of complexifying a real vector space. It involves showing that every projective module over

$k[x_1,\ldots,x_m]$ is a tensor product of a projective module over $k[x_1,\ldots,x_{m-1}]$ with $k[x_1,\ldots,x_m]$. The reason the induction step is successful is that the property of having a basis survives the construction.

**References**

1. D. Quillen, Projective modules over polynomial rings, Invent. Math., 36 (1976) 167–171.
2. A. A. Suslin, Projective modules over a polynomial ring are free, Soviet Math. Dokl., 17 (1976) 1160–1164.
3. T. -Y. Lam, Serre's conjecture, Springer-Verlag, Berlin, 1978.

# THE WORK OF F. RIESZ

P. R. HALMOS*

## PROLOGUE

Frederic Riesz contributed memorable theorems to mathematics, as well as new proofs and improvements of the discoveries of others. By his insights and emphasis he bequeathed to the generation that followed him an attitude of mathematical cleanness and simplicity that might possibly turn out to be just as valuable as his theorems and proofs.

Let me begin this report with some statistics on what Riesz did. The two-volume edition of his complete works lists 95 publications. Many of them (22 to be exact) are in Hungarian, but the mathematician who doesn't read that language wouldn't miss much: most of the Hungarian articles are available in either French or German translation also, and the rare exception is certain to be re-explained, re-stated, or at least summarized elsewhere (possibly in bits and pieces). As far as other languages go: one paper is in Italian, four in English, 25 in German, and 41 in French. The two books (the one on linear equations in infinitely many unknowns, 1913, and the joint one with Sz.-Nagy on functional analysis, 1952) were first published in French. (The second one was subsequently translated into English.) The first English paper was written in collaboration with E. R. Lorch, and it appeared when Riesz was 56 years old; the other three (all on ergodic theory) are solo works. Speaking of collaborations: there aren't very many. In addition to the Lorch paper and the Sz.-Nagy book there are just five: one each with Fejér, Radó, Szegő, Sz.-Nagy, and Marcel Riesz.

The collected works is divided into nine sections: topology, real functions, function spaces, complex functions, harmonic functions, functional analysis, ergodic theory, geometry, plus the inevitable miscellany. Riesz has only two papers on geometry, including his doctor's thesis on quartic curves treated by synthetic projective methods. After 1907 he never returned to the subject.

It is neither possible nor desirable to review here everything that Riesz ever wrote. What I propose to do instead is to give a thumbnail sketch of the high points of each of several directions to which Riesz made a contribution.

*The preparation of this report was supported in part by a grant from the National Science Foundation. The report was given as a lecture at the Conference on Functions, Series, and Operators, in Budapest, Hungary, August 1980.

## TOPOLOGY

The topology section of the collected works consists of ten papers. The first one contains a formulation and proof of a slightly generalized form of Schoenfliess's converse of the Jordan curve theorem, and the last one contains a proof of the Jordan curve theorem. The Schoenflies theorem says that each set that separates the plane the way circles do is homeomorphic to a circle. More precisely, the Riesz version goes as follows. Suppose that $P$ is a bounded perfect set in the plane and $I$ and $E$ (to be thought of as interior and exterior) form a partition of the complement of $P$ such that the following three conditions are satisfied: 1) any two points of $I$ can be joined by a Jordan curve in $I$, and the same is true of $E$; 2) every point of $P$ can be joined to every point of $I$ by a Jordan curve that meets $P$ at its initial point only, and the same is true of $E$; and 3) every Jordan curve that joins a point of $I$ to a point of $E$ meets $P$. Conclusion: $P$ is homeomorphic to a circle.

The other topological papers are minor ones, but some of them played a role in the development of the subject. One of them, for instance, describes the genesis of the concept of topological spaces, and gives an insight into the meaning of connectedness that was far from trite in 1907. In another one there is a proof of what many of us were taught to call the Heine–Borel theorem, i.e., the statement that a closed and bounded interval is compact. According to Riesz, Borel proved the existence of finite subcovers for *countable* open covers only!

## REAL FUNCTIONS

Egoroff's theorem says that if a sequence of measurable functions (on a set of finite measure) converges everywhere except on a set of measure zero, then it converges uniformly except on a set of arbitrarily small measure. The theorem seems to have annoyed Riesz. He said that, striking as it may appear, it does not lead to any essentially new conclusions, and (in a couple of papers) he discussed proofs of the theorem via continuity that do not, I think, come to grips with the combinatorial heart of the matter.

He was an early precursor of the school of thought that prefers to do integration without measure. He noted that the concept of a set of measure zero is more elementary than general quantitative measure theory, and that, with that elementary concept, all of Lebesgue integration theory could be derived from simple functions (in the usual technical sense of that phrase). One of his papers offers a proof of that completely non-measure-theoretic vestigial curiosity called Osgood's theorem (which is nothing but the Lebesgue bounded convergence theorem for continuous functions on a closed bounded interval).

Elsewhere Riesz develops in some detail an approach to integrals via derivatives. Basic idea: call a positive function (on a closed bounded interval) integrable if it is the derivative almost everywhere of a monotone function. Basic lemma: for each integrable $f$ on $[a, b]$ there exists a monotone $F$ such that $f = F'$ almost everywhere and such that $F$ is minimal in the following sense: if $G$ is monotone and $f \leqq G'$ almost everywhere, then $F(d) - F(c) \leqq G(d) - G(c)$ for all intervals $[c, d]$ included in $[a, b]$. Definition:

$$\int_c^d f(x)\,dx = F(d) - F(c),$$

and the theory is off and running.

## ERGODIC THEORY

Riesz's most individual contribution to real function theory is the rising sun lemma and its applications. The rising sun lemma thinks of the graph of a continuous function $f$ on $[a, b]$ as the profile of a mountain range, and assumes that we are looking at it at sunrise (when the sun is at $x = +\infty$). The assertion is that the set of those points in $[a, b]$ over which the mountain is in the shade consists of open intervals $(c, d)$ such that $f(c) = f(d)$ (unless $c = a$, in which case, possibly, $f(c) < f(d)$).

Among the applications of the lemma are a proof of a Hardy–Littlewood maximal theorem, a proof of the almost everywhere differentiability of monotone functions, a proof of Lebesgue's theorem about points of density, and a proof of the Birkhoff ergodic theorem. (In the Hardy–Littlewood–Pólya book on inequalities, the discoverers used Riesz's proof of the maximal theorem instead of their own.)

Riesz deserves to share with Wiener the credit for recognizing the connection between differentiation theory and ergodic theory. In both subjects the central issue is the existence of a certain limit; in one case a basic interval shrinks to zero and in the other it expands to infinity. The ideas and the methods are astonishingly similar, and I don't think we have heard the last of the matter yet. It could well be that there is a comprehensive and elegant theory that includes both, and in which the Lebesgue theorem about derivatives and the Birkhoff theorem about averages of iterates appear as easy corollaries.

There are two kinds of ergodic theory: one talks about functions and the other talks about elements of abstract structures (such as Banach spaces). Riesz's main interest was in the second kind, whose discussion might better be included under the heading of functional analysis, but, as long as the subject has already come up, I shall briefly describe what Riesz did with it.

The first ergodic theorem was von Neumann's; it is about convergence properties of the averages of iterates of unitary operators on Hilbert space. Riesz's contribution to the subject was to extend the result in two ways: the space can be $L^p$ (not just for $p = 2$ but for any $p$, $1 < p < \infty$) and the operator can be any contraction (not just the isometric kind). He obtains a result for

$p = 1$ also, but in that case the unit ball is not weakly compact and that causes trouble; an added hypothesis is necessary (namely, the uniform absolute continuity of the averages that enter).

Riesz returned to ergodic theory several times, polishing and extending, and the result of one of his returns (a joint work with Sz.–Nagy) contains what is surely the ultimate proof of the mean ergodic theorem for unitary operators on Hilbert space. (Summary: the asserted convergence is obvious in both the range and the kernel of $1 - U$, and, if $U$ is unitary, the entire space is the closed direct sum of those two subspaces.) The method applies, with no change, to any operator $T$ that is a contraction and that has the same fixed points as $T^*$. A pretty and pleasant surprise of mathematics, first proved in the Riesz–Nagy paper, is that the latter condition is implied by the former. (Proof: if $Tf = f$, then $\|f - T^*f\|^2 = \|f - T^*Tf\|^2 = \|f\|^2 - 2\|Tf\|^2 + \|T^*Tf\|^2 \leqq -\|f\|^2 + \|f\|^2 = 0$.)

## ANALYTIC FUNCTIONS

Two of the several papers that Riesz wrote on classical ("hard") analysis are outstanding; they continue to be widely quoted and applied. One is about analytic functions and the other about subharmonic functions.

The principal Riesz result about analytic functions was obtained in collaboration by the two brothers, Frederic and Marcel Riesz, and it is always referred to as "the F. and M. Riesz theorem".

Consider, to begin with, a function in $H^\infty$ (i.e., a bounded analytic function in the open unit disk). It follows from the work of Fatou that the radial limit of such a function exists almost everywhere on the perimeter; the first F. and M. Riesz result (conjectured by Fatou) is that except when the original function is identically zero, the radial limit function can vanish on a set of measure zero only. So much for $H^\infty$. As the Riesz brothers immediately remark, the result is easy to extend to $H^p$ for all the usual values of $p$ $(1 < p \leqq \infty)$; the depth of the matter is in the case $p = 1$.

If $F'(z) = f(z)$ for $|z| < 1$, where $f \in H^p$, $1 < p \leqq \infty$, then $F \in H^\infty$, and, moreover, by Fatou and Pringsheim, the function $F$ (extended by radial limits) is continuous on the closed disk and of bounded variation on the perimeter. The preceding results imply that if $F$ is not a constant and if $M$ is a Borel subset of the perimeter, then a necessary and sufficient condition that $F(M)$ have measure zero is that $M$ have measure zero. (To avoid worries about self-intersections, the function $F$ is to be thought of as mapping the disk onto a Riemann surface.) The general F. and M. Riesz theorem drops the assumption on $f$: it says that if $F$ is any non-constant continuous function on the closed disk that is analytic in the interior and of bounded variation on the perimeter, then, for Borel subsets $M$ of the perimeter, either both $M$ and $F(M)$ have measure zero or else neither one does. (In this formulation the derivative $F'$ is in $H^1$ instead of some higher $H^p$.)

The theorem can be (and nowadays usually is) stated in somewhat different language. In one formulation the conclusion is that $F$ (on the perimeter) is absolutely continuous and has a derivative that is different from zero almost everywhere. A more familiar formulation is that an analytic measure in the perimeter (i.e., a Borel measure $\mu$ such that $\int z^n \, d\mu(z) = 0$ for $n = 1, 2, 3, \ldots$) is equivalent to Lebesgue measure (in the sense of mutual absolute continuity).

## SUBHARMONIC FUNCTIONS

As for subharmonic functions, Riesz (and now we're back to F. Riesz only) contributed even to their definition. According to one definition, a function $u$ (defined on a domain $G$ in the plane) is subharmonic if, for every bounded subdomain $G_0$ with $\overline{G}_0 \subset G$ and for every harmonic function $U$ such that $u \leq U$ on the boundary of $G_0$, the same inequality holds in the interior of $G_0$. A conceptual disadvantage of this definition is that $u$ must be compared with every conceivable harmonic $U$ on every conceivable subdomain $G_0$. Riesz's apparently less stringent definition, which turns out to be equivalent, requires only this: the value of $u$ at each point of $G$ must be dominated by the mean value of $u$ on the perimeter of each sufficiently small circle with center at that point. (Incidentally Riesz noticed and emphasized that subharmonic functions are the natural way to generalize to several variables the concept of convex functions of one variable. For one real variable the harmonic functions are just the linear ones, and, correspondingly and consequently, the subharmonic functions are the convex ones.)

The Riesz theorem about subharmonic functions says that each such function is the sum of a potential and a harmonic function. More precisely: if $u$ is a subharmonic function (not identically $-\infty$) on a domain $G$ in the plane, then there exists a unique Borel measure $\mu$ in $G$ such that, for each compact subset $E$ of $G$, the difference

$$u(x) - \int_E \log\|x - y\| \, d\mu(y)$$

is harmonic in the interior of $E$. (Here $x$ and $y$ are vectors in the plane.) This is a representation theorem, of course, quite similar in detail and in spirit to other more general and more famous representation theorems associated with the name of F. Riesz. It has higher-dimensional versions (Riesz stated the 2-dimensional one only) whose statement and proof are almost exactly the same; all that has to be remembered is that if $n > 2$, then the appropriate potential in $\mathbb{R}^n$ is defined in terms of $-1/\|x\|^{n-2}$ (instead of $\log\|x\|$).

## THE RIESZ–FISCHER THEOREM

While some might remember F. Riesz as a general topologist, or as an early developer of integration theory, or as a profound contributor to ergodic theory, to complex function theory, and to the theory of harmonic functions, first and foremost he was a functional analyst who might even lay claim to having singlehandedly founded large portions of modern functional analysis. He was among the first to emphasize $L^p$ spaces (and to note the singular importance of $L^1$, $L^2$, and $L^\infty$), to study weak convergence and weak compactness, and to define and use what is now known as a functional calculus. His name is attached to three major parts of the subject: the Riesz–Fischer theorem, the Riesz representation theorem, and the Riesz theory of compact operators.

Riesz's version of the Riesz–Fischer theorem is that $l^2$ and $L^2$ are isomorphic. More precisely, he showed that if $\{e_n\}$ is an orthonormal basis in $L^2(0,1)$, then the mapping that associates with each $f$ in $L^2$ the sequence $\alpha$ in $l^2$ defined by $\alpha_n = \int_0^1 f(x)e_n(x)\,dx$ is a bijection. Riesz's first proof used the classical trigonometric series and arguments involving continuity and differentiability, and then passed to an arbitrary orthonormal basis by using the appropriate unitary operator on $l^2$. (I was able to formulate this telegraphic summary in terminology more compressed than the one that was available to Riesz; I took advantage of many decades of hindsight.)

Fischer's version is the one more commonly used today: $L^2$ is complete. Fischer's proof is slightly more modern in spirit than Riesz's, but it too uses indefinite integrals and derivatives. Fischer mentions that his version has Riesz's as an immediate corollary.

The Riesz–Fischer theorem is not the result of collaboration; it deserves its double name because the two men discovered it simultaneously. Riesz presented his note to the Paris academy on March 18 (1907), and Fischer lectured on his in Brünn (now Brno) on March 5.

[I have always wondered what else Fischer did, and I looked it up. He was apparently an active mathematician of some substance, five years older than F. Riesz, the same age as M. Riesz. He was born in Vienna, taught first in Brno, then briefly in Erlangen, and then, from 1920 to 1938, he was professor in Cologne; he died in 1954. He wrote mainly on invariant theory and determinants, with only an occasional excursion toward analytic matters.]

Riesz referred to the theorem often. Once or twice he called it Fischer's theorem, sometimes he said something like "the theorem named after Fischer and myself", but most often he described it as "the so-called Riesz–Fischer theorem". Late in his career (in a 1949 lecture) he dropped an interesting hint about how he discovered it. He said that the idea of applying Lebesgue integration to the problems that had been concerning him, and the courage to do so, came from reading Fatou's thesis, and, in particular, from the result that has come to be known as Fatou's lemma. (With typically sharp insight Riesz describes that lemma as the assertion that integration is lower semicontinuous.)

## THE RIESZ REPRESENTATION THEOREM

There are several theorems known as the Riesz representation theorem, but whether they deal with $C$, $L^2$, $L^1$, $L^\infty$, or the most general $L^p$, they all come to the same thing. The most famous and perhaps the most typical one is the one for $C$.

It is obvious that if $g$ is a fixed function in $C$ and if $\varphi$ is defined on $C$ by $\varphi(f)=\int_0^1 f(x)g(x)\,dx$, then $\varphi$ is a bounded linear functional. An early important theorem of Hadamard's (1903) went this far in the converse direction: if $\varphi$ is a bounded linear functional on $C$, then there exists a sequence $\{g_n\}$ of functions in $C$ such that $\varphi(f)=\lim_n \int_0^1 f(x)g_n(x)\,dx$. Riesz's incisive step (1909) was to replace the infinite number of Riemann integrals by a single Stieltjes integral: there exists a function $\alpha$ of bounded variation such that $\varphi(f)=\int_0^1 f(x)\,d\alpha(x)$. The preferred language nowadays is in terms of Borel measures, but the difference is more terminological than conceptual.

All manifestations of the Riesz representation theorem have the same form: every bounded linear functional is one of the obvious kind that the context suggests. The statement for $L^2$, for instance, is that if $\varphi$ is a bounded linear functional, then there exists a function $g$ in $L^2$ such that $\varphi(f)=\int_0^1 f(x)g(x)\,dx$ for every $f$ in $L^2$. The obvious approach to the proof uses some non-trivial measure theory (the Radon–Nikodým theorem). The usual geometric proof of the corresponding theorem in abstract Hilbert space (every bounded linear functional is an inner product) uses nothing. How can that be?

What goes on is that the Riesz representation theorem and the Riesz–Fischer theorem are nearly equally deep, and once one of them is known it is easy to recapture the other. The geometric proof of the representation theorem in Hilbert space depends on the space being complete; to apply it to $L^2$ is to use the Riesz–Fischer theorem. The reverse direction is easier (and less important): the set of all bounded linear functionals on a normed vector space is always complete, and the Riesz representation theorem says, in effect, that the set of all bounded linear functionals on $L^2$ coincides with $L^2$.

## COMPACT OPERATORS

Much of what is now called functional analysis received its first treatment in Riesz's first book (1913) and in his major Acta paper (1918). The book foreshadows, in particular, what is sometimes called the Riesz functional calculus. The idea is to define a function $f$ of an operator $T$ by a Cauchy-like integral such as $(1/2\pi i)\int_C f(\lambda)(\lambda-T)^{-1}d\lambda$, and to establish other contacts between operator theory and the methods of analytic functions.

The main emphasis in both the book and the Acta paper is on the Riesz theory of compact operators. That theory treats operators of the form $1-C$, with $C$ compact, and describes how such an operator is related to its adjoint and how the kernels and the ranges of such operators and their adjoints are related to each other. The culmination of the study is complete information about the spectral properties of compact operators.

Riesz's contributions to functional analysis include studies of Hilbert-Schmidt integral equations, the spectral theorem for both bounded and unbounded Hermitian operators, and the vector lattices that have come to be called Riesz spaces; the results I described above are the most outstanding ones among them.

## EPILOGUE

The work of a mathematician can be original and it can be polished; it is rare that one and the same piece of work deserves both adjectives. Riesz certainly did much original work (and sometimes left it to others to do the polishing), and he did much polishing. In a report such as this it seems proper to emphasize the original work, but I cannot resist the temptation to give an example of the sort of pretty polishing that Riesz did so well. In a letter to Hardy (1930) he offered the following proof of the famous arithmetic-geometric-mean inequality. Since $\log t \leqq t-1$ for all positive $t$, it follows that if $f$ is a positive integrable function on $(0,1)$, then

$$\log \frac{f(x)}{A(f)} \leqq \frac{f(x)}{A(f)} - 1$$

for all $x$. (Here $A(f)$ is the arithmetic mean $\int_0^1 f(x)\,dx$.) Integrate to get $A(\log f) - \log A(f) \leqq 0$, and then exponentiate to get the desired result in its usual form.

The purpose of this report was to describe Riesz's chief original contributions, and thereby to try to get an insight into the history of a part of mathematics that is alive and lively. The main reason for such looks into the past is not to adjudicate priority disputes and ascribe credits, but to increase our understanding of the present and of the future. When I report that Riesz did something, I do not mean to detract from the work of others and to say that he and only he did it or even that he was the first to do it. I do mean to say that he was certainly one of the doers, that without him the subject would have grown more slowly, or in a different direction, and that his contribution was a valuable step forward that was not and could not have been taken by just any of us ordinary mortals.

Indiana University
Bloomington, IN 47405

# CHAPTER III

Extrait de *L'Enseignement mathématique*, T. XVI, fasc. 2, 1970

# HOW TO WRITE MATHEMATICS

P. R. Halmos

## 0. Preface

This is a subjective essay, and its title is misleading; a more honest title might be HOW I WRITE MATHEMATICS. It started with a committee of the American Mathematical Society, on which I served for a brief time, but it quickly became a private project that ran away with me. In an effort to bring it under control I asked a few friends to read it and criticize it. The criticisms were excellent; they were sharp, honest, and constructive; and they were contradictory. "Not enough concrete examples" said one; "don't agree that more concrete examples are needed" said another. "Too long" said one; "maybe more is needed" said another. "There are traditional (and effective) methods of minimizing the tediousness of long proofs, such as breaking them up in a series of lemmas" said one. "One of the things that irritates me greatly is the custom (especially of beginners) to present a proof as a long series of elaborately stated, utterly boring lemmas" said another.

There was one thing that most of my advisors agreed on; the writing of such an essay is bound to be a thankless task. Advisor 1: "By the time a mathematician has written his second paper, he is convinced he knows how to write papers, and would react to advice with impatience." Advisor 2: "All of us, I think, feel secretly that if we but bothered we could be really first rate expositors. People who are quite modest about their mathematics will get their dander up if their ability to write well is questioned." Advisor 3 used the strongest language; he warned me that since I cannot possibly display great intellectual depth in a discussion of matters of technique, I should not be surprised at "the scorn you may reap from some of our more supercilious colleagues".

My advisors are established and well known mathematicians. A credit line from me here wouldn't add a thing to their stature, but my possible misunderstanding, misplacing, and misapplying their advice might cause them annoyance and embarrassment. That is why I decided on the unscholarly procedure of nameless quotations and the expression of nameless thanks. I am not the less grateful for that, and not the less eager to acknowledge that without their help this essay would have been worse.

"Hier stehe ich; ich kann nicht anders."

## 1. There is no recipe and what it is

I think I can tell someone how to write, but I can't think who would want to listen. The ability to communicate effectively, the power to be intelligible, is congenital, I believe, or, in any event, it is so early acquired that by the time someone reads my wisdom on the subject he is likely to be invariant under it. To understand a syllogism is not something you can learn; you are either born with the ability or you are not. In the same way, effective exposition is not a teachable art; some can do it and some cannot. There is no usable recipe for good writing.

Then why go on? A small reason is the hope that what I said isn't quite right; and, anyway, I'd like a chance to try to do what perhaps cannot be done. A more practical reason is that in the other arts that require innate talent, even the gifted ones who are born with it are not usually born with full knowledge of all the tricks of the trade. A few essays such as this may serve to "remind" (in the sense of Plato) the ones who want to be and are destined to be the expositors of the future of the techniques found useful by the expositors of the past.

The basic problem in writing mathematics is the same as in writing biology, writing a novel, or writing directions for assembling a harpsichord: the problem is to communicate an idea. To do so, and to do it clearly, you must have something to say, and you must have someone to say it to, you must organize what you want to say, and you must arrange it in the order you want it said in, you must write it, rewrite it, and re-rewrite it several times, and you must be willing to think hard about and work hard on mechanical details such as diction, notation, and punctuation. That's all there is to it.

## 2. Say something

It might seem unnecessary to insist that in order to say something well you must have something to say, but it's no joke. Much bad writing, mathematical and otherwise, is caused by a violation of that first principle.

Just as there are two ways for a sequence not to have a limit (no cluster points or too many), there are two ways for a piece of writing not to have a subject (no ideas or too many).

The first disease is the harder one to catch. It is hard to write many words about nothing, especially in mathematics, but it can be done, and the result is bound to be hard to read. There is a classic crank book by Carl Theodore Heisel [5] that serves as an example. It is full of correctly spelled words strung together in grammatical sentences, but after three decades of looking at it every now and then I still cannot read two consecutive pages and make a one-paragraph abstract of what they say; the reason is, I think, that they don't say anything.

The second disease is very common: there are many books that violate the principle of having something to say by trying to say too many things. Teachers of elementary mathematics in the U.S.A. frequently complain that all calculus books are bad. That is a case in point. Calculus books are bad because there is no such subject as calculus; it is not a subject because it is many subjects. What we call calculus nowadays is the union of a dab of logic and set theory, some axiomatic theory of complete ordered fields, analytic geometry and topology, the latter in both the "general" sense (limits and continuous functions) and the algebraic sense (orientation), real-variable theory properly so called (differentiation), the combinatoric symbol manipulation called formal integration, the first steps of low-dimensional measure theory, some differential geometry, the first steps of the classical analysis of the trigonometric, exponential, and logarithmic functions, and, depending on the space available and the personal inclinations of the author, some cook-book differential equations, elementary mechanics, and a small assortment of applied mathematics. Any one of these is hard to write a good book on; the mixture is impossible.

Nelson's little gem of a proof that a bounded harmonic function is a constant [7] and Dunford and Schwartz's monumental treatise on functional analysis [3] are examples of mathematical writings that have something to say. Nelson's work is not quite half a page and Dunford-Schwartz is more than four thousand times as long, but it is plain in each case that the authors had an unambiguous idea of what they wanted to say. The subject is clearly delineated; it is a subject; it hangs together; it is something to say.

To have something to say is by far the most important ingredient of good exposition—so much so that if the idea is important enough, the work has a chance to be immortal even if it is confusingly misorganized and awkwardly expressed. Birkhoff's proof of the ergodic theorem [1] is almost maximally confusing, and Vanzetti's "last letter" [9] is halting and awkward, but surely anyone who reads them is glad that they were written. To get by on the first principle alone is, however, only rarely possible and never desirable.

## 3. Speak to someone

The second principle of good writing is to write for someone. When you decide to write something, ask yourself who it is that you want to reach. Are you writing a diary note to be read by yourself only, a letter to a friend, a research announcement for specialists, or a textbook for undergraduates? The problems are much the same in any case; what varies is the amount of motivation you need to put in, the extent of informality you may allow yourself, the fussiness of the detail that is necessary, and the number of times things have to be repeated. All writing is influenced by the audience, but, given the audience, an author's problem is to communicate with it as best he can.

Publishers know that 25 years is a respectable old age for most mathematical books; for research papers five years (at a guess) is the average age of obsolescence. (Of course there can be 50-year old papers that remain alive and books that die in five.) Mathematical writing is ephemeral, to be sure, but if you want to reach your audience now, you must write as if for the ages.

I like to specify my audience not only in some vague, large sense (e.g., professional topologists, or second year graduate students), but also in a very specific, personal sense. It helps me to think of a person, perhaps someone I discussed the subject with two years ago, or perhaps a deliberately obtuse, friendly colleague, and then to keep him in mind as I write. In this essay, for instance, I am hoping to reach mathematics students who are near the beginning of their thesis work, but, at the same time, I am keeping my mental eye on a colleague whose ways can stand mending. Of course I hope that (a) he'll be converted to my ways, but (b) he won't take offence if and when he realizes that I am writing for him.

There are advantages and disadvantages to addressing a very sharply specified audience. A great advantage is that it makes easier the mind reading that is necessary; a disadvantage is that it becomes tempting to indulge in snide polemic comments and heavy-handed "in" jokes. It is surely obvious what I mean by the disadvantage, and it is obviously bad; avoid it. The advantage deserves further emphasis.

The writer must anticipate and avoid the reader's difficulties. As he writes, he must keep trying to imagine what in the words being written may tend to mislead the reader, and what will set him right. I'll give examples of one or two things of this kind later; for now I emphasize that keeping a specific reader in mind is not only helpful in this aspect of the writer's work, it is essential.

Perhaps it needn't be said, but it won't hurt to say, that the audience actually reached may differ greatly from the intended one. There is nothing that guarantees that a writer's aim is always perfect. I still say it's better to have a definite aim and hit something else, than to have an aim that is too inclusive or too vaguely specified and have no chance of hitting anything. Get ready, aim, and fire, and hope that you'll hit a target: the target you were aiming at, for choice, but some target in preference to none.

## 4. Organize first

The main contribution that an expository writer can make is to organize and arrange the material so as to minimize the resistance and maximize the insight of the reader and keep him on the track with no unintended distractions. What, after all, are the advantages of a book over a stack of reprints? Answer: efficient and pleasant arrangement, emphasis where emphasis is needed, the indication of interconnections, and the description of the examples and counterexamples on which the theory is based; in one word, organization.

The discoverer of an idea, who may of course be the same as its expositor, stumbled on it helter-skelter, inefficiently, almost at random. If there were no way to trim, to consolidate, and to rearrange the discovery, every student would have to recapitulate it, there would be no advantage to be gained from standing "on the shoulders of giants", and there would never be time to learn something new that the previous generation did not know.

Once you know what you want to say, and to whom you want to say it, the next step is to make an outline. In my experience that is usually impossible. The ideal is to make an outline in which every preliminary heuristic discussion, every lemma, every theorem, every corollary, every remark, and every proof are mentioned, and in which all these pieces occur in an order that is both logically correct and psychologically digestible. In the ideal organization there is a place for everything and everything is in its place. The reader's attention is held because he was told early what to expect, and, at the same time and in apparent contradiction, pleasant surprises keep happening that could not have been predicted from the bare bones of the definitions. The parts fit, and they fit snugly. The lemmas are there when they are needed, and the interconnections of the theorems are visible; and the outline tells you where all this belongs.

I make a small distinction, perhaps an unnecessary one, between organization and arrangement. To organize a subject means to decide what the main headings and subheadings are, what goes under each, and what are the connections among them. A diagram of the organization is a graph, very likely a tree, but almost certainly not a chain. There are many ways to organize most subjects, and usually there are many ways to arrange the results of each method of organization in a linear order. The organization is more important than the arrangement, but the latter frequently has psychological value.

One of the most appreciated compliments I paid an author came from a fiasco; I botched a course of lectures based on his book. The way it started was that there was a section of the book that I didn't like, and I skipped it. Three sections later I needed a small fragment from the end of the omitted section, but it was easy to give a different proof. The same sort of thing happened a couple of times more, but each time a little ingenuity and an ad hoc concept or two patched the leak. In the next chapter, however, something else arose in which what was needed was not a part of the omitted section but the fact that the results of that section were applicable to two apparently very different situations. That was almost impossible to patch up, and after that chaos rapidly set in. The organization of the book was tight; things were there because they were needed; the presentation had the kind of coherence which makes for ease in reading and understanding. At the same time the wires that were holding it all together were not obtrusive; they became visible only when a part of the structure was tampered with.

Even the least organized authors make a coarse and perhaps unwritten outline; the subject itself is, after all, a one-concept outline of the book. If you know that you are writing about measure theory, then you have a two-word outline, and that's something. A tentative chapter outline is something better. It might go like this: I'll tell them about sets, and then measures, and then functions, and then integrals. At this stage you'll want to make some decisions, which, however, may have to be rescinded later;

you may for instance decide to leave probability out, but put Haar measure in.

There is a sense in which the preparation of an outline can take years, or, at the very least, many weeks. For me there is usually a long time between the first joyful moment when I conceive the idea of writing a book and the first painful moment when I sit down and begin to do so. In the interim, while I continue my daily bread and butter work, I daydream about the new project, and, as ideas occur to me about it, I jot them down on loose slips of paper and put them helter-skelter in a folder. An "idea" in this sense may be a field of mathematics I feel should be included, or it may be an item of notation; it may be a proof, it may be an aptly descriptive word, or it may be a witticism that, I hope, will not fall flat but will enliven, emphasize, and exemplify what I want to say. When the painful moment finally arrives, I have the folder at least; playing solitaire with slips of paper can be a big help in preparing the outline.

In the organization of a piece of writing, the question of what to put in is hardly more important than what to leave out; too much detail can be as discouraging as none. The last dotting of the last i, in the manner of the old-fashioned Cours d'Analyse in general and Bourbaki in particular, gives satisfaction to the author who understands it anyway and to the helplessly weak student who never will; for most serious-minded readers it is worse than useless. The heart of mathematics consists of concrete examples and concrete problems. Big general theories are usually afterthoughts based on small but profound insights; the insights themselves come from concrete special cases. The moral is that it's best to organize your work around the central, crucial examples and counterexamples. The observation that a proof proves something a little more general than it was invented for can frequently be left to the reader. Where the reader needs experienced guidance is in the discovery of the things the proof does not prove; what are the appropriate counterexamples and where do we go from here?

## 5. Think about the alphabet

Once you have some kind of plan of organization, an outline, which may not be a fine one but is the best you can do, you are almost ready to start writing. The only other thing I would recommend that you do first is to invest an hour or two of thought in the alphabet; you'll find it saves many headaches later.

The letters that are used to denote the concepts you'll discuss are worthy of thought and careful design. A good, consistent notation can be a tremendous help, and I urge (to the writers of articles too, but especially to the writers of books) that it be designed at the beginning. I make huge tables with many alphabets, with many fonts, for both upper and lower case, and I try to anticipate all the spaces, groups, vectors, functions, points, surfaces, measures, and whatever that will sooner or later need to be baptized. Bad notation can make good exposition bad and bad exposition worse; ad hoc decisions about notation, made mid-sentence in the heat of composition, are almost certain to result in bad notation.

Good notation has a kind of alphabetical harmony and avoids dissonance. Example: either $ax + by$ or $a_1x_1 + a_2x_2$ is preferable to $ax_1 + bx_2$. Or: if you must use $\Sigma$ for an index set, make sure you don't run into $\sum_{\sigma \in \Sigma} a_\sigma$. Along the same lines: perhaps most readers wouldn't notice that you used $|z| < \varepsilon$ at the top of the page and $z \varepsilon U$ at the bottom, but that's the sort of near dissonance that causes a vague non-localized feeling of malaise. The remedy is easy and is getting more and more nearly universally accepted: $\in$ is reserved for membership and $\varepsilon$ for ad hoc use.

Mathematics has access to a potentially infinite alphabet (e.g., $x$, $x'$, $x''$, $x'''$, ...), but, in practice, only a small finite fragment of it is usable. One reason is that a human being's ability to distinguish between symbols is very much more limited than his ability to conceive of new ones; another reason is the bad habit of freezing letters. Some old-fashioned analysts would speak of "*xyz*-space", meaning, I think, 3-dimensional Euclidean space, plus the convention that a point of that space shall always be denoted by "$(x,y,z)$". This is bad: it "freezes" $x$, and $y$, and $z$, i.e., prohibits their use in another context, and, at the same time, it makes it impossible (or, in any case, inconsistent) to use, say, "$(a,b,c)$" when "$(x,y,z)$" has been temporarily exhausted. Modern versions of the custom exist, and are no better. Example: matrices with "property L"—a frozen and unsuggestive designation.

There are other awkward and unhelpful ways to use letters: "CW complexes" and "CCR groups" are examples. A related curiosity that is probably the upper bound of using letters in an unusable way occurs in Lefschetz [6]. There $x_i^p$ is a chain of dimension $p$ (the subscript is just an index), whereas $x_p^i$ is a co-chain of dimension $p$ (and the superscript is an index). Question: what is $x_3^2$?

As history progresses, more and more symbols get frozen. The standard examples are $e$, $i$, and $\pi$, and, of course, 0, 1, 2, 3, .... (Who would dare write "Let 6 be a group."?) A few other letters are almost frozen: many readers would feel offended if "$n$" were used for a complex number, "$\varepsilon$" for a positive integer, and "$z$" for a topological space. (A mathematician's nightmare is a sequence $n_\varepsilon$ that tends to 0 as $\varepsilon$ becomes infinite.)

Moral: do not increase the rigid frigidity. Think about the alphabet. It's a nuisance, but it's worth it. To save time and trouble later, think about the alphabet for an hour now; then start writing.

## 6. Write in spirals

The best way to start writing, perhaps the only way, is to write on the spiral plan. According to the spiral plan the chapters get written and rewritten in the order 1, 2, 1, 2, 3, 1, 2, 3, 4, etc. You think you know how to write Chapter 1, but after you've done it and gone on to Chapter 2, you'll realize that you could have done a better job on Chapter 2 if you had done Chapter 1 differently. There is no help for it but to go back, do Chapter 1 differently, do a better job on Chapter 2, and then dive into Chapter 3. And, of course, you know what will happen: Chapter 3 will show up the weaknesses of Chapters 1 and 2, and there is no help for it ... etc., etc., etc. It's an obvious idea, and frequently an unavoidable one, but it may help a future author to know in advance what he'll run into, and it may help him to know that the same phenomenon will occur not only for chapters, but for sections, for paragraphs, for sentences, and even for words.

The first step in the process of writing, rewriting, and re-rewriting, is writing. Given the subject, the audience, and the outline (and, don't forget, the alphabet), start writing, and let nothing stop you. There is no better incentive for writing a good book than a bad book. Once you have a first draft in hand, spiral-written, based on a subject, aimed at an audience, and backed by as detailed an outline as you could scrape together, then your book is more than half done.

The spiral plan accounts for most of the rewriting and re-rewriting that a book involves (most, but not all). In the first draft of each chapter I recommend that you spill your heart, write quickly, violate all rules, write with hate or with pride, be snide, be confused, be "funny" if you must, be unclear, be ungrammatical—just keep on writing. When you come to rewrite, however, and however often that may be necessary, do not edit but rewrite. It is tempting to use a red pencil to indicate insertions, deletions, and permutations, but in my experience it leads to catastrophic blunders. Against human impatience, and against the all too human partiality everyone feels toward his own words, a red pencil is much too feeble a weapon. You are faced with a first draft that any reader except yourself would find all but unbearable; you must be merciless about changes of all kinds, and, especially, about wholesale omissions. Rewrite means write again—every word.

I do not literally mean that, in a 10-chapter book, Chapter 1 should be written ten times, but I do mean something like three or four. The chances are that Chapter 1 should be re-written, literally, as soon as Chapter 2 is finished, and, very likely, at least once again, somewhere after Chapter 4. With luck you'll have to write Chapter 9 only once.

The description of my own practice might indicate the total amount of rewriting that I am talking about. After a spiral-written first draft I usually rewrite the whole book, and then add the mechanical but indispensable reader's aids (such as a list of prerequisites, preface, index, and table of contents). Next, I rewrite again, this time on the typewriter, or, in any event, so neatly and beautifully that a mathematically untrained typist can use this version (the third in some sense) to prepare the "final" typescript with no trouble. The rewriting in this third version is minimal; it is usually confined to changes that affect one word only, or, in the worst case, one sentence. The third version is the first that others see. I ask friends to read it, my wife reads it, my students may read parts of it, and, best of all, an expert junior-grade, respectably paid to do a good job, reads it and is encouraged not to be polite in his criticisms. The changes that become necessary in the third version can, with good luck, be effected with a red pencil; with bad luck they will cause one third of the pages to be retyped. The "final" typescript is based on the edited third version, and, once it exists, it is read, reread, proofread, and reproofread. Approximately two years after it was started (two working years, which may be much more than two calendar years) the book is sent to the publisher. Then begins another kind of labor pain, but that is another story.

Archimedes taught us that a small quantity added to itself often enough becomes a large quantity (or, in proverbial terms, every little bit helps). When it comes to accomplishing the bulk of the world's work, and, in particular, when it comes to writing a book, I believe that the converse of Archimedes' teaching is also true: the only way to write a large book is to keep writing a small bit of it, steadily every day, with no exception, with no holiday. A good technique, to help the steadiness of your rate of production, is to stop each day by priming the pump for the next day. What will you begin with tomorrow? What is the content of the next section to be; what is its title ? (I recommend that you find a possible short title for each section, before or after it's written, even if you don't plan to print section titles. The purpose is to test how well the section is planned: if you cannot find a title, the reason may be that the section doesn't have a single unified subject.) Sometimes I write tomorrow's first sentence today; some authors begin today by revising and rewriting the last page or so of yesterday's work. In any case, end each work session on an up-beat; give your subconscious something solid to feed on between sessions. It's surprising how well you can fool yourself that way; the pump-priming technique is enough to overcome the natural human inertia against creative work.

## 7. Organize always

Even if your original plan of organization was detailed and good (and especially if it was not), the all-important job of organizing the material does not stop when the writing starts; it goes on all the way through the writing and even after.

The spiral plan of writing goes hand in hand with the spiral plan of organization, a plan that is frequently (perhaps always) applicable to mathematical writing. It goes like this. Begin with whatever you have chosen as your basic concept—vector spaces, say—and do right by it: motivate it, define it, give examples, and give counterexamples. That's Section 1. In Section 2 introduce the first related concept that you propose to study—linear dependence, say—and do right by it: motivate it, define it, give examples, and give counterexamples, and then, this is the important point, review Section 1, as nearly completely as possible, from the point of view of Section 2. For instance: what examples of linearly dependent and independent sets are easily accessible within the very examples of vector spaces that Section 1 introduced ? (Here, by the way, is another clear reason why the spiral plan of writing is necessary: you may think, in Section 2, of examples of linearly dependent and independent sets in vector spaces that you forgot to give as examples in Section 1.) In Section 3 introduce your next concept (of course just what that should be needs careful planning, and, more often, a fundamental change of mind that once again makes spiral writing the right procedure), and, after clearing it up in the customary manner, review Sections 1 and 2 from the point of view of the new concept. It works, it works like a charm. It is easy to do, it is fun to do, it is easy to read, and the reader is helped by the firm organizational scaffolding, even if he doesn't bother to examine it and see where the joins come and how they support one another.

The historical novelist's plots and subplots and the detective story writer's hints and clues all have their mathematical analogues. To make the point by way of an example: much of the theory of metric spaces could be developed as a "subplot" in a book on general topology, in unpretentious comments, parenthetical asides, and illustrative exercises. Such an organization would give the reader more firmly founded motivation and more insight than can be obtained by inexorable generality, and with no visible extra effort. As for clues: a single word, first mentioned several chapters earlier than its definition, and then re-mentioned, with more and more detail each time as the official treatment comes closer and closer, can serve as an inconspicuous, subliminal preparation for its full-dress introduction. Such a procedure can greatly help the reader, and, at the same time, make the author's formal work much easier, at the expense, to be sure, of greatly increasing the thought and preparation that goes into his informal prose writing. It's worth it. If you work eight hours to save five minutes of the reader's time, you have saved over 80 man-hours for each 1000 readers, and your name will be deservedly blessed down the corridors of many mathematics buildings. But remember: for an effective use of subplots and clues, something very like the spiral plan of organization is indispensable.

The last, least, but still very important aspect of organization that deserves mention here is the correct arrangement of the mathematics from the purely logical point of view. There is not much that one mathematician can teach another about that, except to warn that as the size of the job increases, its complexity increases in frightening proportion. At one stage of writing a 300-page book, I had 1000 sheets of paper, each with a mathematical statement on it, a theorem, a lemma, or even a minor comment, complete with proof. The sheets were numbered, any which way. My job was to indicate on each sheet the numbers of the sheets whose statement must logically come before, and then to arrange the sheets in linear order so that no sheet comes after one on which it's mentioned. That problem had, apparently, uncountably many solutions; the difficulty was to pick one that was as efficient and pleasant as possible.

## 8. Write good english

Everything I've said so far has to do with writing in the large, global sense; it is time to turn to the local aspects of the subject.

Why shouldn't an author spell "continuous" as "continous"? There is no chance at all that it will be misunderstood, and it is one letter shorter, so why not ? The answer that probably everyone would agree on, even the most libertarian among modern linguists, is that whenever the "reform" is introduced it is bound to cause distraction, and therefore a waste of time, and the "saving" is not worth it. A random example such as this one is probably not convincing; more people would agree that an entire book written in reformed spelling, with, for instance, "izi" for "easy" is not likely to be an effective teaching instrument for mathematics. Whatever the merits of spelling reform may be, words that are misspelled according to currently accepted dictionary standards detract from the good a book can do: they delay and distract the reader, and possibly confuse or anger him.

The reason for mentioning spelling is not that it is a common danger or a serious one for most authors, but that it serves to illustrate and emphasize a much more important point. I should like to argue that it is important that mathematical books (and papers, and letters, and lectures) be written in good English style, where good means "correct" according to currently and commonly accepted public standards. (French, Japanese, or Russian authors please substitute "French", "Japanese", or "Russian" for "English".) I do not mean that the style is to be pedantic, or heavy-handed, or formal, or bureaucratic, or flowery, or academic jargon. I do mean that it should be completely unobtrusive, like good background music for a movie, so that the reader may proceed with no conscious or unconscious blocks caused by the instrument of communication and not its content.

Good English style implies correct grammar, correct choice of words, correct punctuation, and, perhaps above all, common sense. There is a difference between "that" and "which", and "less" and "fewer" are not the same, and a good mathematical author must know such things. The reader may not be able to define the difference, but a hundred pages of colloquial misusage, or worse, has a cumulative abrasive effect that the author surely does not want to produce. Fowler [4], Roget [8], and Webster [10] are next to Dunford-Schwartz on my desk; they belong in a similar position on every author's desk. It is unlikely that a single missing comma will convert a correct proof into a wrong one, but consistent mistreatment of such small things has large effects.

The English language can be a beautiful and powerful instrument for interesting, clear, and completely precise information, and I have faith that the same is true for French or Japanese or Russian. It is just as important for an expositor to familiarize himself with that instrument as for a surgeon to know his tools. Euclid can be explained in bad grammar and bad diction, and a vermiform appendix can be removed with a rusty pocket knife, but the victim, even if he is unconscious of the reason for his discomfort, would surely prefer better treatment than that.

All mathematicians, even very young students very near the beginning of their mathematical learning, know that mathematics has a language of its own (in fact it is one), and an author must have thorough mastery of the grammar and vocabulary of that language as well as of the vernacular. There is no Berlitz course for the language of mathematics; apparently the only way to learn it is to live with it for years. What follows is not, it cannot be, a mathematical analogue of Fowler, Roget, and Webster, but it may perhaps serve to indicate a dozen or two of the thousands of items that those analogues would contain.

## 9. Honesty is the best policy

The purpose of using good mathematical language is, of course, to make the understanding of the subject easy for the reader, and perhaps even pleasant. The style should be good not in the sense of flashy brilliance, but good in the sense of perfect unobtrusiveness. The purpose is to smooth the reader's way, to anticipate his difficulties and to forestall them. Clarity is what's wanted, not pedantry; understanding, not fuss.

The emphasis in the preceding paragraph, while perhaps necessary, might seem to point in an undesirable direction, and I hasten to correct a possible misinterpretation. While avoiding pedantry and fuss, I do not want to avoid rigor and precision; I believe that these aims are reconcilable. I do not mean to advise a young author to be ever so slightly but very very cleverly dishonest and to gloss over difficulties. Sometimes, for instance, there may be no better way to get a result than a cumbersome computation. In that case it is the author's duty to carry it out, in public; the best he can do to alleviate it is to extend his sympathy to the reader by some phrase such as "unfortunately the only known proof is the following cumbersome computation".

Here is the sort of thing I mean by less than complete honesty. At a certain point, having proudly proved a proposition $p$, you feel moved to say: "Note, however, that $p$ does not imply $q$", and then, thinking that you've done a good expository job, go happily on to other things. Your motives may be perfectly pure, but the reader may feel cheated just the same. If he knew all about the subject, he wouldn't be reading you; for him the non-implication is, quite likely, unsupported. Is it obvious? (Say so.) Will a counterexample be supplied later? (Promise it now.) Is it a standard but for present purposes irrelevant part of the literature? (Give a reference.) Or, horribile dictu, do you merely mean that you have tried to derive $q$ from $p$, you failed, and you don't in fact know whether $p$ implies $q$? (Confess immediately!) In any event: take the reader into your confidence.

There is nothing wrong with the often derided "obvious" and "easy to see", but there are certain minimal rules to their use. Surely when you wrote that something was obvious, you thought it was. When, a month, or two months, or six months later, you picked up the manuscript and re-read it, did you still think that that something was obvious? (A few months' ripening always improves manuscripts.) When you explained it to a friend, or to a seminar, was the something at issue accepted as obvious? (Or did someone question it and subside, muttering, when you reassured him? Did your assurance consist of demonstration or intimidation?) The obvious answers to these rhetorical questions are among the rules that should control the use of "obvious". There is another rule, the major one, and everybody knows it, the one whose violation is the most frequent source of mathematical error: make sure that the "obvious" is true.

It should go without saying that you are not setting out to hide facts from the reader; you are writing to uncover them. What I am saying now is that you should not hide the status of your statements and your attitude toward them either. Whenever you tell him something, tell him where it stands: this has been proved, that hasn't, this will be proved, that won't. Emphasize the important and minimize the trivial. There are many good reasons for making obvious statements every now and then; the reason for saying that they are obvious is to put them in proper perspective for the uninitiate. Even if your saying so makes an occasional reader angry at you, a good purpose is served by your telling him how you view the matter. But, of course, you must obey the rules. Don't let the reader down; he wants to believe in you. Pretentiousness, bluff, and concealment may not get caught out immediately, but most readers will soon sense that there is something wrong, and they will blame neither the facts nor themselves, but, quite properly, the author. Complete honesty makes for greatest clarity.

## 10. Down with the irrelevant and the trivial

Sometimes a proposition can be so obvious that it needn't even be called obvious and still the sentence that announces it is bad exposition, bad because it makes for confusion, misdirection, delay. I mean something like this: "If $R$ is a commutative semisimple ring with unit and if $x$ and $y$ are in $R$, then $x^2 - y^2 = (x - y)(x + y)$." The alert reader will ask himself what semisimplicity and a unit have to do with what he had always thought was obvious. Irrelevant assumptions wantonly dragged in, incorrect emphasis, or even just the absence of correct emphasis can wreak havoc.

Just as distracting as an irrelevant assumption and the cause of just as much wasted time is an author's failure to gain the reader's confidence by explicitly mentioning trivial cases and excluding them if need be. Every complex number is the product of a non-negative number and a number of modulus 1. That is true, but the reader will feel cheated and insecure if soon after first being told that fact (or being reminded of it on some other occasion, perhaps preparatory to a generalization being sprung on him) he is not told that there is something fishy about 0 (the trivial case). The point is not that failure to treat the trivial cases separately may sometimes be a mathematical error; I am not just saying "do not make mistakes". The point is that insistence on legalistically correct but insufficiently explicit explanations ("The statement is correct as it stands—what else do you want ?") is misleading, bad exposition, bad psychology. It may also be almost bad mathematics. If, for instance, the author is preparing to discuss the theorem that, under suitable hypotheses, every linear transformation is the product of a dilatation and a rotation, then his ignoring of 0 in the 1-dimensional case leads to the reader's misunderstanding of the behavior of singular linear transformations in the general case.

This may be the right place to say a few words about the statements of theorems: there, more than anywhere else, irrelevancies must be avoided.

The first question is where the theorem should be stated, and my answer is: first. Don't ramble on in a leisurely way, not telling the reader where you are going, and then suddenly announce "Thus we have proved that ...". The reader can pay closer attention to the proof if he knows what you are proving, and he can see better where the hypotheses are used if he knows in advance what they are. (The rambling approach frequently leads to the "hanging" theorem, which I think is ugly. I mean something like: "Thus we have proved

THEOREM 2 ... ".

The indentation, which is after all a sort of invisible punctuation mark, makes a jarring separation in the sentence, and, after the reader has collected his wits and caught on to the trick that was played on him, it makes an undesirable separation between the statement of the theorem and its official label.)

This is not to say that the theorem is to appear with no introductory comments, preliminary definitions, and helpful motivations. All that comes first; the statement comes next; and the proof comes last. The statement of the theorem should consist of one sentence whenever possible: a simple implication, or, assuming that some universal hypotheses were stated before and are still in force, a simple declaration. Leave the chit-chat out: "Without loss of generality we may assume ... " and "Moreover it follows from Theorem 1 that ... " do not belong in the statement of a theorem.

Ideally the statement of a theorem is not only one sentence, but a short one at that. Theorems whose statement fills almost a whole page (or more!) are hard to absorb, harder than they should be; they indicate that the author did not think the material through and did not organize it as he should have done. A list of eight hypotheses (even if carefully so labelled) and a list of six conclusions do not a theorem make; they are a badly expounded theory. Are all the hypotheses needed for each conclusion ? If the answer is no, the badness of the statement is evident; if the answer is yes, then the hypotheses probably describe a general concept that deserves to be isolated, named, and studied.

## 11. Do and do not repeat

One important rule of good mathematical style calls for repetition and another calls for its avoidance.

By repetition in the first sense I do not mean the saying of the same thing several times in different words. What I do mean, in the exposition of a precise subject such as mathematics, is the word-for-word repetition of a phrase, or even many phrases, with the purpose of emphasizing a slight change in a neighboring phrase. If you have defined something, or stated something, or proved something in Chapter 1, and if in Chapter 2 you want to treat a parallel theory or a more general one, it is a big help to the reader if you use the same words in the same order for as long as possible, and then, with a proper roll of drums, emphasize the difference. The roll of drums is important. It is not enough to list six adjectives in one definition, and re-list five of them, with a diminished sixth, in the second. That's the thing to do, but what helps is to say, in addition: "Note that the first five conditions in the definitions of $p$ and $q$ are the same; what makes them different is the weakening of the sixth."

Often in order to be able to make such an emphasis in Chapter 2 you'll have to go back to Chapter 1 and rewrite what you thought you had already written well enough, but this time so that its parallelism with the relevant part of Chapter 2 is brought out by the repetition device. This is another illustration of why the spiral plan of writing is unavoidable, and it is another aspect of what I call the organization of the material.

The preceding paragraphs describe an important kind of mathematical repetition, the good kind; there are two other kinds, which are bad.

One sense in which repetition is frequently regarded as a device of good teaching is that the oftener you say the same thing, in exactly the same words, or else with slight differences each time, the more likely you are to drive the point home. I disagree. The second time you say something, even the vaguest reader will dimly recall that there was a first time, and he'll wonder if what he is now learning is exactly the same as what he should have learned before, or just similar but different. (If you tell him "I am now saying *exactly* what I first said on p. 3", that helps.) Even the dimmest such wonder is bad. Anything is bad that unnecessarily frightens, irrelevantly amuses, or in any other way distracts. (Unintended double meanings are the woe of many an author's life.) Besides, good organization, and, in particular, the spiral plan of organization discussed before is a substitute for repetition, a substitute that works much better.

Another sense in which repetition is bad is summed up in the short and only partially inaccurate precept: never repeat a proof. If several steps in the proof of Theorem 2 bear a very close resemblance to parts of the proof of Theorem 1, that's a signal that something may be less than completely understood. Other symptoms of the same disease are: "by the same technique (or method, or device, or trick) as in the proof of Theorem 1 ... ", or, brutally, "see the proof of Theorem 1". When that happens the chances are very good that there is a lemma that is worth finding, formulating, and proving, a lemma from which both Theorem 1 and Theorem 2 are more easily and more clearly deduced.

## 12. The editorial we is not all bad

One aspect of expository style that frequently bothers beginning authors is the use of the editorial "we", as opposed to the singular "I", or the neutral "one". It is in matters like this that common sense is most important. For what it's worth, I present here my recommendation.

Since the best expository style is the least obtrusive one, I tend nowadays to prefer the neutral approach. That does *not* mean using "one" often, or ever; sentences like "one has thus proved that ..." are awful. It does mean the complete avoidance of first person pronouns in either singular or plural. "Since $p$, it follows that $q$." "This implies $p$." "An application of $p$ to $q$ yields $r$." Most (all ?) mathematical writing is (should be ?) factual; simple declarative sentences are the best for communicating facts.

A frequently effective and time-saving device is the use of the imperative. "To find $p$, multiply $q$ by $r$." "Given $p$, put $q$ equal to $r$." (Two digressions about "given". (1) Do not use it when it means nothing. Example: "For any given $p$ there is a $q$." (2) Remember that it comes from an active verb and resist the temptation to leave it dangling. Example: Not "Given $p$, there is a $q$", but "Given $p$, find $q$".)

There is nothing wrong with the editorial "we", but if you like it, do not misuse it. Let "we" mean "the author and the reader" (or "the lecturer and the audience"). Thus, it is fine to say "Using Lemma 2 we can generalize Theorem 1", or "Lemma 3 gives us a technique for proving Theorem 4". It is not good to say "Our work on this result was done in 1969" (unless the voice is that of two authors, or more, speaking in unison), and "We thank our wife for her help with the typing" is always bad.

The use of "I", and especially its overuse, sometimes has a repellent effect, as arrogance or ex-cathedra preaching, and, for that reason, I like to avoid it whenever possible. In short notes, obviously in personal historical remarks, and, perhaps, in essays such as this, it has its place.

## 13. Use words correctly

The next smallest units of communication, after the whole concept, the major chapters, the paragraphs, and the sentences are the words. The preceding section about pronouns was about words, in a sense, although, in a more legitimate sense, it was about global stylistic policy. What I am now going to say is not just "use words correctly"; that should go without saying. What I do mean to emphasize is the need to think about and use with care the small words of common sense and intuitive logic, and the specifically mathematical words (technical terms) that can have a profound effect on mathematical meaning.

The general rule is to use the words of logic and mathematics correctly. The emphasis, as in the case of sentence-writing, is not encouraging pedantry; I am not suggesting a proliferation of technical terms with hairline distinctions among them. Just the opposite; the emphasis is on craftsmanship so meticulous that it is not only correct, but unobtrusively so.

Here is a sample: "Prove that any complex number is the product of a non-negative number and a number of modulus 1." I have had students who would have offered the following proof: "$-4i$ is a complex number, and it is the product of 4, which is non-negative, and $-i$, which has modulus 1; q.e.d." The point is that in everyday English "any" is an ambiguous word; depending on context it may hint at an existential quantifier ("have you any wool?", "if anyone can do it, he can") or a universal one ("any number can play"). Conclusion: never use "any" in mathematical writing. Replace it by "each" or "every", or recast the whole sentence.

One way to recast the sample sentence of the preceding paragraph is to establish the convention that all "individual variables" range over the set of complex numbers and then write something like

$$\forall z \exists p \exists u \, [(p = |p|) \wedge (|u| = 1) \wedge (z = pu)].$$

I recommend against it. The symbolism of formal logic is indispensable in the discussion of the logic of mathematics, but used as a means of transmitting ideas from one mortal to another it becomes a cumbersome code. The author had to code his thoughts in it (I deny that anybody thinks in terms of $\exists$, $\forall$, $\wedge$, and the like), and the reader has to decode what the author wrote; both steps are a waste of time and an obstruction to understanding. Symbolic presentation, in the sense of either the modern logician or the classical epsilontist, is something that machines can write and few but machines can read.

So much for "any". Other offenders, charged with lesser crimes, are "where", and "equivalent", and "if ... then ... if ... then". "Where" is usually a sign of a lazy afterthought that should have been thought through before. "If $n$ is sufficiently large, then $|a_n| < \varepsilon$, where $\varepsilon$ is a preassigned positive number"; both disease and cure are clear. "Equivalent" *for theorems* is logical nonsense. (By "theorem" I mean a mathematical truth, something that has been proved. A meaningful statement can be false, but a theorem cannot; "a false theorem" is self-contradictory). What sense does it make to say that the completeness of $L^2$ is equivalent to the representation theorem for linear functionals on $L^2$? What is meant is that the proofs of both theorems are moderately hard, but once one of them has been proved, either one, the other can be proved with relatively much less work. The logically precise word "equivalent" is not a good word for *that*. As for "if ... then ... if ... then", that is just a frequent stylistic bobble committed by quick writers and rued by slow readers. "If $p$, then if $q$, then $r$." Logically all is well ($p \Rightarrow (q \Rightarrow r)$), but psychologically it is just another pebble to stumble over, unnecessarily. Usually all that is needed to avoid it is to recast the sentence, but no universally good recasting exists; what is best depends on what is important in the case at hand. It could be "If $p$ and $q$, then $r$", or "In the presence of $p$, the hypothesis $q$ implies the conclusion $r$", or many other versions.

## 14. Use technical terms correctly

The examples of mathematical diction mentioned so far were really logical matters. To illustrate the possibilities of the unobtrusive use of precise language in the everyday sense of the working mathematician, I briefly mention three examples: function, sequence, and contain.

I belong to the school that believes that functions and their values are sufficiently different that the distinction should be maintained. No fuss is necessary, or at least no visible, public fuss; just refrain from saying things like "the function $z^2 + 1$ is even". It takes a little longer to say "the function $f$ defined by $f(z) = z^2 + 1$ is even", or, what is from many points of view preferable, "the function $z \to z^2 + 1$ is even", but it is a good habit that can sometimes save the reader (and the author) from serious blunder and that always makes for smoother reading.

"Sequence" means "function whose domain is the set of natural numbers". When an author writes "the union of a sequence of measurable sets is measurable" he is guiding the reader's attention to where it doesn't belong. The theorem has nothing to do with the firstness of the first set, the secondness of the second, and so on; the *sequence* is irrelevant. The correct statement is that "the union of a countable set of measurable sets is measurable" (or, if a different emphasis is wanted, "the union of a countably infinite set of measurable sets is measurable"). The theorem that "the limit of a sequence of measurable functions is measurable" is a very different thing; there "sequence" is correctly used. If a reader knows what a sequence is, if he feels the definition in his bones, then the misuse of the word will distract him and slow his reading down, if ever so slightly; if he doesn't really know, then the misuse will seriously postpone his ultimate understanding.

"Contain" and "include" are almost always used as synonyms, often by the same people who carefully coach their students that $\in$ and $\subset$ are not the same thing at all. It is extremely unlikely that the interchangeable use of contain and include will lead to confusion. Still, some years ago I started an experiment, and I am still trying it: I have systematically and always, in spoken word and written, used "contain" for $\in$ and "include" for $\subset$. I don't say that I have proved anything by this, but I can report that (a) it is very easy to get used to, (b) it does no harm whatever, and (c) I don't think that anybody ever noticed it. I suspect, but that is not likely to be provable, that this kind of terminological consistency (with no fuss made about it) might nevertheless contribute to the reader's (and listener's) comfort.

Consistency, by the way, is a major virtue and its opposite is a cardinal sin in exposition. Consistency is important in language, in notation, in references, in typography—it is important everywhere, and its absence can cause anything from mild irritation to severe misinformation.

My advice about the use of words can be summed up as follows. (1) Avoid technical terms, and especially the creation of new ones, whenever possible. (2) Think hard about the new ones that you must create; consult Roget; and make them as appropriate as possible. (3) Use the old ones correctly and consistently, but with a minimum of obtrusive pedantry.

## 15. Resist symbols

Everything said about words applies, mutatis mutandis, to the even smaller units of mathematical writing, the mathematical symbols. The best notation is no notation; whenever it is possible to avoid the use of a complicated alphabetic apparatus, avoid it. A good attitude to the preparation of written mathematical exposition is to pretend that it is spoken. Pretend that you are explaining the subject to a friend on a long walk in the woods, with no paper available; fall back on symbolism only when it is really necessary.

A corollary to the principle that the less there is of notation the better it is, and in analogy with the principle of omitting irrelevant assumptions, avoid the use of irrelevant symbols. Example: "On a compact space every real-valued continuous function $f$ is bounded." What does the symbol "$f$" contribute to the clarity of that statement ? Another example: "If $0 \leqq lim_n \alpha_n{}^{1/n} = \rho \leqq 1$, then $lim_n \alpha_n = 0$." What does "$\rho$" contribute here? The answer is the same in both cases (nothing), but the reasons for the presence of the irrelevant symbols may be different. In the first case "$f$" may be just a nervous habit; in the second case "$\rho$" is probably a preparation for the proof. The nervous habit is easy to break. The other is harder, because it involves more work for the author. Without the "$\rho$" in the statement, the proof will take a half line longer; it will have to begin with something like "Write $\rho = lim_n \, \alpha_n{}^{1/n}$." The repetition (of "$lim_n \, \alpha_n{}^{1/n}$") is worth the trouble; both statement and proof read more easily and more naturally.

A showy way to say "use no superfluous letters" is to say "use no letter only once". What I am referring to here is what logicians would express by saying "leave no variable free". In the example above, the one about continuous functions, "$f$" was a free variable. The best way to eliminate that particular "$f$" is to omit it; an occasionally preferable alternative is to convert it from free to bound. Most mathematicians would do that by saying "If $f$ is a real-valued continuous function on a compact space, then $f$ is bounded." Some logicians would insist on pointing out that "$f$" is still free in the new sentence (twice), and technically they would be right. To make it bound, it would be necessary to insert "for all $f$" at some grammatically appropriate point, but the customary way mathematicians handle the problem is to refer (tacitly) to the (tacit) convention that every sentence is preceded by all the universal quantifiers that are needed to convert all its variables into bound ones.

The rule of never leaving a free variable in a sentence, like many of the rules I am stating, is sometimes better to break than to obey. The sentence, after all, is an arbitrary unit, and if you want a free "$f$" dangling in one sentence so that you may refer to it in a later sentence in, say, the same paragraph, I don't think you should necessarily be drummed out of the regiment. The rule is essentially sound, just the same, and while it may be bent sometimes, it does not deserve to be shattered into smithereens.

There are other symbolic logical hairs that can lead to obfuscation, or, at best, temporary bewilderment, unless they are carefully split. Suppose, for an example, that somewhere you have displayed the relation

$$(*) \qquad \int_0^1 |f(x)|^2 dx < \infty,$$

as, say, a theorem proved about some particular $f$. If, later, you run across another function $g$ with what looks like the same property, you should resist the temptation to say "$g$ also satisfies (*)". That's logical and alphabetical nonsense. Say instead "(*) remains satisfied if $f$ is replaced by $g$", or, better, give (*) a name (in this case it has a customary one) and say "$g$ also belongs to $L^2(0,1)$".

What about "inequality (*)", or "equation (7)", or "formula (iii)"; should all displays be labelled or numbered? My answer is no. Reason: just as you shouldn't mention irrelevant assumptions or name irrelevant concepts, you also shouldn't attach irrelevant labels. Some small part of the reader's attention is attracted to the label, and some small part of his mind will wonder why the label is there. If there is a reason, then the wonder serves a healthy purpose by way of preparation, with no fuss, for a future reference to the same idea; if there is no reason, then the attention and the wonder were wasted.

It's good to be stingy in the use of labels, but parsimony also can be carried to extremes. I do not recommend that you do what Dickson once did [2]. On p. 89 he says: "Then ... we have (1) ... "—but p. 89 is the beginning of a new chapter, and happens to contain no display at all, let alone one bearing the label (1). The display labelled (1) occurs on p. 90, overleaf, and I never thought of looking for it *there*. That trick gave me a helpless and bewildered five minutes. When I finally saw the light, I felt both stupid and cheated, and I have never forgiven Dickson.

One place where cumbersome notation quite often enters is in mathematical induction. Sometimes it is unavoidable. More often, however, I think that indicating the step from 1 to 2 and following it by an airy "and so on" is as rigorously unexceptionable as the detailed computation, and much more understandable and convincing. Similarly, a general statement about $n \times n$ matrices is frequently best proved not by the exhibition of many $a_{ij}$'s, accompanied by triples of dots laid out in rows and columns and diagonals, but by the proof of a typical (say $3 \times 3$) special case.

There is a pattern in all these injunctions about the avoidance of notation. The point is that the rigorous concept of a mathematical proof can be taught to a stupid computing machine in one way only, but to a human being endowed with geometric intuition, with daily increasing experience, and with the impatient inability to concentrate on repetitious detail for very long, that way is a bad way. Another illustration of this is a proof that consists of a chain of expressions separated by equal signs. Such a proof is easy to write. The author starts from the first equation, makes a natural substitution to get the second, collects terms, permutes, inserts and immediately cancels an inspired factor, and by steps such as these proceeds till he gets the last equation. This is, once again, coding, and the reader is forced not only to learn as he goes, but, at the same time, to decode as he goes. The double effort is needless. By spending another ten minutes writing a carefully worded paragraph, the author can save each of his readers half an hour and a lot of confusion. The paragraph should be a recipe for action, to replace the unhelpful code that merely reports the results of the act and leaves the reader to guess how they were obtained. The paragraph would say something like this: "For the proof, first substitute $p$ for $q$, then collect terms, permute the factors, and, finally, insert and cancel a factor $r$."

A familiar trick of bad teaching is to begin a proof by saying: "Given $\varepsilon$, let $\delta$ be $\left(\frac{\varepsilon}{3M^2 + 2}\right)^{1/2}$". This is the traditional backward proof-writing of classical analysis. It has the advantage of being easily *verifiable* by a machine (as opposed to *understandable* by a human being), and it has the dubious advantage that something at the end comes out to be less than $\varepsilon$, instead of less than, say, $\left(\frac{(3M^2 + 7)\,\varepsilon}{24}\right)^{1/3}$. The way to make the human reader's task less demanding is obvious: write the proof forward. Start, as the author always starts, by putting something less than $\varepsilon$, and then do what needs to be done—multiply by $3M^2 + 7$ at the right time and divide by 24 later, etc., etc.—till you end up with what you end up with. Neither arrangement is elegant, but the forward one is graspable and rememberable.

## 16. Use symbols correctly

There is not much harm that can be done with non-alphabetical symbols, but there too consistency is good and so is the avoidance of individually unnoticed but collectively abrasive abuses. Thus, for instance, it is good to use a symbol so consistently that its verbal translation is always the same. It is good, but it is probably impossible; nonetheless it's a better aim than no aim at all. How are we to read "$\in$": as the verb phrase "is in" or as the preposition "in"? Is it correct to say: "For $x \in A$, we have $x \in B$," or "If $x \in A$, then $x \in B$"? I strongly prefer the latter (always read "$\in$" as "is in") and I doubly deplore the former (both usages occur in the same sentence). It's easy to write and it's easy to read "For $x$ in A, we have $x \in B$"; all dissonance and all even momentary ambiguity is avoided. The same is true for "$\subset$" even though the verbal translation is longer, and even more true for "$\leqq$". A sentence such as "Whenever a positive number is $\leqq 3$, its square is $\leqq 9$" is ugly.

Not only paragraphs, sentences, words, letters, and mathematical symbols, but even the innocent looking symbols of standard prose can be the source of blemishes and misunderstandings; I refer to punctuation marks. A couple of examples will suffice. First: an equation, or inequality, or inclusion, or any other mathematical clause is, in its informative content, equivalent to a clause in ordinary language, and, therefore, it demands just as much to be separated from its neighbors. In other words: punctuate symbolic sentences just as you would verbal ones. Second: don't overwork a small punctuation mark such as a period or a comma. They are easy for the reader to overlook, and the oversight causes backtracking, confusion, delay. Example: "Assume that $a \in X$. $X$ belongs to the class $C$, ... ". The period between the two $X$'s is overworked, and so is this one: "Assume that $X$ vanishes. $X$ belongs to the class $C$, ... ". A good general rule is: never start a sentence with a symbol. If you insist on starting the sentence with a mention of the thing the symbol denotes, put the appropriate word in apposition, thus: "The set $X$ belongs to the class $C$, ... ".

The overworked period is no worse than the overworked comma. Not "For invertible $X$, $X^*$ also is invertible", but "For invertible $X$, the adjoint $X^*$ also is invertible". Similarly, not "Since $p \neq 0$, $p \in U$", but "Since $p \neq 0$, it follows that $p \in U$". Even the ordinary "If you don't like it, lump it" (or, rather, its mathematical relatives) is harder to digest than the stuffy-sounding "If you don't like it, then lump it"; I recommend "then" with "if" in all mathematical contexts. The presence of "then" can never confuse; its absence can.

A final technicality that can serve as an expository aid, and should be mentioned here, is in a sense smaller than even the punctuation marks, it is in a sense so small that it is invisible, and yet, in another sense, it's the most conspicuous aspect of the printed page. What I am talking about is the layout, the architecture, the appearance of the page itself, of all the pages. Experience with writing, or perhaps even with fully conscious and critical reading, should give you a feeling for how what you are now writing will look when it's printed. If it looks like solid prose, it will have a forbidding, sermony aspect; if it looks like computational hash, with a page full of symbols, it will have a frightening, complicated aspect. The golden mean is golden. Break it up, but not too small; use prose, but not too much. Intersperse enough displays to give the eye a chance to help the brain;

use symbols, but in the middle of enough prose to keep the mind from drowning in a morass of suffixes.

## 17. All communication is exposition

I said before, and I'd like for emphasis to say again, that the differences among books, articles, lectures, and letters (and whatever other means of communication you can think of) are smaller than the similarities.

When you are writing a research paper, the role of the "slips of paper" out of which a book outline can be constructed might be played by the theorems and the proofs that you have discovered; but the game of solitaire that you have to play with them is the same.

A lecture is a little different. In the beginning a lecture is an expository paper; you plan it and write it the same way. The difference is that you must keep the difficulties of oral presentation in mind. The reader of a book can let his attention wander, and later, when he decides to, he can pick up the thread, with nothing lost except his own time; a member of a lecture audience cannot do that. The reader can try to prove your theorems for himself, and use your exposition as a check on his work; the hearer cannot do that. The reader's attention span is short enough; the hearer's is much shorter. If computations are unavoidable, a reader can be subjected to them; a hearer must never be. Half the art of good writing is the art of omission; in speaking, the art of omission is nine-tenths of the trick. These differences are not large. To be sure, even a good expository paper, read out loud, would make an awful lecture—but not worse than some I have heard.

The appearance of the printed page is replaced, for a lecture, by the appearance of the blackboard, and the author's imagined audience is replaced for the lecturer by live people; these are big differences. As for the blackboard: it provides the opportunity to make something grow and come alive in a way that is not possible with the printed page. (Lecturers who prepare a blackboard, cramming it full before they start speaking, are unwise and unkind to audiences.) As for live people: they provide an immediate feedback that every author dreams about but can never have.

The basic problems of all expository communication are the same; they are the ones I have been describing in this essay. Content, aim and organization, plus the vitally important details of grammar, diction, and notation—they, not showmanship, are the essential ingredients of good lectures, as well as good books.

## 18. Defend your style

Smooth, consistent, effective communication has enemies; they are called editorial assistants or copyreaders.

An *editor* can be a very great help to a writer. Mathematical writers must usually live without this help, because the editor of a mathematical book must be a mathematician, and there are very few mathematical editors. The ideal editor, who must potentially understand every detail of the author's subject, can give the author an inside but nonetheless unbiased view of the work that the author himself cannot have. The ideal editor is the union of the friend, wife, student, and expert junior-grade whose contribution to writing I described earlier. The mathematical editors of book series and journals don't even come near to the ideal. Their editorial work is but a small fraction of their life, whereas to be a good editor is a full-time job. The ideal mathematical editor does not exist; the friend-wife-etc. combination is only an almost ideal substitute.

The *editorial assistant* is a full-time worker whose job is to catch your inconsistencies, your grammatical slips, your errors of diction, your misspellings—everything that you can do wrong, short of the mathematical content. The trouble is that the editorial assistant does not regard himself as an extension of the author, and he usually degenerates into a mechanical misapplier of mechanical rules. Let me give some examples.

I once studied certain transformations called "measure-preserving". (Note the hyphen: it plays an important role, by making a single word, an adjective, out of two words.) Some transformations pertinent to that study failed to deserve the name; their failure was indicated, of course, by the prefix "non". After a long sequence of misunderstood instructions, the printed version spoke of a "nonmeasure preserving transformation". That is nonsense, of course, amusing nonsense, but, as such, it is distracting and confusing nonsense.

A mathematician friend reports that in the manuscript of a book of his he wrote something like "$p$ or $q$ holds according as $x$ is negative or positive". The editorial assistant changed that to "$p$ or $q$ holds according as $x$ is positive or negative", on the grounds that it sounds better that way. That could be funny if it weren't sad, and, of course, very very wrong.

A common complaint of anyone who has ever discussed quotation marks with the enemy concerns their relation to other punctuation. There appears to be an international typographical decree according to which a period or a comma immediately to the right of a quotation is "ugly". (As here: the editorial assistant would have changed that to "ugly." if I had let him.) From the point of view of the logical mathematician (and even more the mathematical logician) the decree makes no sense; the comma or period should come where the logic of the situation forces it to come. Thus,

He said: "The comma is ugly."

Here, clearly, the period belongs inside the quote; the two situations are different and no inelastic rule can apply to both.

Moral: there are books on "style" (which frequently means typographical conventions), but their mechanical application by editorial assistants can be harmful. If you want to be an author, you must be prepared to defend your style; go forearmed into the battle.

## 19. Stop

The battle against copyreaders is the author's last task, but it's not the one that most authors regard as the last. The subjectively last step comes just before; it is to finish the book itself—to stop writing. That's hard.

There is always something left undone, always either something more to say, or a better way to say something, or, at the very least, a disturbing vague sense that the perfect addition or improvement is just around the corner, and the dread that its omission would be everlasting cause for regret. Even as I write this, I regret that I did not include a paragraph or two on the relevance of euphony and prosody to mathematical exposition. Or, hold on a minute !, surely I cannot stop without a discourse on the proper naming of concepts (why "commutator" is good and "set of first category" is bad) and the proper way to baptize theorems (why "the closed graph theorem" is good and "the Cauchy-Buniakowski-Schwarz theorem" is bad). And what about that sermonette that I haven't been able to phrase satisfactorily about following a model. Choose someone, I was going to say, whose writing can touch you and teach you, and adapt and modify his style to fit your personality and your subject—surely I must get that said somehow.

There is no solution to this problem except the obvious one; the only way to stop is to be ruthless about it. You can postpone the agony a bit, and you should do so, by proofreading, by checking the computations, by letting the manuscript ripen, and then by reading the whole thing over in a gulp, but you won't want to stop any more then than before.

When you've written everything you can think of, take a day or two to read over the manuscript quickly and to test it for the obvious major points that would first strike a stranger's eye. Is the mathematics good, is the exposition interesting, is the language clear, is the format pleasant and easy to read ? Then proofread and check the computations; that's an obvious piece of advice, and no one needs to be told how to do it. "Ripening" is easy to explain but not always easy to do: it means to put the manuscript out of sight and try to forget it for a few months. When you have done all that, and then re-read the whole work from a rested point of view, you have done all you can. Don't wait and hope for one more result, and don't keep on polishing. Even if you do get that result or do remove that sharp corner, you'll only discover another mirage just ahead.

To sum it all up: begin at the beginning, go on till you come to the end, and then, with no further ado, stop.

## 20 The last word

I have come to the end of all the advice on mathematical writing that I can compress into one essay. The recommendations I have been making are based partly on what I do, more on what I regret not having done, and most on what I wish others had done for me. You may criticize what I've said on many grounds, but I ask that a comparison of my present advice with my past action not be one of them. Do, please, as I say, and not as I do, and you'll do better. Then rewrite this essay and tell the next generation how to do better still.

## REFERENCES

[1] Birkhoff, G. D. Proof of the ergodic theorem, *Proc. N.A.S., U.S.A.* 17 (1931) 656-660.
[2] Dickson, L. E., *Modern algebraic theories*, Sanborn, Chicago (1926).
[3] Dunford N. and Schwartz J. T., *Linear operators*, Interscience, New York (1958, 1963).
[4] Fowler H. W., *Modern English usage* (Second edition, revised by Sir Ernest Gowers), Oxford, New York (1965).
[5] Heisel C. T., *The circle squared beyond refutation*, Heisel, Cleveland (1934).
[6] Lefschetz, S. Algebraic topology, *A.M.S.*, New York (1942).
[7] Nelson E. A proof of Liouville's theorem, *Proc. A.M.S.* 12 (1961) 995.
[8] *Roget's International Thesaurus*, Crowell, New York (1946).
[9] Thurber J. and Nugent E., *The male animal*, Random House, New York (1940).
[10] *Webster's New International Dictionary* (Second edition, unabridged), Merriam, Springfield (1951).

Indiana University

# HOW TO TALK MATHEMATICS

By P. R. Halmos

Apology

The purpose of what follows is to suggest to a young mathematician what he might do (and what he had better not do) the first few times that he gives a public lecture on his subject. By a "public lecture" I mean something like a colloquium talk (to more or less the entire mathematics department at a large university), or an invited address (to more or less the entire membership in attendance at a meeting of the American Mathematical Society); I do not mean a classroom lecture (to reluctant beginners) or a seminar talk (to dedicated experts).

That an article on how to talk mathematics might serve a good purpose was suggested by some of the officers of the American Mathematical Society. It seems that there have been more and more complaints about invited addresses ("they are incomprehensible, and therefore useless"), and that, therefore, it might do some good to let a speaker know about such complaints before he adds to the reason for them.

A genius makes his own rules, but a "how to" article is written by one ordinary mortal for the benefit of another. Harpo Marx, one of the greatest harpists of all times, was never taught how to play; everything he did was "wrong" according to standard teaching. Most things that an article such as this one can say have at least one counterexample in the practice of some natural born genius. Authors of articles such as this one know that, but, in the first approximation, they must ignore it, or nothing would ever get done.

Why lecture?

What is the purpose of a public lecture? Answer: to attract and to inform. We like what we do, and we should like for others to like it too; and we believe that the subject's intrinsic qualities are good enough so that anyone who knows what they are cannot help being attracted to them. Hence, better answer: the purpose of a public lecture is to inform, but to do so in a manner that makes it possible for the audience to absorb the information. An attractive presentation with no content is worthless, to be sure, but a lump of indigestible information is worth no more.

The question then becomes this: what is the best way to describe a subject (or that small part of a subject that has recently been the center of the lecturer's attention) to an audience of mathematicians most of whom are interested in something else? The problem is different from describing a subject to students who, willy nilly, must learn it in usable detail, and it is different from sharing a new discovery with fellow experts who have been thinking about the same sort of thing and are wondering what you know that they don't.

Simplicity

Less is more, said the great architect Mies van der Rohe, and if all lecturers remembered that adage, all audiences would be both wiser and happier.

Have you ever disliked a lecture because it was too elementary? I am sure that there are people who would answer yes to that question, but not many. Every time I have asked the question, the person who answered said no, and then looked a little surprised at hearing the answer. A public lecture should be simple and elementary; it should not be complicated and technical. If you believe and can act on this injunction ("be simple"), you can stop reading here; the rest of what I have to say is, in comparison, just a matter of minor detail.

To begin a public lecture to 500 people with "Consider a sheaf of germs of holomorphic functions..." (I have heard it happen) loses people and antagonizes them. If you mention the Künneth formula, it does no harm to say that, at least as far as Betti numbers go, it is just like what happens when you multiply polynomials. If you mention functors, say that a typical example is the formation of the duals of vector spaces and the adjoints of linear transformations.

Be simple by being concrete. Listeners are prepared to accept unstated (but hinted) generalizations much more than they are able, on the spur of the moment, to decode a precisely stated abstraction and to re-invent the special cases that motivated it in the first place. Caution: being concrete should not lead to concentrating on the trees and missing the woods. In many parts of mathematics a generalization is simpler and more incisive than its special parent. (Examples: Artin's solution of Hilbert's 17th problem about definite forms via formally real fields; Gelfand's proof of Wiener's theorem about absolutely convergent Fourier series via Banach algebras.) In such cases there is always a concrete special case that is simpler than the seminal one and that illustrates the generalization with less fuss; the lecturer who knows his subject will explain the complicated special case, and the generalization, by discussing the simple cousin.

Some lecturers defend complications and technicalities by saying that that's what their subject is like, and there is nothing they can do about it. I am skeptical, and I am willing to go so far as to say that such statements indicate incomplete understanding of the subject and of its place in mathematics. Every subject, and even every small part of a subject, if it is identifiable, if it is big enough to give an hour talk on, has its simple aspects, and they, the simple aspects, the roots of the subject, the connections with more widely known and older parts of mathematics, are what a non-specialized audience needs to be told.



Reprinted from the April 1974 issue of the NOTICES, Volume 21, Number 3

Many lecturers, especially those near the foot of the academic ladder, anxious to climb rapidly, feel under pressure to say something brand new - to impress their elders with their brilliance and profundity. Two comments: (1) the best way to do that is to make the talk simple, and (2) it doesn't really have to be done. It may be entirely appropriate to make the lecturer's recent research the focal point of the lecture, but it may also be entirely appropriate not to do so. An audience's evaluation of the merits of a talk is not proportional to the amount of original material included; the explanation of the speaker's latest theorem may fail to improve his chances of creating a good impression.

An oft-quoted compromise between trying to be intelligible and trying to seem deep is this advice: address the first quarter of your talk to your high-school chemistry teacher, the second to a graduate student, the third to an educated mathematician whose interests are different from yours, and the last to the specialists. I have done my duty by reporting the formula, but I'd fail in my duty if I didn't warn that there are many who do not agree with it. A good public lecture should be a work of art. It should be an architectural unit whose parts reinforce each other in conveying the maximum possible amount of information - not a campaign speech that offers something to everybody and, more likely than not, ends by pleasing nobody.

Make it simple, and you won't go wrong.

Details

Some lecturers, with the best of intentions, striving for simplicity, try to achieve it by being overly explicit and overly detailed; that's a mistake.

"Explicit" refers to computations. If a proof can be carried out by multiplying two horrendous expressions, say so and let it go at that; the logical simplicity of the steps doesn't necessarily make the computation attractive or informative to carry out. Landau, legend has it, never omitted a single epsilon from his lectures, and his lectures were inspiring anyway - but that's the exception, not the rule. If, on an exceptional occasion, you think that a brief computation will be decisive and illuminating, put it in, but the rule for ordinary mortals still stands: do not compute in public. It may be an explicit and honest thing to do, but that's not what makes a lecture simple.

"Detailed" refers to definitions. Some lecturers think that the way to reach an audience of non-experts is to tell them everything. ("To get to the theorem I proved last week, I need, starting from the beginning, 14 definitions and 11 theorems that my predecessors have proved. If I talk and write fast, I can present those 25 nuggets in 25 minutes, and in the rest of the time I can state and prove my own thing.") This, too, is honest, and it makes the lecture self-contained, in some sense - but it is impossible to digest, and its effect is dreadful. If someone told you, in half an hour, the meaning of each ideogram on a page of Chinese, could you then read and enjoy the poem on that page in the next half hour?

Proofs

Some lecturers understand the injunction "be simple" to mean "don't prove anything". That isn't quite right. It is true, I think, that it is not the main purpose of a public lecture to prove things, but to prove nothing at all robs the exposition of an essential part of what mathematicians regard as attractive and informative. I would advise every lecturer to be sure to prove something - one little theorem, one usable and elegant lemma, something that is typical of the words and the methods used in the subject. If the proof is short enough, it almost doesn't matter that it may, perhaps, not be understood. It is of value to the listener to hear the lecturer say that Bernoulli numbers enter the theory of stable homotopy groups, even if the listener has only an approximate idea of what Bernoulli numbers or homotopy groups are.

Something that's even better than a sample proof is the idea of a proof, the intuition that suggested it in the first place, the reason why the theorem is true. To find the right words to describe the central idea of a proof is sometimes hard, but it is worth the trouble; when it can be done, it provides the perfect way to communicate mathematics.

Problems

In the same vein, it is a false concept of simplicity that makes a lecturer concentrate only on what is safe and known; I strongly recommend that every public lecture reach the frontiers of knowledge, and at least mention something that is challenging and unknown. It doesn't have to be, it shouldn't be, the most delicate and newest technicality. Don't be afraid of repeating an old one; remember that many in your audience probably haven't heard of your subject since they took a course in it in graduate school, a long time ago. They will learn something just by hearing today that the unsolved problem they learned about years ago is still unsolved. The discussion of unsolved problems is a valuable part of the process of attracting and informing - it is, I think, an indispensable part. A field is not well described if its boundaries are missing from the description; some knowledge of the boundaries is essential for an understanding of where the field is today as well as for enlarging the area of our knowledge tomorrow. A public lecture must be simple, yes, but not at the cost of being empty, or, not quite that bad but bad enough, it must not be incomplete to the point of being dishonest.

Organization

The organization of a talk is like the skeleton of a man: things would fall apart without it, but it's bad if it shows. Organize your public lecture, plan it, prepare it carefully, and then deliver it impromptu, extemporaneously.

To prepare a talk, the first thing to know is the subject, and a very close second is the audience. It's much more important to adjust the level to fit the audience in a public lecture than it is in a book. ("Adjust the level" is not a euphemism for "talk down". Don't insult the audience, but be realistic. Slightly over the mark, very slightly, doesn't do much harm, but too much over is much worse than somewhat under.) A reader can put down a book and come back to it when he has learned more; an annoyed and antagonized listener will, in spirit, leave you, and, as far as this talk is concerned, he'll never come back.

The right level for a talk is a part of what organization is meant to achieve, but, of course, the first and more important thing to organize is the content. Here I have two recommendations (in addition to "prove something" and "ask something", already mentioned): (1) discuss three or four related topics, and the connections between them, rather than relentlessly pursue one central topic, and (2) break each topic into four or five sub-topics, portable, freely addable or subtractable modules, the omission of any one of which would not wreck the continuity.

As for extemporaneous delivery, there are two reasons for that: it sounds good, and it makes possible an interaction between the speaker and the listeners. The faces in the audience can be revealing and helpful: they can indicate the need to slow down, to speed up, to explain something, to omit something.

Preparation

To prepare a lecture means to prepare the subjects it will cover, the order in which those subjects are to come, and the connections between them that you deem worthy of mention; it does not mean to write down all the words with the intention of memorizing them (or, much worse, reading them aloud). Still: to write it all out is not necessarily a bad idea. "All" means all, including, especially, exactly what is to be put on the blackboard (with a clear idea of when it will be put on and whether it will remain for long or be rubbed out right away). To have it all written out will make it easier to run through it once, out loud, by a blackboard, and thus to get an idea of the timing. (Warning: if the dry run takes an hour, then the actual delivery will take an hour and a half.)

Brevity

Most talks are described as "one-hour lectures", but, by a generally shared tradition, most are meant to last for 50 minutes only. Nobody will reproach you for sitting down after 45 minutes, but the majority of the audience will become nervous after 55, and most of them will glare at you, displeased and uncomfortable, after 65.

To take long, to run over time, is rude. Your theorems, or your proofs, are not all that important in other peoples' lives; that hurried, breathless last five minutes is expendable. If you didn't finish, say so, express your regret if you must, but stop; it's better thus than to give the audience cause for regret.

Techniques

A public lecture usually begins with an introduction by the chairman of the session. Rule of etiquette: give him a chance. Before the lecture begins, sit somewhere by the side of the room, or with the audience, near the front; do not stand by or near the blackboard, or hover near the chairman worrying him.

One good trick to overcome initial stage-fright is to memorize one sentence, the opener. After that, the preparation and your knowledge of the subject will take over.

Try very hard to avoid annoying mannerisms. Definition: an annoying mannerism is anything that's repeated more than twice. A mannerism can be verbal ("in other words", pronounced " 'n 'zer w'rs", meaning nothing), it can be visual (surrounding a part of the material on the blackboard by elaborate fences), or it can be dynamic (teeter-tottering at the edge of the platform).

If you are in mechanical trouble, catch the chairman's eye and say, to him only, "I am out of chalk", or "May I have an eraser?". Do not bumble about your awkwardness and do not keep on apologizing. ("Oh, dear, where can I put this - sorry, I seem to have run out of room - well, let's see, perhaps we don't need this any-more...".) Make the appropriate decision and take the appropriate action, but do so silently. Keep your own counsel, and do not distract the audience with irrelevancies.

Silence is a powerful tool at other times too; the best speakers are also the best non-speakers. A long period of silence (five seconds, say, or ten at most) after an important and crisply stated definition or theorem puts the audience on notice ("this is important") and gives them a chance to absorb what was just said. Don't overdo it, but three or four times during the hour, at the three or four high points, you might very well find that the best way to explain something is to say nothing.

Speak slowly and speak loudly; write large and speak as you write; write slowly and do not write much. Intelligently chosen abbreviations, arrows for implications, and just reminder words, not deathless prose, are what a board is for; their purpose is to aid the audience in following you by giving them something to look at as well as something to listen to. (Example: do not write "semisimple is defined as follows:"; write "semisimple:".) Do not, ever, greet an audience with a carefully prepared blackboard (or overhead projector sheets) crammed with formulas, definitions, and theorems. (An occasionally advisable exception to this rule has to do with pictures - if a picture, or two pictures, would help your exposition but would take too long to draw as you talk, at least with the care it deserves, the audience will forgive you for drawing it before the talk begins.) The audience can take pleasure in seeing the visual presentation grow before its eyes - the growth is part of your lecture, or should be.

Flexibility

Because of the unpredictability of the precise timing (you didn't rehearse enough, the audience asks questions during the talk, the lecture room is reserved for another group at 5:00 sharp, or you just plain get mixed up and waste time trying to get unscrambled), flexibility is an important quality to build into a lecture. You must be prepared to omit (or to add!) material, and you must be prepared to do so under pressure, in public, on the spur of the moment, without saying so, and without seeming to do so. There are probably many ways to make a lecture flexible; I'll mention two that I have found useful.

The first is exercises. Prepare two or three statements whose detailed discussion might well be a part of the lecture but whose omission would not destroy continuity, and, at the proper places during your lecture, "assign" them to the audience as exercises. You run the slight risk of losing the attention of some of the more competitive members of the group for the rest of the hour. What you gain is something else that you can gracefully fill out your time with if (unlikely as that may be) you finish everything else too soon, and, at the same time, something that'll never be missed if you do not discuss the solution. (Exercises in this sense may yield another fringe benefit: they'll give the audience something to ask their courtesy questions about.)

A second way to make a lecture flexible is one I mentioned before and I believe is worth emphasizing again: portable modules. My notes for a lecture usually consist of about 20 telegraphically written paragraphs. The detailed presentation of each paragraph may take between 2 and 4 minutes, and at least half the paragraphs (the last 10) are omittable. These omittable modules often contain material dear to my heart: that clever proof, that ingenious generalization, that challenging question - but no one (except me) will miss them if I keep mum. Knowing that those modules are there, I sail through the first half of the period with no worries: I am sure that I won't run out of things to say, and I am sure that everything that I must say will get said. In the second half, or last third, of the period I keep an eye on the time, and, without saying anything about it, make instantaneous decisions about what to throw overboard.

One disadvantage of this method is that at the end of your time you might sound too abrupt, as if you had stopped in the middle of a sentence. To avoid the abrupt ending, prepare your peroration, and do not omit it. The peroration can be a three-sentence summary of the whole lecture, or it can be the statement of the most important unsolved problem of the subject. Make it whatever you think proper for an ending, and then end with it.

Rule of etiquette: when you stop, sit down. Literally sit down. Do not just stop talking and look helpless, and do not ask for questions; that's the chairman's job.

## Short talks

Short talks are harder to prepare and to deliver than long ones. The lecturer has less time to lay the groundwork, and the audience has less time to catch on; the lecturer feels under pressure to explain quickly, and the audience is under pressure to understand quickly.

In my experience a 20-minute talk can still be both enjoyable and enlightening; all you need to do is prepare a 10-minute talk and present it leisurely. A 10-minute talk is the hardest to do right; the precepts presented above (simple, organized, and short) must be applied again, but this time there is no room for error. Focus on one idea only, and on its simplest nontrivial special case at that, practice the talk and time it carefully, and under no circumstances allow a 10-minute contributed paper to become a 45-minute uninvited address. It has been done, but the results were neither informative nor attractive.

Some experts are willing to relax the rules for a 10-minute talk: it is all right, they say, to dive into the middle of things immediately, and it is all right, they say, to use prepared projection sheets. Others, having in mind the limited velocity and capacity of the human mind to absorb technicalities, disagree.

## Summary

My recommendations amount to this: make it simple, organized, and short. Make your lecture simple (special and concrete); be sure to prove something and ask something; prepare, in detail; organize the content and adjust to the level of the audience; keep it short, and, to be sure of doing so, prepare it so as to make it flexible.

Remember that you are talking in order to attract the listeners to your subject and to inform them about it; and remember that less is more.

Reprinted from the
AMERICAN MATHEMATICAL MONTHLY
Vol. 82, No. 1, pp. 14–17, Jan. 1975

## II. WHAT TO PUBLISH — BY P. R. HALMOS

When I accepted membership on this panel, I talked about it with almost every mathematician I saw and I asked for advice and help — in effect I conducted a sort of one-man opinion research poll. Some of the comments I received were not helpful. "Boy, you've got yourself into a corner!", they said. "That's an impossible subject!"

The first question, clearly, is "What is the question?" A proper understanding of the meaning of "What to publish?" depends on at least an approximate answer to "Why publish?". That answer varies with the self-awareness, the honesty, and the idealism of the one who gives it. It could, for instance, be to improve one's financial, academic, and social standing, or co-operatively to push the frontiers of knowledge ever more forward.

Another preliminary examination of the meaning of the question "What to publish?" should ask "Who is asking?" Is an author at the beginning of his career asking for advice on how to tell wheat from chaff? Is a referee or an editor asking which papers to accept? Is a publisher or a mathematical society asking which subjects are commercially or scientifically profitable? Authors, editors, and publishers do in fact ask these questions, every day, and they answer them; they make a decision in each case, somehow. The somehow seems helter-skelter, unsystematic. Is it perhaps the purpose of panel discussions such as this one to point to some guiding principles that can be used in every case? Or is it wrong to think that there *are* any guiding principles, and is it, if there are any, wrong to try to enforce them?

My opinion research poll revealed some hostility to the very asking of the question. Some mathematicians resent it, even before it leads to any detailed discussion of standards. Emphasis on the question implies to them that standards that they do not share might be applied, that somebody might interfere with their own publication program — it all smacks of elitism.

(I digress to comment how curiously pejorative the word "elitism" has become in recent years. My dictionary defines the "elite" as the "best or most skilled members of a given social group". To be against elitism would seem therefore like saying that "good" means "bad". But that's how it is; to insist nowadays that something — almost anything — be good — be it automobiles, the air we breathe, high school education, supreme court justices, the council of the American Mathematical Society, or published theorems — is elitist, and therefore bad.)

Only a few of my informants were, however, hostile to the question; a somewhat larger number took a pathetically friendly attitude. They were happy about the question; they almost said "I thought you'd never ask!"; and they gave positive answers. What to publish? By all means more expository papers (not a surprising answer); that folk theorem that is "well known" but not accessible anywhere; and more book reviews (I wasn't expecting that one!).

Let me remind you that most laws (with the exception only of the regulatory statutes that govern traffic and taxes) are negative. Consider, as an example, the Ten Commandments. When Moses came back from Mount Sinai, he told us what to be by telling us, eight out of ten times, what *not* to do. It may therefore be considered appropriate to say what *not* to publish. I warn you in advance that all the principles that I was able to distill from interviews and from introspection, and that I'll now tell you about, are a little false. Counterexamples can be found to each one — but as directional guides the principles still serve a useful purpose.

First, then, do not publish fruitless speculations; do not publish polemics and diatribes against a friend's error. Do not publish the detailed working out of a known principle. (Gauss discovered exactly which regular polygons are ruler-and-compass constructible, and he proved, in particular, that the one with 65537 sides — a Fermat prime — is constructible; please do not publish the details of the procedure. It's been tried.)

Do not publish in 1975 the case of dimension 2 of an interesting conjecture in algebraic geometry, one that you don't know how to settle in general, and then follow it by dimension 3 in 1976, dimension 4 in 1977, and so on, with dimension $k - 3$ in 197$k$. Do not, more generally, publish your failures: I tried to prove so-and-so; I couldn't; here it is — see?!

Adrian Albert used to say that a theory is worth studying if it has at least three distinct good hard examples. Do not therefore define and study a new class of functions, the ones that possess left upper bimeasurably approximate derivatives, unless you can, at the very least, fulfill the good graduate student's immediate request: show me some that do and show me some that don't.

A striking criterion for how to decide not to publish something was offered by my colleague John Conway. Suppose that you have just finished typing a paper. Suppose now that I come to you, horns, cloven hooves, forked tail and all, and ask: if I gave you $1000.00, would you tear the paper up and forget it? If you hesitate, your paper is lost — do not publish it. That's part of a more general rule: when in doubt, let the answer be no.

As I went around asking people's opinion about what to publish, the view that I heard expressed most often is that too much is now being published. The active, talented mathematicians, both the beginners and the established ones, complain about the flood of junk in the journals; their estimates of how much of it should have been published vary from a generous 50% to a stringent 2%. The less active, less motivated members of the community complain (in my opinion rightly so) about the great pressure that is used on them to publish, publish, publish. They are frequently people of good taste, and they do not want to contribute to the flood of junk.

Almost everybody's answer to "What to publish?" can be expressed in either one word — "less" — or two words — "good stuff".

The trouble, of course, is how to define "good stuff" — how to establish canons of taste. G. H. Hardy's criteria are easy to say (is it true?, is it new?, is it interesting?), but are they easy to apply?

The newness of a paper can manifest itself in various ways: it can contain a new fact, a new proof, or a new method. Usually, of course, the three are mixed together, but there are striking cases when they occur in pure form. A paper by Morrison and Brillhart on the factorization of $2^{2^7}+1$ in the *Bulletin of the AMS* (1971) is nothing but a brutal fact, but both the authors and the editors of the Bulletin were right to decide that the fact was sufficiently new and that it sufficiently satisfied a pre-existing curiosity for the mathematical world as a whole to receive it with interest. Landau's *tour-de-force* proof of the irreducibility of cyclotomic polynomials (*Math. Z.*, 1929) reveals no new facts and no new methods, but if someone can do in one paragraph what, till then, many others have done in many laborious pages, he is right to think that the rest of us will think his proof interesting. When Cantor proved the existence of transcendental numbers (*Crelle's J.*, 1874), he didn't tell the world any fact that the world didn't already know; and he didn't improve Liouville's proof (*J. de Math.*, 1851) — indeed not — Cantor's result was much less sharp — but, undeniably, he introduced into mathematics a new method, the method of proving the existence of something by transfinite arithmetic, that has become a standard part of every graduate student's mathematical toolkit.

It's not much good to say "publish your *deep* results only", because depth is no easier to define than interest. It is of some use, just the same, because even the rough approximations to a definition may be suggestive of the direction in which the truth lies. Is a theorem deep when its proof is complicated? Yes, often, but by no means always. The theorem that 1 plus aleph null is aleph null, whose proof is short and trivial, is in my opinion one of the deepest in mathematics. Does "deep" mean the same as "surprising"? Yes, sometimes, but by no means always. No one finds the Jordan curve theorem surprising, but if a long and messy proof is any criterion, it sure is deep. One other test of depth deserves mention, namely breadth of contact. If a theorem about number theory uses the methods of complex function theory in its proof and has applications to the topological problem of determining the homotopy groups of spheres, it is probably deep.

Scientific publication has always been a fiercely competitive activity, but nowadays it is much more so than before, perhaps because it has become a question of many people's livelihood — of money. Graduate students compete with each other, assistant professors vie for promotion, and the full professors try to fight off the bright youngsters who are threatening them from below.

In any case the question — "What to publish?" — is not an operational one. If I knew the answer and revealed it, would anyone do what I told him to? What people will publish is not what anyone tells them to, but what the current social, political, military, financial, academic, and perhaps even mathematical atmosphere calls for. The thing that discussion can best clarify is not what should but what will happen — we may not prescribe, but, possibly, we can predict. My own perhaps too optimistic prediction is that the junk will not continue to increase. The flood crest was brought onto us by an unexpected and unresisted increase in the availability of funds, and that is now over. In the coming decades no one, I think, will go into mathematics unless he wants to. Those of us who are in already will not be as rich as we used to be — a lot of the newly founded journals will go broke — the quantity of publication will decrease. The pressure to publish, no matter what, will decrease. The result will not be Utopia, not by any means, but, I predict, the problems we'll face will be different — and, whatever they are, they will not precipitate panel discussions on "Why, What, and How to Publish?".

DEPARTMENT OF MATHEMATICS, INDIANA UNIVERSITY, BLOOMINGTON, IN 47401.

Reprinted from the AMERICAN MATHEMATICAL MONTHLY
Vol. 82, No. 5, May 1975
pp. 466–470

# THE PROBLEM OF LEARNING TO TEACH

## I. THE TEACHING OF PROBLEM SOLVING — BY P. R. HALMOS

The best way to learn is to do; the worst way to teach is to talk.

About the latter: did you ever notice that some of the best teachers of the world are the worst lecturers? (I can prove that, but I'd rather not lose quite so many friends.) And, the other way around, did you ever notice that good lecturers are not necessarily good teachers? A good lecture is usually systematic, complete, precise — and dull; it is a bad teaching instrument. When given by such legendary outstanding speakers as Emil Artin and John von Neumann, even a lecture can be a useful tool — their charisma and enthusiasm come through enough to inspire the listener to go forth and do something — it looks like such fun. For most ordinary mortals, however, who are not so bad at lecturing as Wiener was — nor so stimulating!— and not so good as Artin — and not so dramatic! — the lecture is an instrument of last resort for good teaching.

My test for what makes a good teacher is very simple: it is the pragmatic one of judging the performance by the product. If a teacher of graduate students consistently produces Ph. D.'s who are mathematicians and who create high-quality new mathematics, he is a good teacher. If a teacher of calculus consistently produces seniors who turn into outstanding graduate students of mathematics, or into leading engineers, biologists, or economists, he is a good teacher. If a teacher of third-grade "new math" (or old) consistently produces outstanding calculus students, or grocery store check-out clerks, or carpenters, or automobile mechanics, he is a good teacher.

For a student of mathematics to hear someone talk about mathematics does hardly any more good than for a student of swimming to hear someone talk about swimming. You can't learn swimming technique by having someone tell you where to put your arms and legs; and you can't learn to solve problems by having someone tell you to complete the square or to substitute sin $u$ for $y$.

Can one learn mathematics by reading it? I am inclined to say no. Reading has an edge over listening because reading is more active — but not much. Reading with pencil and paper on the side is very much better — it is a big step in the right direction. The very best way to read a book, however, with, to be sure, pencil and paper on the side, is to keep the pencil busy on the paper and throw the book away.

Having stated this extreme position, I'll rescind it immediately. I know that it is extreme, and I don't really mean it — but I wanted to be very emphatic about not going along with the view that learning means going to lectures and reading books. If we had longer lives, and bigger brains, and enough dedicated expert teachers to have a student/teacher ratio of 1/1, I'd stick with the extreme view — but we don't. Books and lectures don't do a good job of transplanting the facts and techniques of

Talks given at the Annual Meeting in San Francisco, January 17, 1974, at a joint AMS-MAA Panel discussion.

the past into the bloodstream of the scientist of the future — but we must put up with a second best job in order to save time and money. But, and this is the text of my sermon today, if we rely on lectures and books only, we are doing our students, and their students, a grave disservice.

What mathematics is really all about is solving concrete problems. Hilbert once said (but I can't remember where) that the best way to understand a theory is to find, and then to study, a prototypal concrete example of that theory, a root example that illustrates everything that can happen. The biggest fault of many students, even good ones, is that although they might be able to spout correct statements of theorems, and remember correct proofs, they cannot give examples, construct counterexamples, and solve special problems. I have seen many students who could state something they called the spectral theorem for Hermitian operators on Hilbert space but who had no idea how to diagonalize a $3 \times 3$ real symmetric matrix. That's bad — that's bad learning, probably caused, at least in part, by bad teaching. The full-time professional mathematician and the occasional user of mathematics, and the whole spectrum of the scientific community in between — they all need to solve problems, mathematical problems, and our job is to teach them how to do it, or, rather, to teach their future teachers how to teach them to do it.

I like to start every course I teach with a problem. The last time I taught the introductory course in set theory, my first sentence was the definition of algebraic numbers, and the second was a question: are there any numbers that are not algebraic? The last time I taught the introductory course in real function theory, my first sentence was a question: is there a non-decreasing continuous function that maps the unit interval into the unit interval so that length of its graph is equal to 2? For almost every course one can find a small set of questions such as these — questions that can be stated with the minimum of technical language, that are sufficiently striking to capture interest, that do not have trivial answers, and that manage to embody, in their answers, all the important ideas of the subject. The existence of such questions is what one means when one says that mathematics is really all about solving problems, and my emphasis on problem solving (as opposed to lecture attending and book reading) is motivated by them.

A famous dictum of Pólya's about problem solving is that if you can't solve a problem, then there is an easier problem that you can't solve — find it! If you can teach that dictum to your students, teach it so that they can teach it to theirs, you have solved the problem of creating teachers of problem solving. The hardest part of answering questions is to ask them; our job as teachers and teachers of teachers is to teach how to ask questions. It's easy to teach an engineer to use a differential equations cook book; what's hard is to teach him (and his teacher) what to do when the answer is not in the cook book. In that case, again, the chief problem is likely to be "what is the problem?". Find the right question to ask, and you're a long way toward solving the problem you're working on.

What then is the secret — what is the best way to learn to solve problems? The answer is implied by the sentence I started with: solve problems. The method I advocate is sometimes known as the "Moore method," because R. L. Moore developed and used it at the University of Texas. It is a method of teaching, a method of creating the problem-solving attitude in a student, that is a mixture of what Socrates taught us and the fiercely competitive spirit of the Olympic games.

The way a bad lecturer can be a good teacher, in the sense of producing good students, is the way a grain of sand can produce pearl-manufacturing oysters. A smooth lecture and a book entitled "Freshman algebra for girls" may be pleasant; a good teacher challenges, asks, annoys, irritates, and maintains high standards — all that is generally not pleasant. A good teacher may not be a popular teacher (except perhaps with his *ex*-students), because some students don't like to be challenged, asked, annoyed, and irritated — but he produces pearls (instead of casting them in the proverbial manner).

Let me tell you about the time I taught a course in linear algebra to juniors. The first hour I handed to each student a few sheets of paper on which were dittoed the precise statements of fifty theorems. That's all — just the statements of the theorems. There was no introduction, there were no definitions, there were no explanations, and, certainly, there were no proofs.

The rest of the first hour I told the class a little about the Moore method. I told them to give up reading linear algebra (for that semester only!), and to give up consulting with each other (for that semester only). I told them that the course was in their hands. The course was those fifty theorems; when they understood them, when they could explain them, when they could buttress them with the necessary examples and counterexamples, and, of course, when they could prove them, then they would have finished the course.

They stared at me. They didn't believe me. They thought I was just lazy and trying to get out of work. They were sure that they'd never learn anything that way.

All this didn't take as much as a half hour. I finished the hour by giving them the basic definitions that they needed to understand the first half dozen or so theorems, and, wishing them well, I left them to their own devices.

The second hour, and each succeeding hour, I called on Smith to prove Theorem 1, Kovacs to prove Theorem 2, and so on. I encouraged Kovacs and Herrero and all to watch Smith like hawks, and to pounce on him if he went wrong. I myself listened as carefully as I could, and, while I tried not to be sadistic, I too pounced when I felt I needed to. I pointed out gaps, I kept saying that I didn't understand, I asked questions about side issues, I asked for, and sometimes supplied, counterexamples, I told about the history of the subject when I had a chance, and I pointed out connections with other parts of mathematics. In addition I took five minutes or so of most hours to introduce the new definitions needed. Altogether I probably talked 20 minutes out of each of the 50-minute academic hours that we were together. That's a lot — but it's a lot less than 50 (or 55) out of 50.

It worked like a charm. By the second week they were proving theorems and finding errors in the proofs of others, and obviously taking pleasure in the process. Several of them had the grace to come to me and confess that they were skeptical at first, but they had been converted. Most of them said that they spent more time on that course than on their other courses that semester, and learned more from it.

What I just now described is like the "Moore method" as R. L. Moore used it, but it's a much modified Moore method. I am sure that hundreds of modifications could be devised, to suit the temperaments of different teachers and the needs of different subjects. The details don't matter. What matters is to make students ask and answer questions.

Many times when I've used the Moore method, my colleagues commented to me, perhaps a semester or two later, that they could often recognize those students in their classes who had been exposed to a "Moore class" by those students' attitude and behavior. The distinguishing characteristics were greater mathematical maturity than that of the others (the research attitude), and greater inclination and ability to ask penetrating questions.

The "research attitude" is a tremendous help to all teachers, and students, and creators, and users of mathematics. To illustrate, for instance, how it is a help to me when I teach elementary calculus (to a class that's too large to use the Moore method on), I must first of all boast to you about my wonderful memory. Wonderfully bad, that is. If I don't teach calculus, say, for a semester or two, I forget it. I forget the theorems, the problems, the formulas, the techniques. As a result, when I prepare next week's lecture, which I do by glancing at the prescribed syllabus, or, if there is none, at the table of contents of the text, but never at the text itself, I start almost from scratch — I do research in calculus. The result is that I have more fun than if I had it all by rote, that time after time I am genuinely surprised and pleased by some student's re-discovery of what Leibniz probably knew when he was a teenager, and that my fun, surprise, pleasure, and enthusiasm is felt by the class, and is taken as an accolade by each discoverer.

To teach the research attitude, every teacher should do research and should have had training in doing research. I am not saying that everyone who teaches trigonometry should spend half his time proving abstruse theorems about categorical teratology and joining the publish-or-perish race. What I am saying is that everyone who teaches, even if what he teaches is high-school algebra, would be a better teacher if he thought about the implications of the subject outside the subject, if he read about the connections of the subject with other subjects, if he tried to work out the problems that those implications and connections suggest — if, in other words, he did research in and around high-school algebra. That's the only way to keep the research attitude, the question-asking attitude, alive in himself, and thus to keep it in a condition suitable for transmitting it to others.

Here it is, summed up, in a few nut shells:

The best way to learn is to do — to ask, and to do.

The best way to teach is to make students ask, and do. Don't preach facts — stimulate acts.

The best way to teach teachers is to make them ask and do what they, in turn, will make their students ask and do.

Good luck, and happy teaching, to us all.

DEPARTMENT OF MATHEMATICS, INDIANA UNIVERSITY, BLOOMINGTON, IN 47401.

Reprinted from the MATHEMATICS MAGAZINE
Vol. 50, No. 1, January 1977
pp. 5–11

# Logic from A to G

## *A sketch for a mathematician's mechanical helper, flippantly annotated by a working mathematician.*

PAUL R. HALMOS
*University of Calfornia, Santa Barbara*

### What logic is and is not

Originally "logic" meant the same as "the laws of thought" and logicians studied the subject in the hope that they could discover better ways of thinking and surer ways of avoiding error than their forefathers knew, and in the hope that they could teach these arts to all mankind. Experience has shown, however, that this is a wild-goose chase. A normal healthy human being has built in him all the "laws of thought" anybody has ever invented, and there is nothing that logicians can teach him about thinking and avoiding error. This is not to say that he knows *how* he thinks and it is not to say that he never makes errors. The situation is analogous to the walking equipment all normal healthy human beings are born with. I don't know how I walk, but I do it. Sometimes I stumble. The laws of walking might be of interest to physiologists and physicists; all I want to do is to keep on walking.

The subject of mathematical logic, which is the subject of this paper, makes no pretense about discovering and teaching the laws of thought. It is called *mathematical* logic for two reasons. One reason is that it is concerned with the kind of activity that mathematicians engage in when they prove things. Mathematical logic studies the nature of a proof and tries to forecast in a general way all possible types of things that mathematicians ever will prove, and all that they never can. Another reason for calling the subject *mathematical* logic is that it itself is a part of mathematics. It attacks its subject in a mathematical way and proves things exactly the same way as do the other parts of mathematics whose methods it is concerned with. The situation is like that of a factory that makes machines whose purpose is to make machines. A worker at such an establishment is no different from a worker at any other machine factory, except perhaps that he understands a little better what makes machines in general work as they do (and a little less well how any particular machine works).

The history of logic, like the history of most subjects, developed all in the wrong order. If I told it to you straight, you'd get completely confused. I propose to tell you a little bit of the history of logic in the "right" order, that is, in the order in which it *should* have happened.

### First Boole and propositions

According to my version of history it all begins with George Boole about 100 years ago. Boole's contribution was a systematic study of the innocuous little words that we all use every day to tie propositions together, the so-called propositional connectives. These connectives are, in English,

*and, or, not, implies,* and *if-and-only-if.*

It is convenient to have abbreviations for these words. The customary mathematical symbols used to abbreviate them are

$$\&,\ \vee,\ \neg,\ \Rightarrow,\ \text{and}\ \Leftrightarrow.$$

Thus, for example, if $P$ and $Q$ are propositions, then so also is $P \& Q$. If $P$ says "the sun's shining" and $Q$ says "it's hot", then $P \& Q$ says "the sun's shining and it's hot".

### Next Peano and numbers

The next figure in our revised history is the 19th-century Italian mathematician Peano who studied the foundations of arithmetic. He studied, in particular, the properties of the basic numbers *zero* and *one*, the basic operations of *addition* and *multiplication*, and the basic relation of *equality*. The symbols for these things are, of course, known to everybody: they are

$$0,\ 1,\ +,\ \times,\ \text{and}\ =.$$

Thus, for example, the popular proposition that "two and two make four" can be written in the unabbreviated form

$$(1+1)+(1+1)=1+(1+(1+1)).$$

### Then Aristotle and a quantifier

Immediately after Peano comes Aristotle, who lived well over 2000 years ago, and to whom we are indebted for the first analysis of the crucial words *all* and *some*. (Incidentally, we have now also reached the beginning of the alphabet: the "A" in "Logic from A to G" is, of course, Aristotle.) The abbreviations are $\forall$ and $\exists$. To illustrate the use of these symbols, consider a well-known sentence such as "He who hesitates is lost". In pedantic mathematese this can be said as follows: "For all $X$, if $X$ hesitates, then $X$ is lost". Using $H(X)$ and $L(X)$ as abbreviations for "$X$ hesitates" and "$X$ is lost", we may write

$$(\forall X)(H(X) \Rightarrow L(X)).$$

If we doubt this assertion, if, that is, we are inclined to believe it quite likely that hesitation is possible without subsequent perdition, we may express our skepticism in the form

$$(\exists X)(H(X) \& (\neg L(X)).$$

### Finally Frege and many quantifiers

An essential part of such Aristotelean abbreviations is the use of auxiliary symbols such as the "$X$" above; symbols such as that play the role of pronouns in the language of logic. Recall that, in the example, "$X$" took the place of "he". Symbols used in this way are called *individual variables*, and the next historical figure to be mentioned is the first one to face them courageously. His name is Frege, and he too lived in the 19th century. His role for us today is to emphasize that *one* variable, that is one pronoun, is much too meager equipment for most scientific and mathematical purposes. Thus, for instance, if we want to express the very modest assertion that there are more than two numbers, that is, that there exist at least three different ones, we would do it this way:

$$(\exists X)(\exists Y)(\exists Z)(\neg(X=Y) \& \neg(Y=Z) \& \neg(X=Z)).$$

For this purpose, pretty clearly, we need at least three variables. To say that there are more than ten numbers, we need at least eleven variables. To do any non-trivial amount of mathematics, we need an (at least potentially) infinite supply of variables. An efficient way of getting such a supply (since ordinary alphabets are much too finite) is to use one letter, say "$X$", and one extra symbol, say a dash ′, and then use in the role of variables the symbols

$$X, X', X'', X''', X'''', \text{etc.}$$

In addition to all the symbols I have mentioned so far, there are two others that I have already used, and compulsive honesty now forces me to adjoin them to the list. The symbols I mean are the *left parenthesis* and *right parenthesis* denoted, of course, by

$$($$

and

$$).$$

**How many symbols?**

It turns out that a considerably large body of mathematics, namely, all arithmetic, can be expressed by means of the symbols I have listed so far. (Thus, for instance, Euler's celebrated theorem that every positive integer is the sum of four squares can be written as follows:

$$(\forall X)(\exists X')(\exists X'')(\exists X''')(\exists X'''')(X = (X' \times X') + ((X'' \times X'') + ((X''' \times X''') + (X'''' \times X''''))))).)$$

In fact certain savings can be made; some symbols are naturally and easily definable in terms of others. Our list can easily be cut down to an even dozen, namely to

$$\& \ \neg \ + \ \times \ 0 \ 1 \ \exists \ X \ ' \ = \ ( \ ).$$

To recapture $\vee$ observe that $P \vee Q$ is the same as $\neg(\neg P \ \& \ \neg Q)$. To recapture $\Rightarrow$ observe that $P \Rightarrow Q$ is the same as $\neg P \vee Q$. To get $\forall$ note that $(\forall X)P(X)$ is the same as $\neg(\exists X)(\neg P(X))$. In words: to say that everybody hates spinach is the same as to deny that there is somebody who loves it. There is some technical advantage in such abbreviations, but there's no sense in tying ourselves down. Whenever convenient I'll still use $\vee$ and $\Rightarrow$ and $\forall$, and I'll even use $Y$, $Z$, $U$, $V$, and such, as variables. The proper way to interpret these irregularities is clear by now: replace $\vee$, $\Rightarrow$, $\forall$, and the like by their definitions, and replace $Y$, $Z$, $U$, $V$, and such, by $X'$, $X''$, $X'''$, $X''''$, etc.

It might be fun to make a couple of side observations here. One of them is that what we are doing for elementary arithmetic can just as well be done for *all* extant mathematics. The technical machinery, that is the symbolism and the rules governing it, doesn't even get more complicated: the only difference is that we have to think a little harder. Since that is clearly undesirable, I am going to stick to elementary arithmetic. The second observation is that there is nothing magic about the number twelve. A dozen symbols are adequate for arithmetic, and in fact for all mathematics (though we might have to find a different dozen for that). A little care and stinginess, however, can easily reduce the dozen to still fewer, and the best possible result that you would hope for in your wildest dreams is true, namely, that two symbols are enough. A complete exposition of all mathematics written with, say, the dots and dashes of the Morse code wouldn't make particularly thrilling reading, but in principle it is perfectly feasible.

**The mechanical mathematician**

Let us now set about designing the mechanical mathematician to end all mathematicians. The virtues of this imaginary machine are purely conceptual; it will not be claimed that there is any practical advantage in building it. It will never replace the live mathematician.

The first step is to build into the machine a typewriter ribbon, a dozen typewriter keys (one bearing each one of the dozen basic symbols), and a potentially infinitely long piece of typing paper. The idea is that when a certain crucial button is pushed the machine is to start printing, and to print one after another all things that it could conceivably ever print. One way to program this would be to arrange the dozen symbols in an arbitrary order (call it alphabetical), and then to direct the machine to print the first twelve symbols ("letters") of the alphabet, then to print the 144 two-letter "words", next the 1728 three-letter "words", and so on *ad infinitum.* A machine so designed would print a lot of nonsense (for example

$$= \; = \; = \; ( \; ( \; ( \; 0 \; + \; ' \; ),$$

and it would print a lot of lies (for example

$$(\exists X)((0 \times X) = 1)),$$

but it would also, sooner or later, get around to printing all arithmetical statements.

**Insist on grammar**

In order to make the machine more like a flesh-and-blood mathematician, the next thing we must do is to arrange matters so that the output of the machine, while possibly false, should never be arrant nonsense. This is in principle quite easy. There is no point in listing here all the restrictions to which the machine must be subjected, but let us look at a few samples. First, teach the machine some "grammar". Let us say that a *noun* is any sequence of symbols built up from 0's and 1's by successively sandwiching $+$'s and $\times$'s between them and separating the results by the appropriate parentheses. Examples:

$$0,$$

$$1 + 1,$$

$$((1 + 1) \times (1 + (1 + (1 + 1))))$$

are nouns in this sense. Let us, similarly, say that a *pronoun* is anything obtained from $X$ by successively appending dashes. Thus the pronouns are

$$X, X', X'', X''', \text{etc.}$$

To go on in the same spirit, a *substantive* shall be a string of nouns and pronouns put together by means of addition and multiplication (and the auxiliary use of parentheses) the same way as nouns were originally put together from 0's and 1's. It is not at all difficult to design the machine so that it is able to recognize a substantive when it sees one. Once that is done, we can direct the machine as follows: "Start printing substantives (in some systematic order). After printing one, print an equal sign and then print another substantive. Learn to recognize the strings you have printed in this manner (substantive, equal, substantive), and call each such string a *clause.*" Our machine can now print sensible clauses, and recognize them as such. It is just one step from here to teach the machine to print (and to recognize) compound *phrases.* The idea is to put clauses together by suitably restricted use of the logical operators *or, not,* and *some.* A machine that prints only phrases is thus within conceptual reach. Such a machine might still print incomplete phrases (for example $X = 0$) and lies (for example $1 = 0$), but it will no longer print gibberish.

The incidental mention of "incomplete phrases" suggests another look at what we want the mechanical mathematician to do. An incomplete phrase, in the sense I want that expression to have now, is something like "he is lost". The natural reaction upon seeing or hearing those words is to ask "Who is lost?" Similarly, "$X = 0$" should evoke the reaction "What is $X$?". Phrases with "dangling pronouns" like the "he" in "he is lost" and like the "$X$" in "$X = 0$" are the ones I mean to call incomplete here. A phrase that has no such dangling pronouns shall be called a *sentence.* The next step in perfecting the mechanical mathematician is to teach it to recognize a sentence when it sees one, and

to instill in it an inhibition that permits it to print complete sentences only (that is, no incomplete phrases). Now if the button is pushed, the machine will start printing sensible sentences. It will never present the machine operator with "$X = 0$". It might say something uninteresting (for example $(\exists X)(X = 0)$) or false (for example $(\forall X)(X = 0)$), but at any rate it will always say something.

**Establish the axioms**

The machine now knows how to talk; the next step is to teach it how to prove things. Neither the machine nor its flesh-and-blood prototype, however, can prove something from nothing. A live mathematician has his axioms; the machine must have built into it certain sentences that it is instructed to start from. Let us call those sentences *axioms*. Once again I shall not define here exactly which sentences are to be the axioms of arithmetic, but I shall indicate some examples. (The complete definition of "axiom" for elementary arithmetic is not even very long or complicated. This is just not the time and place to get technical.) Very well then: the machine might be taught that whenever $P$ and $Q$ are sentences, then the sentence

(A) $$P \Rightarrow (P \vee Q)$$

shall be printed in red ink. The idea, of course, is that every such sentence is to be regarded as an axiom. Similarly, we might say that if $P(X)$ and $Q(X)$ are phrases (containing the dangling pronoun "$X$", but no other), then the sentence

$$(\exists X)(P(X) \vee Q(X)) \Rightarrow (\exists X)P(X) \vee (\exists X)Q(X) \tag{B}$$

shall be printed in red. Final sample: a sentence such as

$$(\forall X)(\forall Y)(X + Y = Y + X) \tag{C}$$

might be an axiom of elementary arithmetic. (Interpretive examples: (A) If it's hot, then either the sun's shining or else it's hot or both. (B) If somebody likes either spinach or broccoli, then either somebody likes spinach or somebody likes broccoli. (C) $2+3=3+2$.)

The main point is this: in some sensible and systematic manner a certain collection of sentences is singled out from among all the sentences that the machine can print, and the machine is directed to print those special sentences in red. The red sentences are called the axioms for the machine.

**Program the procedure**

I said before that nobody can prove something from nothing, and, for that reason, I endowed the mechanical mathematician with some starting sentences. But it is just as true that nobody can prove something *with* nothing. The machine can print black and red sentences, and, if the axioms were chosen in accordance with the wishes of a reasonable being, the red sentences (the axioms) will be true. The machine might nevertheless still print a lot of false stuff, and it still has no means of producing new true sentences out of old ones.

The process of educating the machine has now reached its final step. In that step the machine must learn to recognize certain patterns of sentences and is rewarded by being permitted to use more red ink. The total list of such *rules of procedure* is not long, and the list of two examples from among them, that I propose actually to give, is even shorter. Possible rule number one: if $P$ is a red sentence, and if $P \Rightarrow Q$ is a red sentence, then print $Q$ in red. (Interpretation: if "The sun's shining" is an axiom or has already been proved, and if the same is true of "If the sun's shining, then it's hot", then we may consider to have proved "It's hot".) Possible rule number two: if an incomplete phrase $P(X)$ containing the dangling pronoun "$X$" (and no others) is such that the sentence $P(0)$ is red ($P(0)$ is the sentence obtained from $P(X)$ by substituting 0 for $X$), then print the sentence $(\exists X)P(X)$ in red. (Interpretation: if we substitute 0 for $X$ in "$1 + X = 1$", we obtain "$1 + 0 = 1$". If this sentence is an axiom or has already been proved, then we may consider to have proved "$1 + X = 1$ for some $X$".)

With the last modification we can now rest on our laurels. We may if we wish change the internal design of the machine so that it no longer deigns to print anything but red sentences. When the button is pushed, the machine starts printing axioms, and, by means of the rules of procedure, it goes on to print *theorems* that it can "derive" from the axioms. It can do this in some systematic (say, alphabetic) order.

The millennium is come; the mechanical mathematician is complete. Push one button and sit back. One after another the theorems of elementary arithmetic will appear on the tape. If you wait long enough, sooner or later you will see all theorems pass before your eyes. The machine never talks nonsense, and the machine never tells lies. If you find the machine somewhat boring, possibly repetitious, and much much too slow for your merely human patience, that is not its fault.

**Are there contradictions?**

We have incorporated into the machine all that we ourselves, its builders, know about elementary arithmetic. The internal workings of the machine *are* elementary arithmetic. The external theory of the machine, its design and its structural properties, are part of another discipline often called *metamathematics* (or, in this case, metaarithmetic). The following question is typical of the ones that can be asked in metamathematics: "Will the machine ever print both a sentence $P$ and its negation $\neg P$?" Should it ever do so, we would probably express our displeasure at this state of affairs by saying that elementary arithmetic is inconsistent. Fortunately this is not so: arithmetic is consistent. The proof of consistency depends on a very sophisticated, definitely non-elementary study of the structure of the "machine" we've been describing. The study is "non-elementary" in several senses of the word. In the most precise technical sense the fact is that the proof that the machine is consistent is not one of the theorems that the machine itself is able to prove. To say it again: the machine will never contradict itself, but it is not able to prove that it won't.

**Can everything be proved?**

Another interesting metamathematical question is this: "Is the machine complete, in the sense that it either proves or disproves every sentence of elementary arithmetic?" I've already told you the answer to this question, but the point will bear repetition. As the machine now stands, everything it prints it proves. If I am interested in a particular arithmetical sentence, I may write that sentence on a piece of paper and then, after setting the machine into operation, compare each successive output of the machine with my prepared slip. If the slip says $P$ and if, at some stage, the machine also says $P$, I retire victorious: my $P$ is proved. If the slip says $P$, and if, at some stage, the machine says $\neg P$, I retire in ignominy: my $P$ is disproved. Isn't there, however, a third possibility? Couldn't it happen that the machine will never print $P$ and neither will it ever print $\neg P$? Couldn't it happen that the machine will never decide the $P$ versus $\neg P$ controversy? It could, and it does, and we have thereby reached the end of the alphabet. $G$ is for Gödel, the brilliant twentieth-century logician. In the early 1930's Gödel proved, by a delicate and ingenious analysis of the arithmetic machine, that there are sentences (many of them) that the machine never decides. His proof is quite explicit: he gives a complete set of directions for writing down an undecidable sentence. The proof that the sentence obtained by following his directions is undecidable depends on a detailed examination of those very directions themselves. There is nothing wrong, there is no paradox, and it all hangs together. The fact that no one has ever bothered actually to write down Gödel's undecidable sentence is, once again, the fault of human impatience and the brevity of human life.

I said when I raised the question of completeness that I had already answered it. Indeed, consider a sentence (written out formally in the dozen formal symbols of arithmetic) that says that arithmetic is consistent. It is not at all clear that the apparently meager formal apparatus of arithmetic is capable of expressing such a sentence; it is one of Gödel's accomplishments to have shown that it is capable of doing so. If we take that for granted, and if we call one such sentence $P$, then what we know is that $P$ is not provable in elementary arithmetic. What about $\neg P$? Well, clearly, $\neg P$ cannot be provable either. Reason: everything that is provable is true, as we already know from our earlier thoughts on the subject. (This depends on the fact that arithmetic is consistent.) The sentence $\neg P$ is certainly *not* true. (Recall that $\neg P$ denies the consistency of arithmetic.) Conclusion: neither $P$ nor $\neg P$ is provable. (Note: of the two, $P$ is the one that is true.)

**There is more**

That is the end of the road, for us, for now. It is by no means the end of the road for mathematical logic. What I've been reporting to you happened in the 1930's, and science has not stood still since then. Gödel himself has contributed several other striking results to our knowledge of formal logic. Many others have taken up the field and opened up unexpected applications and complications. Who, for instance, could have expected formal mathematical logic to turn out to be one of the most important tools in the design of honest-to-goodness circuits-and-printouts electronic computing machines? Mathematical logic is alive and well; much remains to be done; it'll be a long time before anyone can describe mathematical logic from $A$ to $Z$.

Reprinted from *The American Mathematical Monthly,* Vol. 87, No. 7, Aug. - Sept. 1980.

# THE HEART OF MATHEMATICS

P. R. HALMOS

**Introduction.** What does mathematics *really* consist of? Axioms (such as the parallel postulate)? Theorems (such as the fundamental theorem of algebra)? Proofs (such as Gödel's proof of undecidability)? Concepts (such as sets and classes)? Definitions (such as the Menger definition of dimension)? Theories (such as category theory)? Formulas (such as Cauchy's integral formula)? Methods (such as the method of successive approximations)?

Mathematics could surely not exist without these ingredients; they are all essential. It is nevertheless a tenable point of view that none of them is at the heart of the subject, that the mathematician's main reason for existence is to solve problems, and that, therefore, what mathematics *really* consists of is problems and solutions.

"Theorem" is a respected word in the vocabulary of most mathematicians, but "problem" is not always so. "Problems," as the professionals sometimes use the word, are lowly exercises that are assigned to students who will later learn how to prove theorems. These emotional overtones are, however, not always the right ones.

The commutativity of addition for natural numbers and the solvability of polynomial equations over the complex field are both theorems, but one of them is regarded as trivial (near the basic definitions, easy to understand, easy to prove), and the other as deep (the statement is not obvious, the proof comes via seemingly distant concepts, the result has many surprising applications). To find an unbeatable strategy for tic-tac-toe and to locate all the zeroes of the Riemann zeta function are both problems, but one of them is trivial (anybody who can understand the definitions can find the answer quickly, with almost no intellectual effort and no feeling of accomplishment, and the answer has no consequences of interest), and the other is deep (no one has found the answer although many have sought it, the known partial solutions require great effort and provide great insight, and an affirmative answer would imply many non-trivial corollaries). Moral: theorems can be trivial and problems can be profound. Those who believe that the heart of mathematics consists of problems are not necessarily wrong.

**Problem Books.** If you wanted to make a contribution to mathematics by writing an article or a book on mathematical problems, how should you go about it? Should the problems be elementary (pre-calculus), should they be at the level of undergraduates or graduate students, or should they be research problems to which no one knows the answer? If the solutions are known, should your work contain them or not? Should the problems be arranged in some systematic order (in which case the very location of the problem is some hint to its solution), or should they be arranged in some "random" way? What should you expect the reader to get from your work: fun, techniques, or facts (or some of each)?

All possible answers to these questions have already been given. Mathematical problems have quite an extensive literature, which is still growing and flowering. A visit to the part of the stacks labeled QA43 (Library of Congress classification) can be an exciting and memorable revelation, and there are rich sources of problems scattered through other parts of the stacks too. What follows is a quick review of some not quite randomly selected but probably typical problem collections that even a casual library search could uncover.

---

The author received his Ph.D. from the University of Illinois; he has held positions at (consecutively) Illinois, Syracuse, Chicago, Michigan, Hawaii, Indiana, Santa Barbara, and Indiana, with visiting positions for various periods at the Institute for Advanced Study, Montevideo, Miami, Harvard, Tulane, University of Washington (Seattle), Berkeley, Edinburgh, and Perth. He has published eight or more books and many articles; he has held a Guggenheim Fellowship and is a member of the Royal Society of Edinburgh and the Hungarian Academy of Sciences. The MAA has given him a Chauvenet Prize and two Ford awards. He has been active in the affairs of both the AMS and the MAA, and will become editor of this MONTHLY on Jan. 1, 1982.

His mathematical interests are in measure and ergodic theory, algebraic logic, and operators on Hilbert space, with excursions to probability, statistics, topological groups, and Boolean algebras.—*Editors*

**Hilbert's Problems.** The most risky and possibly least rewarding kind of problem collection to offer to the mathematical public is the one that consists of research problems. Your problems could become solved in a few weeks, or months, or years, and your work would, therefore, be out of date much more quickly than most mathematical exposition. If you are not of the stature of Hilbert, you can never be sure that your problems won't turn out to be trivial, or impossible, or, perhaps worse yet, just orthogonal to the truth that we all seek—wrongly phrased, leading nowhere, and having no lasting value.

A list of research problems that has had a great effect on the mathematical research of the twentieth century was offered by Hilbert in the last year of the nineteenth century at the International Congress of Mathematicians in Paris [3]. The first of Hilbert's 23 problems is the continuum hypothesis: is every uncountable subset of the set $\mathbb{R}$ of real numbers in one-to-one correspondence with $\mathbb{R}$? Even in 1900 the question was no longer new, and although great progress has been made since then and some think that the problem is solved, there are others who feel that the facts are far from fully known yet.

Hilbert's problems are of varying depths and touch many parts of mathematics. Some are geometric (if two tetrahedra have the same volume, can they always be partitioned into the same finite number of smaller tetrahedra so that corresponding pieces are congruent?—the answer is no), and some are number-theoretic (is $2^{\sqrt{2}}$ transcendental?—the answer is yes). Several of the problems are still unsolved. Much of the information accumulated up to 1974 was brought up to date and collected in one volume in 1976 [5], but the mathematical community's curiosity did not stop there—a considerable number of both expository and substantive contributions has been made since then.

**Pólya-Szegő.** Perhaps the most famous and still richest problem book is that of Pólya and Szegő [6], which first appeared in 1925 and was republished (in English translation) in 1972 and 1976. In its over half a century of vigorous life (so far) it has been the mainstay of uncountably many seminars, a standard reference book, and an almost inexhaustible source of examination questions that are both inspiring and doable. Its level stretches from high school to the frontiers of research. The first problem asks about the number of ways to make change for a dollar, the denominations of the available coins being 1, 5, 10, 25, and 50, of course; in the original edition the question was about Swiss francs, and the denominations were 1, 2, 5, 10, 20, and 50. From this innocent beginning the problems proceed, in gentle but challenging steps, to the Hadamard three circles theorem, Tchebychev polynomials, lattice points, determinants, and Eisenstein's theorem about power series with rational coefficients.

**Dörrie.** "The triumph of mathematics" is the original title (in German) of Dörrie's book [1]. This is a book that deserves to be much better known than it seems to be. It is eclectic, it is spread over 2000 years of history, and it ranges in difficulty from elementary arithmetic to material that is frequently the subject of graduate courses.

It contains, for instance, the following curiosity attributed to Newton (Arithmetica Universalis, 1707). If "$a$ cows graze $b$ fields bare in $c$ days, $a'$ cows graze $b'$ fields bare in $c'$ days, $a''$ cows graze $b''$ fields bare in $c''$ days, what relation exists between the nine magnitudes $a$ to $c''$? It is assumed that all fields provide the same amount of grass, that the daily growth of the fields remains constant, and that all the cows eat the same amount each day." Answer:

$$\det\begin{pmatrix} b & bc & ac \\ b' & b'c' & a'c' \\ b'' & b''c'' & a''c'' \end{pmatrix} = 0.$$

This is Problem 3, out of a hundred.

The problems lean more toward geometry than anything else, but they include also Catalan's question about the number of ways of forming a product of $n$ prescribed factors in a multiplicative system that is totally non-commutative and non-associative ("how many different ways can a product of $n$ different factors be calculated by pairs?," Problem 7), and the Fermat-Gauss impossibility theorem ("the sum of two cubic numbers cannot be a cubic number," Problem 21).

Two more examples should give a fair idea of the flavor of the collection as a whole: "every quadrilateral can be considered as a perspective image of a square" (Problem 72), and "at what point of the earth's surface does a perpendicularly suspended rod appear the longest?"(Problem 94). The style and the attitude are old-fashioned, but many of the problems are of the eternally interesting kind; this is an excellent book to browse in.

**Steinhaus.** My next mini-review is of a Polish contribution, Steinhaus [7], which (like Dörrie's) has exactly 100 problems, and they are genuinely elementary and good solid fun. When someone says "problem book" most people think of something like this one, and, indeed, it is an outstanding exemplar of the species. The problems are, however, not equally interesting or equally difficult. They illustrate, moreover, another aspect of problem solving: it is sometimes almost impossible to guess how difficult a problem is, or, for that matter, how interesting it is, till after the solution is known.

Consider three examples. (1) Does there exist a sequence $\{x_1, x_2, \ldots, x_{10}\}$ of ten numbers such that (a) $x_1$ is contained in the closed interval $[0, 1]$, (b) $x_1$ and $x_2$ are contained in different halves of $[0, 1]$, (c) each of $x_1, x_2$, and $x_3$ is contained in a different third of the interval, and so on up through $x_1, x_2, \ldots, x_{10}$? (2) If 3000 points in the plane are such that no three lie on a straight line, do there exist 1000 triangles (meaning interior and boundary) with these points as vertices such that no two of the triangles have any points in common? (3) Does there exist a disc in the plane (meaning interior and boundary of a circle) that contains exactly 71 lattice points (points both of whose coordinates are integers)?

Of course judgments of difficulty and interests are subjective, so all I can do is record my own evaluations. (1) is difficult and uninteresting, (2) is astonishingly easy and mildly interesting, and (3) is a little harder than it looks and even prima facie quite interesting. In defense of these opinions, I mention one criterion that I used: if the numbers (10, 1000, 71) cannot be replaced by arbitrary positive integers, I am inclined to conclude that the corresponding problem is special enough to be dull. It turns out that the answer to (1) is yes, and Steinhaus proves it by exhibiting a solution (quite concretely: $x_1 = .95$, $x_2 = .05$, $x_3 = .34$, $x_4 = .74$, etc.). He proves (the same way) that the answer is yes for 14 instead of 10, and, by three pages of unpleasant looking calculation, that the answer is no for 75. He mentions that, in fact, the answer is yes for 17 and no for every integer greater than 17. I say that's dull. For (2) and (3) the answers are yes (for all $n$ in place of 1000, or in place of 71).

**Glazman-Ljubič.** The book of Glazman and Ljubič [2] is an unusual one (I don't know of any others of its kind), and, despite some faults, it is a beautiful and exciting contribution to the problem literature. The book is, in effect, a new kind of textbook of (finite-dimensional) linear algebra and linear analysis. It begins with the definitions of (complex) vector spaces and the concepts of linear dependence and indepedence; the first problem in the book is to prove that a set consisting of just one vector $x$ is linearly independent if and only if $x \neq 0$. The chapters follow one another in logical dependence, just as they do in textbooks of the conventional kind: Linear operators, Bilinear functionals, Normed spaces, etc.

The book is not expository prose, however; perhaps it could be called expository poetry. It gives definitions and related explanatory background material with some care. The main body of the book consists of problems; they are all formulated as assertions, and the problem is to prove them. The proofs are not in the book. There are references, but the reader is told that he will not need to consult them.

The really new idea in the book is its sharp focus: this is really a book on functional analysis, written for an audience who is initially not even assumed to know what a matrix is. The ingenious idea of the authors is to present to a beginning student the easy case, the transparent case, the motivating case, the finite-dimensional case, the purely algebraic case of some of the deepest analytic facts that functional analysts have discovered. The subjects discussed include spectral theory, the Toeplitz-Hausdorff theorem, the Hahn-Banach theorem, partially ordered vector spaces, moment problems, dissipative operators, and many other such analytic sounding results. A beautiful course could be given from this book (I would love to give it), and a student brought up in such a course could become an infant prodigy functional analyst in no time.

(A regrettable feature of the book, at least in its English version, is the willfully unorthodox terminology. Example: the (canonical) projection from a vector space to a quotient space is called a "contraction", and what most people call a contraction is called a "compression". Fortunately the concept whose standard technical name is compression is not discussed.)

**Klambauer.** The last addition to the problem literature to be reviewed here is Klambauer's **[4]**. Its subject is real analysis, and, although it does have some elementary problems, its level is relatively advanced. It is an excellent and exciting book. It does have some faults, of course, including some misprints and some pointless repetitions, and the absence of an index is an exasperating feature that makes the book much harder to use than it ought to be. It is, however, a great source of stimulating questions, of well known and not so well known examples and counterexamples, and of standard and not so standard proofs. It should be on the bookshelf of every problem lover, of every teacher of analysis (from calculus on up), and, for that matter, of every serious student of the subject.

The table of contents reveals that the book is divided into four chapters: Arithmetic and combinatorics, Inequalities, Sequences and series, and Real functions. Here are some examples from each that should serve to illustrate the range of the work, perhaps to communicate its flavor, and, I hope, stimulate the appetite for more.

The combinatorics chapter asks for a proof of the "rule for casting out nines" (is that expression for testing the divisibility of an integer by 9 via the sum of its decimal digits too old-fashioned to be recognized?), it asks how many zeroes there are at the end of the decimal expansion of 1000!, and it asks for the coefficient of $x^k$ in $(1+x+x^2+\cdots+x^{n-1})^2$. Along with such problems there are also unmotivated formulas that probably only their father could love, and there are a few curiosities (such as the problem that suggests the use of the well ordering principle to prove the irrationality of $\sqrt{2}$). A simple but striking oddity is this statement: if $m$ and $n$ are distinct positive integers, then

$$m^{n^m} \neq n^{m^n}.$$

The chapter on inequalities contains many of the famous ones (Hölder, Minkowski, Jensen), and many others that are analytically valuable but somewhat more specialized and therefore somewhat less famous. A curiosity the answer to which very few people are likely to guess is this one: for each positive integer $n$, which is bigger

$$\sqrt{n}^{\sqrt{n+1}} \quad \text{or} \quad \sqrt{n+1}^{\sqrt{n}}?$$

The chapter on sequences has the only detailed and complete discussion that I have ever seen of the fascinating (and non-trivial) problem about the convergence of the infinite process indicated by the symbol

$$x^{x^{x^{\cdot^{\cdot^{\cdot}}}}}$$

Students might be interested to learn that the result is due to Euler; the reference given is to the article *De formulis exponentialibus replicatis*, Acta Academica Scientiarum Imperialis Petropolitanae, 1777. One more teaser: what is the closure of the set of all real numbers of the form $\sqrt{n}-\sqrt{m}$ (where $n$ and $m$ are positive integers)?

The chapter on real functions is rich too. It includes the transcendentality of $e$, some of the basic properties of the Cantor set, Lebesgue's example of a continuous but nowhere differentiable function, and F. Riesz's proof (via the "rising sun lemma") that every continuous monotone function is differentiable almost everywhere. There is a discussion of that vestigial curiosity called Osgood's theorem, which is the Lebesgue bounded convergence theorem for continuous functions on a closed bounded interval. The Weierstrass polynomial approximation theorem is here (intelligently broken down into bite-size lemmas), and so is one of Gauss's proofs of the fundamental theorem of algebra. For a final example I mention a question that should be asked much more often that it probably is: is there an example of a series of functions, continuous on a closed bounded interval, that converges absolutely and uniformly, but for which the Weierstrass $M$-test fails?

**Problem Courses.** How can we, the teachers of today, use the problem literature? Our assigned task is to pass on the torch of mathematical knowledge to the technicians, engineers, scientists, humanists, teachers, and, not least, research mathematicians of tomorrow: do problems help?

Yes, they do. The major part of every meaningful life is the solution of problems; a considerable part of the professional life of technicians, engineers, scientists, etc., is the solution of mathematical problems. It is the duty of all teachers, and of teachers of mathematics in particular, to expose their students to problems much more than to facts. It is, perhaps, more satisfying to stride into a classroom and give a polished lecture on the Weierstrass $M$-test than to conduct a fumble-and-blunder session that *ends* in the question: "Is the boundedness assumption of the test necessary for its conclusion?" I maintain, however, that such a fumble session, intended to motivate the student to search for a counterexample, is infinitely more valuable.

I have taught courses whose entire content was problems solved by students (and then presented to the class). The number of theorems that the students in such a course were exposed to was approximately half the number that they could have been exposed to in a series of lectures. In a problem course, however, exposure means the acquiring of an intelligent questioning attitude and of some technique for plugging the leaks that proofs are likely to spring; in a lecture course, exposure sometimes means not much more than learning the name of a theorem, being intimidated by its complicated proof, and worrying about whether it would appear on the examination.

**Covering Material.** Many teachers are concerned about the amount of material they must cover in a course. One cynic suggested a formula: since, he said, students on the average remember only about 40% of what you tell them, the thing to do is to cram into each course 250% of what you hope will stick. Glib as that is, it probably would not work.

Problem courses do work. Students who have taken my problem courses were often complimented by their subsequent teachers. The compliments were on their alert attitude, on their ability to get to the heart of the matter quickly, and on their intelligently searching questions that showed that they understood what was happening in class. All this happened on more than one level, in calculus, in linear algebra, in set theory, and, of course, in graduate courses on measure theory and functional analysis.

Why must we cover everything that we hope students will ultimately learn? Even if (to stay with an example already mentioned) we think that the Weierstrass $M$-test is supremely important, and that every mathematics student must know that it exists and must understand how to apply it—even then a course on the pertinent branch of analysis might be better for omitting it. Suppose that there are 40 such important topics that a student *must* be exposed to in a term. Does it follow that we must give 40 complete lectures and hope that they will all sink in? Might it not be better to give 20 of the topics just a ten-minute mention (the name, the statement, and an indication of one of the directions in which it can be applied), and to treat the other 20 in depth, by student-solved problems, student-constructed counterexamples, and student-discovered applications? I firmly believe that the latter method teaches more and teaches better. Some of the material doesn't get *covered* but a lot of it gets *discovered* (a telling old pun that deserves to be kept alive), and the method thereby opens doors whose very existence might never have been suspected behind a solidly built structure of settled facts. As for the Weierstrass $M$-test, or whatever else was given short shrift in class—well, books and journals do exist, and students have been known to read them in a pinch.

**Problem Seminars.** While a problem course might be devoted to a sharply focused subject, it also might not—it might just be devoted to fostering the questioning attitude and improving technique by discussing problems widely scattered over several fields. Such technique courses, sometimes called Problem Seminars, can exist at all levels (for beginners, for Ph.D. candidates, or for any intermediate group).

The best way to conduct a problem seminar is, of course, to present problems, but it is just as bad for an omniscient teacher to do all the asking in a problem seminar as it is for an omniscient teacher to do all the talking in a lecture course. I strongly recommend that students in a problem seminar be encouraged to discover problems on their own (at first perhaps by slightly modifying problems that they have learned from others), and that they should be given public praise (and grade credit) for such discoveries. Just as you should not tell your students all the answers, you should also not ask them all the questions. One of the hardest parts of problem solving is to ask the right question, and the only way to learn to do so is to practice. On the research level, especially, if I pose a definite thesis problem to a candidate, I am not doing my job of teaching him to do research. How will he find his next problem, when I am no longer supervising him?

There is no easy way to teach someone to ask good questions, just as there is no easy way to teach someone to swim or to play the cello, but that's no excuse to give up. You cannot swim for someone else; the best you can do is to supervise with sympathy and reinforce the right kind of fumble by approval. You can give advice that sometimes helps to make good questions out of bad ones, but there is no substitute for repeated trial and practice.

An obvious suggestion is: generalize; a slightly less obvious one is: specialize; a moderately sophisticated one is: look for a non-trivial specialization of a generalization. Another well-known piece of advice is due to Pólya: make it easier. (Pólya's dictum deserves to be propagated over and over again. In slightly greater detail it says: if you cannot solve a problem, then there is an easier problem that you cannot solve, and your first job is to find it!) The advice I am fondest of is: make it sharp. By that I mean: do not insist immediately on asking the natural question ("what is ...?", "when is ...?", "how much is ...?"), but focus first on an easy (but nontrivial) yes-or-no question ("is it ...?").

**Epilogue.** I do believe that problems are the heart of mathematics, and I hope that as teachers, in the classroom, in seminars, and in the books and articles we write, we will emphasize them more and more, and that we will train our students to be better problem-posers and problem-solvers than we are.

**References**

**1.** H. Dörrie, 100 Great Problems of Elementary Mathematics, Dover, New York, 1965.

**2.** I. M. Glazman and Ju. I. Ljubič, Finite-Dimensional Linear Analysis: A Systematic Presentation in Problem Form, MIT, Cambridge, 1974.

**3.** D. Hilbert, Mathematical problems, Bull. Amer. Math. Soc., 8 (1902) 437–479.

**4.** G. Klambauer, Problems and Propositions in Analysis, Dekker, New York, 1979.

**5.** Mathematical Developments Arising from Hilbert Problems, AMS, Providence, 1976.

**6.** G. Pólya and G. Szegö, Problems and Theorems in Analysis, Springer, Berlin, 1972, 1976.

**7.** H. Steinhaus, One Hundred Problems in Elementary Mathematics, Basic Books, New York, 1964.

DEPARTMENT OF MATHEMATICS, INDIANA UNIVERSITY, BLOOMINGTON, IN 47405.

Reprinted from the
MATHEMATICAL INTELLIGENCER
Vol. 3, No. 4, pp. 147–153, 1981

# Does Mathematics Have Elements?

P. R. Halmos

## Prologue

For Empedocles, a little over 2400 years ago, there were four chemical elements, fire, water, earth, and air, and they were continually brought together and torn apart by two opposing forces, harmony and discord. For Aristotle, a hundred years later, two binary classification schemes took the place of one: instead of harmony-discord, he had wet-dry and hot-cold. The serious alchemists of the middle ages found that nature was even more complicated than that that; they classified matter by its luster, heaviness, combustibility, solubility, etc. Boyle, in the 1600's, offered a definition similar to that of a prime number – an element is a substance from which other substances can be made but which cannot be separated into different substances – and analytical chemistry was off and running. A hundred years later Lavoisier (inspired in part by Newton's insight into the paramount importance of weight) could formulate a modified definition that led for the first time to quantitative tables of chemical elements similar to our modern ones.

A chemist asks what things are made of and how they are put together. A systematic listing of chemical elements, incomplete and awkward at first, is not by itself a good answer, but it helps to organize what's known and to indicate what may still be missing.

No doubt many mathematicians have noted that there are some basic ideas that keep cropping up in widely different parts of their subject, combining and re-combining with one another in a way faintly reminiscent of how all matter is made up of elements. A subconscious intuitive awareness of these "elements" of mathematics probably contributes to (possibly it constitutes) the research insight that distinguishes great mathematicians from ordinary mortals. What are they – what are the elements of mathematics?

## Examples

Here are three possible examples of the sort of basic concepts that might be considered mathematical elements: geometric series, quotient structures, and eigenvectors. They seem to me to be of three different kinds, perhaps at three different levels of depth; as a first step toward classification the words computational, categorical, and conceptual (respectively) could be used to describe them.

*Geometric series.* In a not necessarily commutative ring with unit (e.g., in the set of all $3 \times 3$ square matrices with real entries), if $1 - ab$ is invertible, then $1 - ba$ is invertible. However plausible this may seem, few people can see their way to a proof immediately; the most revealing approach belongs to a different and distant subject.

Every student knows that

$$1 - x^2 = (1 + x)(1 - x),$$

and some even know that

$$1 - x^3 = (1 + x + x^2)(1 - x).$$

The generalization

$$1 - x^{n+1} = (1 + x + \ldots + x^n)(1 - x)$$

is not far away. Divide by $1 - x$ and let $n$ tend to infinity; if $|x| < 1$, then $x^{n+1}$ tends to 0, and the conclusion is that

$$\frac{1}{1 - x} = 1 + x + x^2 + \ldots .$$

This simple classical argument begins with easy algebra, but the meat of the matter is analysis: numbers, absolute values, inequalities, and convergence are needed not only for the proof but even for the final equation to make sense.

In the general ring theory question there are no numbers, no absolute values, no inequalities, and no limits – those concepts are totally inappropriate and cannot be brought to bear. Nevertheless an impressive-sounding classical phrase, "the principle of permanence of functional form", comes to the rescue and yields an analytically inspired proof in pure algebra. The idea is to pretend that $\dfrac{1}{1-ba}$ can be expanded in a geometric series (which is utter nonsense), so that

$$(1-ba)^{-1} = 1 + ba + baba + bababa + \ldots .$$

It follows (it doesn't really, but it's fun to kepp pretending) that

$$(1-ba)^{-1} = 1 + b(1 + ab + abab + ababab + \ldots)a.$$

and, after one more application of the geometric series pretense, this yields

$$(1-ba)^{-1} = 1 + b(1-ab)^{-1}a.$$

Now stop the pretense and verify that, despite its unlawful derivation, the formula works. If, that is, $c = (1-ab)^{-1}$, so that $(1-ab)c = c(1-ab) = 1$, then $1 + bca$ is the inverse of $1-ba$. Once the statement is put this way, its proof becomes a matter of (perfectly legal) mechanical computation.

Why does it all this work? What goes on here? Why does it seem that the formula for the sum of an infinite geometric series is true even for an abstract ring in which convergence is meaningless? What general truth does the formula embody? I don't know the answer, but I note that the formula is applicable in other situations where it ought not to be, and I wonder whether it deserves to be called one of the (computational) elements of mathematics.

*Quotient structures.* The symmetric difference of two subsets $A$ and $B$ of, say, the plane, sometimes denoted by $A \,\Delta\, B$, is the set of those points that belong either to $A$ or to $B$ but not to both. Question: is the operation $\Delta$ associative?

The answer is accessible by just plain work, but it's not pleasant that way: too many sets have to be kept track of and too many cases must be considered. There is a much better way to guess the answer and, at the same time, to prove that the guess is true, a way that embeds the question into a powerful general theory of structures.

Integers can be added and subtracted; better said, with respect to addition they form an abelian group. Integer-valued functions on an arbitrary non-empty set can be added and subtracted just as well, and they too form an abelian group. There is a well-known and useful correspondence between the subsets of a set (and, in particular, the subsets of the plane) and certain special integer-valued functions: to each set $A$ there corresponds the function $\chi_A$, its characteristic function, that takes the value 1 on $A$ and 0 on the complement of $A$. When two sets are combined by $\Delta$, what happens to the corresponding functions? The best way to find out is to calculate their sum:

$$\chi_A(x) + \chi_B(x) = \begin{cases} 0 \text{ if } x \notin A, x \notin B, \\ 1 \text{ if } x \in A \,\Delta\, B, \\ 2 \text{ if } x \in A, x \in B. \end{cases}$$

Otherwise expressed: $\chi_A(x) + \chi_B(x)$ is odd or even according as $x \in A \,\Delta\, B$ or not. In still other words: $\chi_{A \Delta B}(x)$ is congruent modulo 2 to the sum of $\chi_A(x)$ and $\chi_B(x)$.

Sudden insight and complete victory: the formation of symmetric differences for sets is the same as addition modulo 2 for their characteristic functions. Since addition modulo 2 is associative, so is the formation of symmetric differences.

In abstract terms the argument is one in the category of abelian groups. What the argument shows is that the collection of subsets of the plane endowed with the operation $\Delta$ is (isomorphic to) the quotient of the abelian group of all integer-valued functions on the plane modulo the subgroup of all even-integer-valued functions. (It is obvious, isn't it? , that the plane plays no special role; any non-empty set can be used similarly to construct a "Boolean group".)

The mathematical element that clamors for attention here is the formation of quotient structures. Integers modulo even integers are a familiar manifestation of the concept, and so are integers modulo multiples of 12. Many other manifestations lead to deep theories; a couple of famous analytic examples are measurable sets modulo sets of measure zero, and operators on Hilbert space modulo compact operators.

*Eigenvalues.* After gathering a pile of coconuts one day, five sailors on a desert island agree to divide them evenly next morning. During the night one sailor secretly gets up and tries to divide the coconuts into five equal piles. Finding that there is a remainder of one coconut, he tosses the odd one to a monkey, and, secreting his pile, mixes up the others and goes back to sleep. The second sailor does the same thing, with the same result, and so do the third, fourth, and fifth. In the morning the remaining pile of coconuts (less one) is again divisible by 5. What is the smallest number of coconuts that the original pile could have contained?

Anybody can solve this old puzzle with nothing but paper, pencil, and patience; just "work backwards". An alternative is to examine, for each possible number $x$ of coconuts, the number $S(x)$ that a typical sailor's activities leave in the pile. The formula is simple enough: $S(x) = \frac{4}{5}(x - 1)$. Call an integer $x$ a "solution" if the result of starting with $x$ and performing the operation $S$ six times is an integer; in this language the problem is to find the smallest positive solution. Since $S(x) = \frac{4}{5}x - \frac{4}{5}$, so that

$$S^6(x) = \left(\frac{4}{5}\right)^6 x - \left(\left(\frac{4}{5}\right)^6 + \left(\frac{4}{5}\right)^5 + \ldots + \left(\frac{4}{5}\right)\right),$$

it follows that
questions that are connected with it. Case in point: Enflo's solution of the (strictly infinite) basis problem for Banach spaces showed that the difficulty was really a matter of understanding the structure of finite-dimensional examples.

*Infinity.* Finiteness is an element, and so is infinity. By "infinity" I do not mean something as shallow as "the point at infinity" in projective geometry or on the Riemann sphere; there is nothing infinite about that. Just as the typically finite argument is the pigeonhole principle, the typically infinite one is induction. I say "induction" because it's a single word that points in the right direction. What I really have in mind, however, is the simplest theorem about infinite cardinal numbers that is at the same time one of the deepest theorems of mathematics, namely the statement that $1 + \aleph_0 = \aleph_0$. That's what is really at the heart of mathematical induction and is the basis of Dedekind's definition of infinity. It enters in analysis (including ergodic theory and, of course, operator theory), in algebra (for example in the classification of infinite abelian groups), in topology, and in logic – its connotations embrace everything that anyone could properly call infinite.

[Here is a curious aside, parallel to the statement that in some sense all mathematics is finite: in another sense all mathematics is infinite. This is more nearly the classical view. Even an uninteresting mathematical assertion, such as

29 + 54 = 83, is a general theorem with an uncountably infinite set of special cases (or do I mean applications?). There is no contradiction: each statement uses a slightly dramatized stylistic ellipsis to make a point, and with an adequate amount of pedantic padding both statements become true and dull.]

It may be worthy of note that although in natural languages "infinite" is a negative word (not finite), whose definition must presumably come after the positive one, in mathematics infinity is the free concept, defined in positive terms (something exists), and finiteness is the restrictive, negative one (something cannot be done).

Dedekind defined a set to be infinite if it can be put in one-to-one correspondence with a proper subset, and finite if it cannot. The idea, transplanted to the context of category theory, can be used to define generalizations of mere set-theoretic infinity and finiteness – generalizations whose study could possibly shed some light on those elements. What I have in mind can be illustrated in the category of groups by these definitions: a group is *infinitary* if it is isomorphic to a proper subgroup and *finitary* otherwise. Example: the infinite group of all those roots of unity whose exponent is a power of 2 is finitary. Similar definitions in other categories (e.g., metric spaces) yield other suggestive examples of finitary objects (e.g., the compact ones).

## Composition

A difficult aspect of trying to discover the elements of mathematics is nomenclature. "Infinity" is probably a good name, and "geometric series" is probably a bad one. I have no good name for the next cluster of ideas I should like to recall; till a better name comes along, I'll leave them grouped under the heading *composition.* Here they are, my next batch of possible elements, and these are of the conceptual kind.

*Iteration.* It all starts from the trivial observation the mappings can be composed. In other words, if we can perform each of two operations (and if their domains and ranges match properly), then there is nothing to stop us from performing them both, one after the other, and, in particular, there is nothing to stop us from performing one operation over and over again. This concept is at the root of Archimedes' theorem (if $K$ and $\epsilon$ are positive numbers, then some positive integral multiple of $\epsilon$ is greater than $K$, or, informally, every little bit helps). It is at the root of the idea of integration also (the sum of many "infinitesimals" can be large), and at the root of greatly generalized integrals such as arise in spectral theory.

Another part of analysis where the notion of iteration plays a central role is the method of successive approximations. The "method" is also called Banach's fixed point theorem. Note that the element under discussion makes contact with the element of invariance: fixed points are good to look for and to find, whether the search is iterative or not.

*Cross section.* Composition is an important mathematical element even when it is not done infinitely often; one reason is its connection with the element that may be called transversality. The most famous instance of what I want to discuss here is the axiom of choice. Possible formulation: for every disjoint set of sets there exists a set that has exactly one member in common with each of them, or, in other words, every partition has a transversal set. (Disjointness is not always a part of the statement, but it might as well be; the versions with it and, suitably rephrased, without it, are equivalent.) It doesn't take much thought to see that the axiom can be expressed as follows: whenever $f$ is a function from a set $Y$ onto a set $X$, then there exists a function $g$ from $X$ into $Y$ such that $f(g(x)) = x$ for all $x$ in $X$. (The idea is that if $Y$ is the disjoint union of sets $Y_x$, with $x$ in $X$, and if $f(y) = x$ for each $y$ in $Y_x$, then $g$ is a choice function that assigns to each $x$ in $X$ a member $y_x$ of $Y_x$.) In other words, every function (better: every surjection) has a right inverse, or, in a different but frequently used terminology, every function has a cross section.

The element "cross section" is a part of many mathematical compounds, subject, of course, in each case, to the

$$S^6(x) - S^6(y) = \left(\frac{4}{5}\right)^6 (x - y)$$

for all $x$ and $y$. This implies that if both $x$ and $y$ are solutions, then $x$ and $y$ are congruent modulo $5^6$; if, conversely, $x$ is a solution and $x$ and $y$ are congruent modulo $5^6$, then $y$ also is a solution.

The simplest conceivable "solution" is the "eigenvector" belonging to the "eigenvalue equation" $S(x) = x$; that solution is $-4$. The smallest positive solution, therefore, is $-4 + 5^6 = 15{,}621$. Moral: the concept of invariance (fixed point, eigenvalue, eigenvector) is a possible element of mathematics.

Whatever their provenance, all three of these examples ($(1 - ba)^{-1}$. $A \triangle B$, $5^6$) are parts of mathematical folklore by now. The geometric series one is usually ascribed to N. Jacobson, and the eigenvalue one to P. A. M. Dirac.

## Universal algebra

I chose the three examples above not only because they illustrate the possiblility of three different kinds of elements, but, to be honest, because they have also a qualitiy that is as rare as it is striking: in each of them the element can be used to solve a problem in a manner that is efficient and at least slightly surprising. All other candidates that I have been able to think of are of a more obvious and less spectacular kind. The commonest ones, the most visible ones, are on the level of universal algebra. The purpose of what follows is to describe (or at least to mention) some of the most conspicuous ones, first on the categorical and then on the conceptual level.

*Structure.* Mathematicians often (always?) traffic in sets with structures. The structure may come from an internal operation or two (as in groups and fields), or an external function or two (as in metric spaces or analytic manifolds). The structure may involve more than one set (as the scalars that act on and the vectors that belong to a vector space), and it may be defined in terms of classes of subsets (as in topological spaces and in measure theory).

A pertinent observation (which is a friend of every effective teacher and every productive research mathematician, but which seems never to have received official recognition) is that the constituents of a structure cannot, must not, dangle separately, but must be subjected to appropriate structural compatibility conditions. A ring is not just a set in which both addition and multiplication can be performed – it is vitally important that they be connected by the distributive law. A topological group is not just a set that has a topological structure as well as a multiplicative one – it is vitally important that they be connected by continuity.

*Categories.* It has been viscerally obvious to the working mathematician for a long time that homomorphisms between groups, continuous functions between topological spaces, and linear transformations between vector spaces are all the same sort of thing, playing the same roles, and behaving in many respect the same way. The concept of a "structure" might be an element of mathematics, or, perhaps better, might be an appropriate classification heading for some of the elements. The subject that discusses structures in varying degrees of generality has been called various names. It can be called universal algebra, and, under that name, it has been loved by some and hated by others; some of the people who profess to despise it have put salt on its tail and called the result category theory. Once that's done, the abstract versions of the mappings that keep arising are dubbed (homo)morphisms, and, motivated by set-theoretic notions such as injection, surjection, and bijection, a large Greek lexicon of mono-, epi-, iso-, as well as endo- and auto-morphisms goes to work. Arguably, each one of the concepts here named is an element of mathematics.

*Isomorphism.* The recognition that two structures are "the same in some sense" – isomorphic – (recall the eigenvalue solution of the coconut problem) is surely a reasonable candidate for an element. There is a general notion in the background here; a vague way to describe it is a manifestation of the number 1, or uniqueness. (Single-generated structures, e.g., cyclic groups, are also suggested by this circle of ideas.)

The rare and always exciting insight that something is "really" something else – e.g., that probability is really measure theory, or that Fourier series are really a part of the study of locally compact abelian groups – all that belongs to the one-ness element.

[I permit myself an autobiographical disgression. When I was studying David Berg's proof that every normal operator on a Hilbert space of dimension $\aleph_0$ is the sum of a diagonal operator and a compact one, I kept thinking that I had seen something like this before, that the same permutation of ideas belonged to another proof, somewhere else. The insight was slow in coming, but it was worth working for. What I finally remembered was that that "other proof, somewhere else" proved the classical theorem about every compact metric space being a continuous image of the Cantor set. As soon as I saw that, I understood Berg's theorem completely, and, incidentally, I became convinced that the right road to follow was to use the theorem about the Cantor set rather than imitate its proof.]

*Quotients.* Isomorphism is probably the structural element that occurs most often, that has the greatest visibility, but the one that I think has the greatest depth and least obvious manifestations is epimorphism. The set-theoretic language in which surjections (which is what epimorphisms usually are) are most frequently discussed is that of quotient structures (the concept that solved the associativity problem for symmetric differences). Quotient structures occur everywhere in mathematics and play a vital role: witness quotient groups in algebra, identification spaces in topology (that give us, for instance, circles by bending closed intervals), modular arithmetic in number theory, and the (already mentioned) $L^p$ spaces in analysis and Calkin algebra in operator theory.

## Size

Isomorphism, as an element, contrasts with epimorphism; the distinction is between recognizing that two things are the same and identifying them so as to make them the same. From a different point of view isomorphism can be looked at as the emphasis on one-ness, or prime-ness, and from that point of view the proper contrast is two-ness, or (finite) many-ness, or infinity. These words are meant to point to a few of the less universal (and therefore more important? ) categorical elements.

*Primes.* Primes occur in many parts of mathematics. Their first appearance for most of us is in number theory, followed soon afterward by their appearance in algebra (as, for instance, irreducible polynomials), and something like them crops up almost everywhere. The connected components of a topological space may be viewed as its prime constituents; search for "primes" among Beurling's inner functions plays an important role in operator theory; and the determination of all the primes among finite groups (that is, of the simple groups) was a gigantic job that occupied many algebraists for many years.

*Duality.* The most familiar antonym of "one" is "two", and, correspondingly, two-ness or duality should probably be viewed as an element. The word "duality" is used in projective geometry (points and lines in the plane), in category theory, in logic (memento also Boolean algebras and totally disconnected compact Hausdorff spaces), in topology (memento Alexander), in harmonic analysis (Pontrjagin, Tanaka), in Banach space theory (memento reflexivity) – and while it means something different in each of these cases, the underlying element is recognizably the same. Related but different manifestations of two-ness occur in the theory of involutions in groups and algebras, and in the study of order, equivalence, and other types of binary relations.

[A different level at which two-ness enters mathematics is visible in the two-way pull of many well-known examples and theorems. The best theorems don't just say that a strong assumption implies a desirable conclusion; they are more likely to say that despite a strong pull downward something goes upward. The theorem that $\Sigma_{n=1} \frac{1}{n^{1+\epsilon}}$ converges whenever $\epsilon > 0$ is improved by the knowledge that the series diverges when $\epsilon = 0$; the wonder of the Cantor function is that while it remains constant almost all the time it manages to grow from 0 to 1.]

In their role in primality and duality the numbers 1 and 2 are, apparently, different from all other positive integers. Bilinear forms can be generalized to trilinear ones, and the achievement is in many ways quite successful; ternary relations exist as well as binary ones (although no good general theory of them has been found yet); and in a group the elements of order three can be looked at as well as the elements of order two – but somehow it all looks unnatural and forced, like a generalization that does nothing but generalize.

*Pigeonhole.* The concept of finiteness is the next element of mathematics that I should like to call attention to. When finiteness enters mathematical reasoning it usually does so in the guise of the famous pigeonhole principle: ten bills delivered to nine mailboxes are bound to make at least one debtor extra grumpy.

The pigeonhole principle is the quintessence of "finite mathematics"; it is at the heart of the very definition (Dedekind's) of the concept of finiteness, and, in particular, at the heart of the burgeoning activity called combinatorics. Its most subtle generalizations are the Ramsey theorems, some of which have shed surprising light on the theory of sentences that are true but unprovable.

Finiteness has echoes in the most infinite parts of mathematics – it is a defensible thesis that in some sense all mathematics is finite. Compactness, for instance, a concept associated with topological spaces (usually infinite), is thought of by many mathematicians as a cleverly generalized kind of finiteness. [Incidentally, it has always seemed to me that general topology is the right infinite generalization of combinatorics – I might go so far as to say that it *is* infinite combinatorics. As a curious piece of authoritarian evidence for this view I cite Norman Steenrod: he disliked both subjects equally, and thought lowly of them, apparently in the same way and for the same reasons.]

As another example of the finite in infinity I mention my faith (surely it is not mine alone? ) that if we knew *all* about operator theory in spaces of *finite* dimension, then we could answer all questions about operators, even the ones (such as the invariant subspace problem) that are incontrovertibly infinite. I do not mean something as absurd as a one-to-one correspondence: the solution of the invariant subspace problem for finite-dimensional spaces does not solve the problem in general. What I mean, what I must mean, is that the solution of every operator problem can be approached by finding, formulating, and answering the right, the appropriate, the truly central finite-dimensional questions that are connected with it. Case in point: Enflo's solution of the (strictly infinite) basis problem for Banach spaces showed that the difficulty was really a matter of understanding the structure of finite-dimensional examples.

*Infinity.* Finiteness is an element, and so is infinity. By "infinity" I do not mean something as shallow as "the point at infinity" in projective geometry or on the Riemann sphere; there is nothing infinite about that. Just as the typically finite argument is the pigeonhole principle, the typically infinite one is induction. I say "induction" because it's a single word that points in the right direction. What I really have in mind, however, is the simplest theorem about infinite cardinal numbers that is at the same time one of the deepest theorems of mathematics, namely the statement that $1 + \aleph_0 = \aleph_0$. That's what is really at the heart of mathematical induction and is the basis of Dedekind's definition of infinity. It enters in analysis (including ergodic theory and, of course, operator theory), in algebra (for example in the classification of infinite abelian groups), in topology, and in logic – its connotations embrace everything that anyone could properly call infinite.

[Here is a curious aside, parallel to the statement that in some sense all mathematics is finite: in another sense all mathematics is infinite. This is more nearly the classical view. Even an uninteresting mathematical assertion, such as 29 + 54 = 83, is a general theorem with an uncountably infinite set of special cases (or do I mean applications?). There is no contradiction: each statement uses a slightly dramatized stylistic ellipsis to make a point, and with an adequate amount of pedantic padding both statements become true and dull.]

It may be worthy of note that although in natural languages "infinite" is a negative word (not finite), whose definition must presumably come after the positive one, in mathematics infinity is the free concept, defined in positive terms (something exists), and finiteness is the restrictive, negative one (something cannot be done).

Dedekind defined a set to be infinite if it can be put in one-to-one correspondence with a proper subset, and finite if it cannot. The idea, transplanted to the context of category theory, can be used to define generalizations of mere set-theoretic infinity and finiteness – generalizations whose study could possibly shed some light on those elements. What I have in mind can be illustrated in the category of groups by these definitions: a group is *infinitary* if it is isomorphic to a proper subgroup and *finitary* otherwise. Example: the infinite group of all those roots of unity whose exponent is a power of 2 is finitary. Similar definitions in other categories (e.g., metric spaces) yield other suggestive examples of finitary objects (e.g., the compact ones).

### Composition

A difficult aspect of trying to discover the elements of mathematics is nomenclature. "Infinity" is probably a good name, and "geometric series" is probably a bad one. I have no good name for the next cluster of ideas I should like to recall; till a better name comes along, I'll leave them grouped under the heading *composition.* Here they are, my next batch of possible elements, and these are of the conceptual kind.

*Iteration.* It all starts from the trivial observation the mappings can be composed. In other words, if we can perform each of two operations (and if their domains and ranges match properly), then there is nothing to stop us from performing them both, one after the other, and, in particular, there is nothing to stop us from performing one operation over and over again. This concept is at the root of Archimedes' theorem (if $K$ and $\epsilon$ are positive numbers, then some positive integral multiple of $\epsilon$ is greater than $K$, or, informally, every little bit helps). It is at the root of the idea of integration also (the sum of many "infinitesimals" can be large), and at the root of greatly generalized integrals such as arise in spectral theory.

Another part of analysis where the notion of iteration plays a central role is the method of successive approximations. The "method" is also called Banach's fixed point theorem. Note that the element under discussion makes contact with the element of invariance: fixed points are good to look for and to find, whether the search is iterative or not.

*Cross section.* Composition is an important mathematical element even when it is not done infinitely often; one reason is its connection with the element that may be called transversality. The most famous instance of what I want to discuss here is the axiom of choice. Possible formulation: for every disjoint set of sets there exists a set that has exactly one member in common with each of them, or, in other words, every partition has a transversal set. (Disjointness is not always a part of the statement, but it might as well be; the versions with it and, suitably rephrased, without it, are equivalent.) It doesn't take much thought to see that the axiom can be expressed as follows: whenever $f$ is a function from a set $Y$ onto a set $X$, then there exists a function $g$ from $X$ into $Y$ such that $f(g(x)) = x$ for all $x$ in $X$. (The idea is that if $Y$ is the disjoint union of sets $Y_x$, with $x$ in $X$, and if $f(y) = x$ for each $y$ in $Y_x$, then $g$ is a choice function that assigns to each $x$ in $X$ a member $y_x$ of $Y_x$.) In other words, every function (better: every surjection) has a right inverse, or, in a different but frequently used terminology, every function has a cross section.

The element "cross section" is a part of many mathematical compounds, subject, of course, in each case, to the additional conditions that the enveloping structure naturally imposes. (The cross section must be not only a function, but an appropriate kind of continuous, or differentiable, or algebraically well-behaved function.) Cross-sections enter, for instance, in differential geometry (connections), in combinatorics (the marriage theorem), in algebra (semidirect products), in analysis (Stone and von Neumann had occasion to study cross sections of the mapping that associates with each measurable set its equivalence class modulo sets of measure zero), and in topology (which continuous mappings have continuous cross sections? , do they all have Borel cross sections? ).

*Exponential.* The idea of composition leads naturally to iteration and thence exponentiation, and that suggests one more element associated with this circle of ideas, namely the exponential function. We all learn about $e^x$ relatively early in calculus, and many of us are surprised and pleased by its complex properties (e.g., its periodicity). A concept that plays a fundamental role not only in interest rates, but also in Banach algebras and Lie theory, is surely deserving of being considered an element of mathematics.

## Analogy

There are patterns that occur and recur in mathematics that seem to lie deeper than the first ones (such as morphism) that category theory calls attention to, but that are vaguer, less well known, and less studied (so far) than some of the classical elements (such as geometric series). Worthy of mention among them are commutativity, symmetry, and continuity.

*Commutativity.* Is the concept of commutativity an element? Could be. Surely the insight that what classical double limit theorems assert is in some respects similar to the relation between two translations acting on Euclidean space, and that both of them resemble the behavior of the paths in certain arrow diagrams – surely that is an insight worth having, and, perhaps, the common feature, the concept of commutativity, is an element of mathematics.

*Symmetry.* The word "symmetry" is not usually defined; mathematicians are more likely to use it informally, as they use words such as "analysis". "Proof by symmetry" is a popular phrase in which the word occurs; is there an element hiding there?

The commonly agreed context to which symmetry is assigned is group theory, and there too an element may be hiding. I do not just mean the well-established concept of group, but I mean the more lately arrived tendency to make a group out of everything; phrases such as Grothendieck group and $K$-theory will indicate the direction I am looking in.

*Continuity.* Continuity is another word that seems to want to mean something even when it shouldn't, even when there is no topology present (or, in any event, no obvious one). Mathematicians seem to be disposed to believe that everything is continuous. The tendency is visible in the Kodaira-Spencer theory of deformation of manifolds, in the Riesz-Thorin interpolation theorem, and in the Kadison-Kastler study of perturbation of von Neumann algebras. (The least precise and oldest manifestation of the phenomenon is the principle of permanence of functional form already referred to.)

All these phenomena – the element candidates of commutativity, symmetry, and continuity, are instances of what might be called illegitimate generalization, or conclusion jumping by analogy. However shady their logical status, however lacking in rigor they may be, they can be valuable weapons in the arsenal of the truth-hunting mathematician: long before all else fails, be sure to ask "do they commute? ", "do they have inverses? ", and "do they converge to the right limit? ".

## Epilogue

Is what I have been saying mathematical mysticism, or is it possible that there really are some underlying guiding principles in mathematics that we should try to learn more about? I think there are such principles, but I do not know; I believe, therefore I think.

I feel like a fumbling freshman disciple of Empedocles, picking "elements" almost at random, not by controlled observation and careful analysis, but based on private experience and personal intuition. The elements I have suggested range from dim analogy (everything is continuous), through conventional wisdom (structural constituents must be compatible) and universal algebra (look for invariants, form quotient structures), to computational trickery (sum the geometric series). My own hesitant first steps were suggested by the geometric series trick, and, more generally, by the conviction that Euler's illegal manipulations with divergent series were not worthless: he had the guidance of genius, an unformulated but sharply focused insight into the fundamental elements of truth. My subsequent thoughts led to a more rarefied air than Euler's solid earth, but I continue to feel that the concrete, down-to-earth candidates (exponential function, pigeonhole principle) are more likely to be elements than the abstract, airy ones (duality, infinity).

In any event, I know that, at best, all I have done was to suggest the existence of a question. If that suggests to others wider applications of the elements I have mentioned, or new elements in parts of mathematics I know nothing about, or steps toward a beautiful and useful theory of mathematical elements, then I'll have accomplished my mission.

*P. R. Halmos*
*Indiana University*
*Department of Mathematics*
*Swain Hall East*
*Bloomington, IN 47405*
*U.S.A.*

Reprinted from the
TWO-YEAR COLLEGE MATHEMATICS JOURNAL
Vol. 13, No. 4, pp. 243–251, Sept. 1982

# The Thrills of Abstraction

P. R. Halmos

My wife and I were invited to a party recently, a party attended by four other couples, making a total of ten people. Some of those ten knew some of the others, and some did not, and some were polite, and some were not. As a result a certain amount of handshaking took place in an unpredictable way, subject only to two obvious conditions: no one shook his or her own hand and no husband shook his wife's hand. When it was all over, I became curious and I went around the party asking each person: "How many hands did you shake? . . . And you? . . . And you?" What answers could I have received? Conceivably some people could have said "None", and others could have given me any number between 1 and 8 inclusive. That's right, isn't it? Since self-handshakes and spouse-handshakes were ruled out, 8 is the maximum number of hands that any one of the party of 10 could have shaken.

I asked nine people (everybody, including my own wife), and each answer could have been any one of the nine numbers 0 to 8 inclusive. I was interested to note, and I hereby report, that the nine different people gave me nine different answers; someone said 0, someone said 1, and so on, and, finally, someone said 8. When it was all over, my curiosity was satisfied: I knew all the answers. Next morning I told the story to my colleagues at the office, exactly as I told it now, and I challenged them, on the basis of the information just given, to tell me how many hands my wife shook.

I mention this problem not because it is what I want mainly to say (but I will parenthetically whisper that it is a legitimate, bona fide problem, and my colleagues could have solved it), but because it illustrates, sort of, one of the basic mathematical principles that I do want to discuss. Here is another question, that leads to another basic mathematical principle: is there a number with the property that when it is multiplied by itself five times the result is the same as if we had just added 2 to it? Anyone who did not manage to remain completely innocent of high-school algebra will recognize the question as a non-symbolic way of asking whether the equation $x^5 = x + 2$ has any solutions.

## Numbers, Phonemes, and Species

Here is one more question: what is there in common between the biologist's concept of a species, the linguist's concept of a phoneme, and the mathematician's concept of a number?

I'll turn to the questions in reverse order, but before doing that I'd like to describe the leitmotif of the whole discussion.

What is a black hole? I don't know, but I get a vague idea every now and then from a casual article in a newspaper or a magazine. It seems to be something very very heavy—so heavy that nothing that once enters its domain of gravitational attraction can ever escape from it, not even light, and, as a result, it is something that we can never perceive with any of our senses—something we can never see, hear, smell, taste, or feel. It has a measurable influence on some of the world we can perceive, but it itself is an abstraction. I probably said that all wrong, but I feel that even my vague and erroneous notion of a black hole is worth knowing—when I learned it, my soul grew, I became richer. I saw a vista I had never dreamt of before, my imagination was stimulated in a new way.

All parts of human intellectual endeavor have their abstractions: the economist's utility, the psychologist's id, the chemist's molecule, the biologist's species, the linguist's phoneme, and, of course, the mathematician's number—all these are abstractions, and each is a seminal part of the field it belongs to. When I asked, however, what species, phoneme, and number have in common, I wasn't just leading up to the shallow answer that they are all abstractions. There is more to the question than that.

A dictionary might define a phoneme of a language as "a smallest unit of speech that distinguishes one word from another". That's too quick, too shallow, too simplistic, but it's a beginning of a definition. An example will help to clarify the issue. If I replace *b* by *m* in "bat", I get another English word, "mat", that means something completely different; that's why *b* and *m* belong to two different English phonemes.

Consider, on the other hand, the words "stone" and "tone". Does everybody realize that the *t* sounds different in those words? In "tone" it is aspirated, and in "stone" it is not—by which the linguist means that if I hold a slip of paper two or three inches from my lips and then say "tone", the paper will move, but if I say "stone", it will not. There are languages (I believe Hindi is one of them) in which the replacement of an unaspirated *t* by an aspirated one can change meaning (just as the replacement of *b* by *m* changes the meaning of "bat"). In English, however, although phoneticians and their machines can distinguish between the two *t*'s, there is no context in which the replacement of one by the other changes the meaning. If a person who is not a native speaker of English uses an unaspirated *t* where he shouldn't, we feel that there is something slightly off, that he has a foreign accent in some sense, but we don't have any trouble understanding him. As far as English is concerned, the two *t*'s are "isosemantic". There is no such word—I just made it up —but everybody can probably guess what it would mean if it existed: it would mean that the replacement of one by the other preserves meaning.

What then is a phoneme? Or, better asked, what is the phoneme of a sound? Answer: the collection of all sounds isosemantic with it. Since *b* and *m* are not isosemantic, *b* does not belong to the phoneme of *m*, but the *t* in "tone" does belong to the phoneme of the *t* in "stone".

A similar analysis of the concept of species is possible, but I will not enter into it now. A dictionary might define a species as "a collection of organisms capable of interbreeding", but before we could discuss the pertinent analogue of "isosemantic", we would need to sort out the sexes, and that digression, while possibly interesting, would take too long.

The concept of number is nearer at hand, and, at least in mathematical circles, very well known. We all use words such as "five" every day, but do many people ask themselves what "five" is? And, by the way, shouldn't they be ashamed of themselves? We wouldn't use words such as "grandfather", or "tax", or "lawnmower" without being able to define them—without, to be specific, being able to tell a ten-year old child exactly what a grandfather, or a tax, or a lawnmower is, but the challenge is to tell him exactly what a number is. I don't mean what a number *does*, or how a number can be used—I mean what it *is*.

All right: what is "5"? We may not know that, but we know that if it's the answer to "How many fingers are there on your right hand", then it's also the answer to "How many players are there on a basketball team?" In other words, while we may not know what "number" is, we do know when two sets of objects (be they fingers, or whatever) are "equinumerous". They are that just when we can establish a correspondence between them (for example, by pointing to each basketball player on the team with a different finger) that is a one-to-one correspondence—each object in each set corresponds to a unique object in the other set.

What then is a number? Or, better asked, what is the number of objects in a set? Answer: the collection of all sets equinumerous with it.

## Abstraction and Attitude: Equivalence Relations and Extensionalism

This is an abstract definition, it is a frightening definition, it's an ingenious definition. It is due to Bertrand Russell, and it leads me now to comment on two things: one, an abstraction, a basic mathematical concept, that includes the way species, phonemes, numbers, and many other concepts in many parts of life are best thought of, and, two, an attitude, a philosophical stand, that some mathematicians embrace, and that contributes greatly to the clarity and precision of mathematics. The name of the concept is "equivalence relation", and it is well known and standard; the name of the attitude is "extensionalism", and, while the attitude is not uncommon, the name, as far as I know, is something that I've been using for some time in private, but no one else has ever heard of.

An equivalence relation is a relation that has three properties in common with the relations of being isosemantic and equinumerous, namely that it is reflexive, symmetric, and transitive. The replacement of an utterance by itself (which is, of course, no replacement at all) surely preserves meaning, and each set has a one-to-one correspondence with itself—that's what "reflexive" means. Officially: a relation is reflexive in case every object in its realm does bear that relation to itself. So, for instance, fatherhood is not reflexive—no one can be his own father—and whether brotherhood among, let us say, human males is or is not is a small hairsplitting debate about how you want to use words. Am I my own brother?

To say that a relation is "symmetric" means that the roles of two objects in the relation can always be reversed. Example: if the initial sound in"pit" is isosemantic with the initial sound in "pendulum", then, vice versa, the initial sound in "pendulum" is isosemantic with the initial sound in "pit". Similarly: if a basketball team is equinumerous with the fingers on my right hand, then, vice versa, the fingers on my right hand are equinumerous with a basketball team. Here are a couple of non-examples: fatherhood is not symmetric (my father bears that relation to me, but I do not bear the same relation to him), and fondness is not symmetric—while it may often happen that someone I am fond of is fond of me, it is not guaranteed, and one single exception disproves the universality of the property.

"Transitivity" is just as easy a concept, but it takes a little longer to say. If three sounds are isosemantic in order, that is, the first and the second are isosemantic, and the second and the third are, then it follows that the first and the third also are.

A well-known non-example is friendship: it isn't always true that if Tom is Dick's friend and Dick is Harry's, then Tom and Harry also bear the relation of friendship to each other.

So, that's what an equivalence relation is: one that is reflexive, symmetric, and transitive. And any time we run across an equivalence relation, the objects to which it applies can be split up into what are called equivalence classes—and, using that language, I can now say that a phoneme is an equivalence class of the relation of being isosemantic, and a number is an equivalence class of being equinumerous.

That's one of my main points, and when I learned of it, I felt that I gained a thrilling insight—that's what I mean by a thrill of abstraction. The notion of equivalence relation is one of the basic building blocks out of which all mathematical thought is constructed. It is simple, it is general, it is widely applicable, and it is 100% explicit and precise. And, what's more, it has nothing to do with columns of numbers or triangles or electronic computers or whatever mathematics is sometimes thought to be about—it is abstract pure thought.

Now, about "extensionalism"—there I am not sure I can explain what I feel. In a short sentence what I am trying to say is that to a mathematician—well, in any event, to me—a concept IS its extension. Consider, for an example, the number 5. What is it? Not "What does it do?", "How can it be used?", or "How can I tell it apart from others?", but "What IS it?" Mathematicians usually ask such questions: their insistence on definitions and their insistence on complete precision in the definitions and complete consistency in their use is one of the distinguishing features of their art. The "extension" of a property (an old, established philosophical term) is the class of all objects that possess it. Thus, the extension of "blue" is the class of all blue things—the sky, the Danube, the books, the neckties, whatever—whatever—everything that happens to be blue. The extension of "5" is the class of all quintuples—basketball team, fingers of a hand, whatever. A cautious lexicographer might be willing to go this far with the mathematician: very well, he might say, 5 is the property that is common to all quintuples. The rigorous mathematician would consider that pussyfooting, however. Just what, pray, is a "property"?, he would ask. And how dare we speak of "the" common property of the set of all quintuples—how do we know there is only one? No, sir!, he would say: all I really know about fiveness is that I am willing to assert it of the fingers on my right hand, and, extending from there, of any other set that I can put in one-to-one correspondence with those fingers. In other words, he would say, I know the extension of fiveness, and that's all I know about it. The only courageous way to define 5, therefore, is to follow the principle that a concept IS its extension—and, as a religiously observant extensionalist, I therefore *define* 5 to be the equivalence class of equinumerousness to which the set of fingers on my right hand belongs.

There is something cold and forbidding, something impersonal and frightening, about this definition—one might feel that while it is intellectually, legalistically defensible, it somehow misses the essence of the concept being defined. It reminds me of the classical, and equally unsatisfying definition of a man as a "featherless biped". When I first heard that I objected. Surely, I thought, there is more to humanity than that. What about soul, what about humor, what about art, culture, technology, war, friendship, motherhood—what about all these "essential" characteristics of humanity—doesn't a cold-blooded definition such as "featherless biped" miss them all, and therefore miss the point? After many years of becoming used to the idea, I no longer feel that discomfort in the presence of an extensionalist definition. If it is indeed true (I repeat: *if* it is indeed true—I am not asserting that it is) that humanity is coextensive with the class of featherless bipeds, then humanity is the class of featherless bipeds. And, similarly, since "fiveness" jolly well is coextensive with the class of all quintuples, I happily embrace the definition according to which 5 *is* that class.

## Dreamers and Non-Constructive Proofs

It's about time I turned to the second of the three questions that I raised, in order to describe a second, very different, basic mathematical belief and behavior. Is there, I asked, a number that when multiplied by itself 5 times gives the same result as adding 2 to it? There are those, both among dreamers and among very practical people, who would answer that question by yes only if they could explicitly produce a number with the property described, or, at the very least, in the worst case, if they could explicitly describe an algorithm, a procedure of calculation, that will produce such a number. Thus, for instance, if I change the number 2 in the problem to 240, and if I go on to observe that $3^5 = 243$, which is the same as $3 + 240$, then, I think, we would all agree that the changed question has been answered, and answered in the affirmative.

There is, however, another way to answer such questions, the way of non-constructive proofs, of which I'll give a modest example. Imagine that I have an ultra-efficient but not especially intelligent computer, programmed to tell me instantaneously which is greater, $x^5$ or $x + 2$, whenever I ask it about any particular whole number $x$, but that knows about whole numbers only. All right, I say to the computer, let's go: $x = 0$. It says: $x + 2$ is greater. I say: $x = 1$. It says: $x + 2$ is greater. I say: $x = 2$. It says: $x^5$ is greater. I say: Hurray!—the game is over, and the answer is yes. That's right, isn't it? If I imagine myself moving along the line, scanning all the numbers from 0 on up, and if I know that somewhere (say, when $x = 1$) $x + 2$ is the larger of $x^5$ and $x + 2$, and somewhat later (when $x = 2$) $x^5$ is the larger, then, by an intuitively obvious and rigorously provable property of continuity I can rest assured that somewhere in between $x^5$ and $x + 2$ will be exactly equal.

What do I know now that I didn't know before? Do I know a number $x$ such that $x^5 = x + 2$? No, I don't. All I know, but that I know for sure, is that although I am not (not yet!) able to construct one, such a number does exist.

I have just given, as I promised, a modest example of a non-constructive existence proof. I call it a "modest" example, because, as a matter of fact, with a little trouble it can be converted into a concrete algorithm that will produce a number of the kind that is wanted as accurately as desired: for the benefit of the reader who is just dying of curiosity, I'll put on record that rounded off to five decimals the answer is 1.26717.

Genuine non-constructive existence proofs, the kind that cannot be converted into a computational procedure, are sometimes a source of heated debates in the mathematical family. They are impressive demonstrations of human ingenuity and of the depth of mathematical thought. Sometimes, for instance, in order to prove that a certain set (such as the set of points on the number line) contains at least one object of a particular kind (such as a number $x$ for which $x^5 = x + 2$), a mathematician might use a "stochastic" method. That's a complicated concept whose detailed description would take us too far afield, but in qualitative terms it means something like this. Design a gambling game, a dice game, say, whose possible outcomes are the objects in the set under consideration. Using the properties that are demanded of the particular objects whose existence is in question, compute the probability that the gambling game will produce one of those objects. If that probability turns out to be a positive number (in other words, not 0), then we can be sure that the set of desired objects is not empty—objects like that must exist—even though the method of proof doesn't even yield a hope this side of heaven of ever concretely exhibiting one.

The stochastic method is a much fairer example of a non-constructive existence proof than the "modest" one based on continuity. Many non-constructive existence proofs use some notion of the "size" of a set (such as probability, or dimension, or even just cardinal number), and achieve their end by proving that the size of the set of objects not known to exist is large—large enough to guarantee that it is not zero!

## Schubfachprinzip

The very first question that I asked (remember?—the handshake question?) can be used to illustrate a third basic principle of thrilling, pure, abstract mathematical thought, the so-called Schubfachprinzip, or pigeonhole principle, but I think I'll yield to my congenital tendency to mathematical sadism, and let that question stand as a puzzle for you—I'll use a different question to explain the Schubfachprinzip.

Suppose that a bunch of us are together in a room, 100 of us, say, and we form temporarily a small society of our own. In this closed society there are a certain number of acquaintanceships: some of us are acquainted with some others. I don't know which ones of us are acquainted with which others, but I'm sure of one thing: I'll bet that there are at least two of us that have the same number of acquaintances.

Believe it? Let's see if I can make it convincing. Suppose that someone asked us, each of us, myself included, "How many other people in this closed society are you acquainted with?" We could all tell him an answer, somewhere between 0 and 100. No, wait a minute. If there are exactly 100 of us, then nobody is acquainted with 100 *other* people; the largest number can be no larger than 99. As far as 0 is concerned, that's all right, there could well be some hermits among us, but it's not likely, and, in any event I can easily settle that case. If there are two or more hermits, then I've already won my bet: any two hermits have the same number of acquaintances. If there is only one hermit, then let's ostracize him—let's not count him—let's go so far as to pretend that he isn't here. I must still prove that among the remaining 99 there are two of us with the same number of acquaintances, and I'll do so—but because 100 is easier to say than 99, let me assume that even if the possible hermit is not counted there are still 100 of us left.

So then, what possible numbers will each of the 100 of us give to the questioner? Answer: any number between 1 and 99 inclusive. What is it that I am betting? Answer: that two of us will give the questioner the same number. Indeed: how could we fail? There are only 99 numbers to tell him and there are 100 of us telling: there must be at least one repetition.

Isn't that pretty? I think it is, and, by the way, it is an application of the impressive sounding but childishly easy Schubfachprinzip. The principle says that if we have a number of pigeonholes, and if there are more letters than pigeonholes, then at least one pigeonhole will end up with more than one letter in it. That childishly easy principle is still another basic building block of mathematics—it occurs over and over again, sometimes in very sophisticated contexts, and it is the backbone of all so-called finite or combinatorial mathematics.

Note that the three basic principles that I have described so far are of three different kinds. "Equivalence relation" is a concept; "non-constructive existence proof" is a technique (and an attitude); and the Schubfachprinzip is a theorem, a fact (with, to be sure, many millions of applications and very different-sounding special cases). I could have, and for greater clarity I feel sure I should have, given other examples of the domains of application of the three basic principles already mentioned, and, by the same token, I could and should have given other principles, that arise in other problems. Anything like completeness in a discussion such as this is impossible in a few pages—but perhaps I could do more justice to both the subject and the reader by at least mentioning what else could have been said.

Thus, for instance, is it obvious that the face of a clock is, in effect, a picture of an equivalence relation? (I have in mind the relation between two numbers that holds when one is obtained from the other by adding 12 to it, or, for that matter, any multiple of 12—so that the 13 o'clock is the "same" as 1 o'clock.) Or is it obvious that round-off downward (permitted by the Internal Revenue Service, or so I am informed, when we calculate our income tax) defines an equivalence relation? (In this sense a tax of $317.23 is equivalent to $317.00; more generally two possible calculated taxes are equivalent if ignoring the pennies, any number of them from 1 to 99, makes them equal.)

As for examples of non-constructive existence proofs: many of them depend on the famous (for some people infamous) law of excluded middle. Do we want to prove that a certain mathematical construct "exists"? Very well—let us assume that there is no number, or triangle, or whatever, that satisfies the definition we are working with; let us proceed to reason from that assumption, and, if we're lucky, we shall presently arrive at a contradiction. Conclusion: non-existence is untenable, and at least one instance of the object must indeed exist. This kind of non-constructive existence proof makes the people who don't believe in it angrier than most other kinds.

## Do Normal Numbers Exist?

Other examples of non-constructive proofs can occur in the theory of the so-called "transcendental numbers" (there are, in the sense of Cantor's set theory, "more" transcendental numbers than non-transcendental ones, hence there must be at least one), and in the theory of "normal" numbers (the "length" of the set of normal numbers, or, in other words, the "probability" that a number be normal, is not zero, and hence there must be at least one such number).

The last thing I mentioned is sufficiently interesting that I am strongly tempted to go into a bit of technical detail. I promise it won't last long.

In this discussion the "numbers" I want to consider are the proper fractions—the positive numbers that have no whole number part, such as

.5000000 . . . ,
.3333000 . . . ,
.3333333 . . . ,
.142857142857 . . . ,
.12345678901234567890 . . . .

When we look at the decimal form of such a number, we can ask how often does the digit 8 occur in that form, in the long run average? The answer for the first three numbers is "never"—8 just isn't in the act. The answer for the fourth number is "one sixth of the time". Isn't that clear? There is exactly one 8 in each successive group of six digits; among the first million digits the number of 8's is approximately one sixth, and if we replace "million" by more and more, the approximation to one sixth becomes more and more nearly perfect. For the last number in the list, the answer is "one tenth"; the reasoning is the same as before.

There are ten digits available to us, and we might consider that a number is "fair" if it treats each of them the same as all others—in other words, if each of the ten digits occurs in that number exactly one tenth of the time, in the long run average. In this sense only the fifth of my five sample numbers is fair.

There is a more sophisticated notion of fairness, however, according to which none of my sample numbers is fair. To illustrate what I mean, let me ask this question. Given a number (in decimal form, with no whole number part), how often do the digits 5 and 7 appear in it, next to each other, in that order (in the same long run sense as before)? The answer is "never" in all my examples, except the fourth, and in that case it is "one sixth". To see what I mean, go along the digits, count all "blocks" of length two, and keep track of what proportion of them are "57".

What should the answer be if we are to regard the number as fair, fair not only to each individual digit, but fair to each conceivable pair as well? The answer depends on how many possible pairs there are—and the answer to that is 100. Clear? Sure it is: just count them, from 00, 01, 02, . . . , 09, 10, . . . , to 97, 98, 99. One hundred it is, and, consequently, the only way a number can be "pair fair" is if it has each possible pair in it one hundredth of the time (in the long run average).

Can we write down a number that is fair to each digit and to each pair of digits as well? Sure we could, with paper, pencil, and some time—but as soon as the task was finished, I'd be ready with a new question that demands to be asked. The new question is about triples, such as 293. I would now refuse to call a number fair unless it treated fairly each digit (with frequency one in ten), each pair (with frequency one in a hundred), and each triple (with frequency—surely the answer is guessable—one in a thousand). And once the pattern is clear, I can continue it: in my infinite greed for justice I'll demand an absolutely fair number, by which I mean one in which all blocks of all lengths occur with the "right" frequency (one in ten, or hundred, or thousand, or ten thousand, etc., for singles, doubles, triples, quadruples, etc.). The usual technical, mathematical name is not "absolutely fair" but "normal", and now we've got a question, a bona fide, hard, juicy mathematical question. Can all these, infinitely many, conditions be satisfied simultaneously? In other words: do there exist any normal numbers?

That one I don't think most people can do, not unless they are professional card-carrying dues-paying members of the mathematician's union. But the mathematician who is not afraid of non-constructive existence proofs, and who has a small amount of training in modern probability theory, can sail right through it. All he needs to do is to consider the process of choosing a number at random, by, for instance, throwing an arrow randomly at the segment of the number line that lies between 0 and 1, compute the probability that the number he hits is normal, and observe that the answer is not 0. The computation is not trivial—that's where some mathematical technique is really needed. The probability that it yields is not only different from 0, but it is as different as it could possibly be: it is equal to 1. In other words: it is almost certain that a randomly chosen number will be normal—which surely guarantees that normal numbers do exist.

The number of what I have called "basic mathematical principles" is surprisingly small. No one has ever listed them, and it would be a risky, controversial thing to do, but most mathematicians agree that mathematics is a unit—it all hangs together, with all subjects interwoven, and all concepts applicable everywhere—the number of bricks needed to build such a marvelously compact edifice cannot be very large.

That's one general comment; I'd like to make one more. I have been discussing the thrills of abstraction, and, in particular, the thrills of mathematics, which most people consider very abstract indeed. Would it be a contradiction if I now said that mathematics is an experimental science? I do think that mathematics is abstract, and I do think that mathematics is an experimental science, and I do not think that those two beliefs contradict one another.

To solve a mathematical problem is not a deductive act—it is a matter of guessing, of trial and error, of experiment. To solve the handshake problem, for instance, for five couples, we could do a lot worse than just plain guess. Guess, for instance, that the answer is 7, and then try it out and see what, if anything, is wrong with that guess. Another procedure, a more dignified one that more nearly deserves to be called an experiment, is to vary the conditions and try to solve some related but, we hope, easier problem. What, for instance, happens to the handshake question if we ask it for only four couples? or three? or two? or even just one?

## Abstractions Are Facts

That's the sort of way that a typical working mathematician proceeds—his attitude is not that of creation but of discovery. The answer is there somewhere, and we have no control over what it is—all we are trying to do is find it. The concepts, techniques, and theorems are abstract all right—but our learning about them proceeds the same way as our learning about the boiling point of a chemical and the acceleration of a falling body. The abstractions are *facts*, facts outside of us, facts that we do not "invent" but that are there for us to find if we can.

Some readers will recognize, of course, that the position I have thereby "proved" is that of an unreconstructed die-hard Platonist, but they won't, I hope, hold that against me. My convictions (please do not call them prejudices) took a long time to grow and I would hate to have to give them up. I am convinced that mathematics is infinite in its extent and applications, yet a unity in its conceptual way of looking at things and describing them; the facts of mathematics are there waiting for us to guess at, experiment on, and finally stumble across; the concepts, the techniques, and the facts are abstract, and, in their very abstraction, one of the most thrilling phenomena of the universe.

P.S. The answer to the handshaking problem is 4.

*Paul R. Halmos is Distinguished Professor of Mathematics at Indiana University, and Editor of the "American Mathematical Monthly." He received his Ph.D. from the University of Illinois and has held positions at Illinois, Syracuse, Chicago, Michigan, Hawaii, and Santa Barbara. He has published numerous books and nearly 100 articles and has been the editor of many journals and several book series. The Mathematical Association of America has given him the Chauvenet Prize and (twice) the Lester Ford award for mathematical exposition. His main mathematical interests are in measure and ergodic theory, algebraic logic, and operators on Hilbert space.*

# CHAPTER IV

Reprinted From
*Scientific American*
Vol. 196, pp. 88–89, 1957

# NICOLAS BOURBAKI[1]

PAUL R. HALMOS

His name is Greek, his nationality is French, and his history is curious. He is one of the most influential mathematicians of the 20th century. The legends about him are many, and they are growing every day. Almost every mathematician knows a few stories about him and is likely to have made up a couple more. His works are read and extensively quoted all over the world. There are young men in Rio de Janeiro almost all of whose mathematical education was obtained from his works, and there are famous mathematicians in Berkeley and in Göttingen who think that his influence is pernicious. He has emotional partisans and vociferous detractors wherever groups of mathematicians congregate. The strangest fact about him, however, is that he does not exist.

This nonexistent Frenchman with the Greek name is Nicolas Bourbaki (rhymes with *Poo-Bah-key*). The fact is that Nicolas Bourbaki is a collective pseudonym used by an informal corporation of mathematicians. (The charming French phrase for corporation, "anonymous society", is quite apt here.) The pseudonymous group is writing a comprehensive treatise on mathematics, starting with the most general basic principles and to conclude, presumably, with the most specialized applications. The project got under way in 1939, and 20 volumes (almost 3,000 pages) of the monumental work have appeared.

Why the authors chose to call themselves Bourbaki is shrouded in mystery. There is reason to think that their choice was inspired by an army officer of some importance in the Franco-Prussian War. General Charles Denis Sauter Bourbaki was quite a colorful character. In 1862, at the age of 46, he was offered a chance to become the King of Greece, but he declined the opportunity. He is remembered now mainly for the unkind way the fortunes of war treated him. In 1871, after fleeing from France to Switzerland with a small remnant of his army, he was interned there and then tried to shoot himself. Apparently he missed, for he is reported to have lived to the venerable age of 83. There is said to be a statue of him in Nancy. This may establish a connection between him and the mathematicians who are using his name, for several of them were at various times associated with the University of Nancy.

[1]An article from *Scientific American*, Vol. 196, No. 5, May, 1957.

One of the legends surrounding the name is that about 25 or 30 years ago first-year students at the Ecole Normale Supérieure (where most French mathematicians get their training) were annually exposed to a lecture by a distinguished visitor named Nicolas Bourbaki, who was in fact an amateur actor disguised in a patriarchal beard, and whose lecture was a masterful piece of mathematical double-talk.

It is necessary to insert a word of warning about the unreliability of most Bourbaki stories. While the members of this cryptic organization have taken no blood oath of secrecy, most of them are so amused by their own joke that their stories about themselves are intentionally conflicting and apocryphal. Outsiders, on the other hand, are not likely to know what they are talking about: they can only report an often-embellished legend. The purpose of this article is to describe Bourbaki's scientific accomplishments and relate a few samples of the stories told about him (them). Some of the stories are unverifiable, to say the least, but that doesn't make them any less entertaining.

Scientific publication under a pseudonym is not, of course, original with this group. The English statistician William Sealy Gosset published his pioneering work on the theory of small samples under the name of "Student", probably to avoid embarrassing his employers (the brewers of Guinness). At about the time Bourbaki was starting up, another group of wags invented E. S. Pondiczery, a purported member of the Royal Institute of Poldavia. The initials (E.S.P., R.I.P.) were inspired by a projected but never-written article on extrasensory perception. Pondiczery's main work was on mathematical curiosa. His proudest accomplishment was the only known use of a second-degree pseudonym. Submitting a paper on the mathematical theory of big-game hunting to *The American Mathematical Monthly*, Pondiczery asked in a covering letter that he be allowed to sign it with a pseudonym, because of the obviously facetious nature of the material. The editor agreed, and the paper appeared (in 1938) under the name of H. Pétard.

Primitive tribes, and occasionally scientists, may find magic in a name. This accounts for a publication which would never have been conceived if the authors' names had been different. George Gamow and his friend Hans Bethe saw and took advantage of a wonderful opportunity when a bright young physicist with an unusual name appeared on the scene. On April 1, 1948, they published in *The Physical Review* a perfectly straight-faced paper on the origin of chemical elements whose only unusual feature was the by-line. It read, of course, Alpher, Bethe and Gamow.

While we are on the subject of articles appearing under strange names, it is appropriate to mention the case of Maurice de Duffahel. This gentleman achieved mathematical immortality by the very simple device of publishing, under his own name, some of the classical papers of the great masters. He made only the feeblest attempt to disguise his activities. In 1936 he republished as his own a paper which had been published only 24 years earlier by Charles Emile Picard. Duffahel's version was identical with Picard's, word for word, symbol for symbol, except for one omission: he left out, for understandable reasons, a footnote in which Picard had referred to one of his earlier papers.

Scholarship eventually caught up with Duffahel. You can fool some editors some of the time, but you can't fool all reviewers all of the time. A reviewer of the Picard–Duffahel paper happened to know the works of Picard well enough to recognize the repetition, and Duffahel's publishing career came to an abrupt end.

The works of Bourbaki do not have to be concealed from the executives of a brewery, they are not mere innocent merriment but serious mathematics, and they are definitely not plagiarized from anyone else. The group originally adopted the pseudonym half in jest and half to avoid a boringly long list of authors on the title page; they continue its use more as a corporate name than as a disguise. The names of the members are an open secret to most mathematicians. The membership of Bourbaki, like that of most corporations, changes from time to time, but the style and the spirit of the work stay the same. It is handy to be able to describe a certain style and spirit by one adjective (the accepted term is Bourbachique) rather than by a reference to the "young French school" or a similar circumlocution.

Bourbaki's first appearance on the scene was in the middle 1930s, when they began to publish notes, reviews, and other papers in the *Comptes Rendus* of the French Academy of Sciences and elsewhere. The major treatise on which they later embarked was explained in a paper which was translated into English and printed (in 1950) in *The American Mathematical Monthly* under the title "The Architecture of Mathematics." A footnote reads:

> Professor N. Bourbaki, formerly of the Royal Poldavian Academy [shades of Pondiczery!], now residing in Nancy, France, is the author of a comprehensive treatise of modern mathematics, in course of publication under the title *Eléments de Mathématique* (Hermann et Cie., Paris, 1939–), of which 10 volumes have appeared so far.

The paper, by the way, is an interesting statement of Bourbaki's view of the concept of "structure" in mathematics; it is a masterful description of the Bourbaki spirit. Another paper, which appeared in *The Journal of Symbolic Logic* for 1949, has the ambitious title "Foundations of Mathematics for the Working Mathematician". It is quite technical, but the authors' personality shows through the symbolism. It concludes:

> On these foundations I state that I can build up the whole of the mathematics of the present day; and if there is anything original in my procedure, it lies solely in the fact that, instead of being content with such a statement, I proceed to prove it in the same way as Diogenes proved the existence of motion; and my proof will become more and more complete as my treatise grows.

This paper gives the author's home institution as the "University of Nancago" (Nancy plus Chicago). The main reason for the combination is that one of the founding fathers is now on the staff of the University of Chicago. His name is André Weil (and he is, by the way, the brother of the well-known religious mystic Simone Weil). Although André Weil is not known to the general public, many of his colleagues are prepared to argue that he is the world's greatest living mathematician. His work on algebraic number theory and algebraic geometry is profound and important; his influence on the development of 20th-century mathematics is great, and even some of his more offhand contributions (for example, uniform structures and harmonic analysis on topological groups) have opened up new directions and inspired further researches. Nancago, incidentally, crops up again in a new series of advanced mathematical books which is being published under the impressive heading *Publications de l'Institut Mathématique de l'Université de Nancago*.

According to one of the Bourbaki legends, their major work, whose general title is *Elements of Mathematics*, owes its origin to a conversation between Weil and Jean Delsarte on how calculus should be taught. Whatever the motivation of the work may originally have been, its present purpose is certainly not elementary pedagogy. It is as if a discussion of the best way to teach an understanding of popular music gave rise to a complete treatise on harmony and musicology. (Mathematicians consider the calculus to be as "trivial" as musicians consider the music of Victor Herbert.) Bourbaki's treatise (written in French) is a survey of all mathematics from a sophisticated point of view.

The whole will presumably consist of several parts, but the 20 volumes that have appeared so far do not even complete Part I, titled *The Fundamental Structures of Analysis*. The names of the six subdivisions of Part I are a mild shock to the layman (or the classical mathematician) who thinks of mathematics in terms of arithmetic, geometry and other such old-fashioned words. They are: (1) Set Theory, (2) Algebra, (3) General Topology, (4) Functions of a Real Variable, (5) Topological Vector Spaces, and (6) Integration.

Each volume comes provided with a loose insert of four pages constituting a set of directions on the proper use of the treatise. They go into detail about the necessary prerequisites for reading the treatise (about two years of university mathematics), describe the organization of the work and specify the "rigorously fixed logical order" in which the chapters, books, and parts must be read. The directions also explain the authors' pedagogical tricks, and some of them are very good tricks indeed. One trick, which other authors could profitably copy, is to warn the reader whenever the subject becomes especially slippery, that is, when he is likely to fall into an error: the slippery passages are flagged by a conspicuous S-curve ("dangerous turn") in the margin.

A less admirable Bourbachique trick is their slightly contemptuous attitude toward the substitution of what they call "abuses of language" for technical terms. It is generally admitted that strict adherence to rigorously correct terminology is likely to end in being pedantic and unreadable. This is especially true of Bourbaki, because their terminology and symbolism are frequently at variance with commonly accepted usage. The amusing fact is that often the "abuse of language" which they employ as an "informal" replacement for a technical term is actually conventional usage: weary of trying to remember their own innovation, the authors slip comfortably into the terminology of the rest of the mathematical world.

Almost every Bourbaki volume contains an excellent set of exercises. Mathematics cannot be learned passively, and the Bourbaki exercises are a challenge to activity. The authors used a lot of ingenuity in inventing new exercises, and in rephrasing and rearranging old ones. As a matter of policy they usually do not give credit to the original authors of the exercises they have revised, but no one seems to mind. A mathematician is even likely to consider it an honor to have one of his papers "stolen" by Bourbaki and used as an exercise.

The Bourbaki gadgetry includes foldout sheets that summarize important definitions and assumptions, a dictionary for each book that serves also as a comprehensive index and a guide to both non-Bourbachique terminology and basic Bourbakese. The only important thing missing is adequate bibliographical guidance. The Bourbaki presentation of each subject is systematic and thorough, and often includes a brilliant historical review of the subject. But the historical essays make only a few grudging references to the classics and fail almost entirely to mention the sources of the modern contributions. No deception is intended (Bourbaki does not claim to have discovered all of modern mathematics), but the practice may have the effect of confusing the future mathematical historian.

These are the external trappings of Bourbaki. The Bourbaki style and spirit, the qualities that attract friends and repel enemies, are harder to describe. Like the qualities of music, they must be felt rather than understood.

One of the things that attracted students to Bourbaki from the start was that they gave the first systematic account of some subjects (for example, general topology and multilinear algebra) which were not available anywhere else in book form. Bourbaki pioneered in reducing to orderly form a large mass of papers which had appeared over several decades in many journals and in several languages. The main features of the Bourbaki approach are a radical attitude about the right order for doing things, a dogmatic insistence on a privately invented terminology, a clean and economical organization of ideas, and a style of presentation which is so bent on saying everything that it leaves nothing to the imagination and has, consequently, a watery, lukewarm effect.

A typical sample of the thoroughness and leisurely pace of the Bourbaki treatment is their approach to defining the number "1". They devote almost 200 pages to preparation before they get to the definition itself. They then define the number 1 in terms of highly condensed abbreviating symbols, explaining in a footnote that the unabbreviated form of the definition in their system of notation would require several tens of thousands of symbols. In all fairness to Bourbaki it must be said that modern mathematical logicians have known for some time that concepts such as the number 1 are not so elementary as they look.

How does a cooperative work of this magnitude ever get written? A large part of the credit goes to Jean Dieudonné (originally from Nancy, now at Northwestern University) who has been Bourbaki's chief scribe almost from the beginning. Since Dieudonné is a prolific writer on mathematics under his own name, there is a certain difficulty about distinguishing his private work from his efforts for Bourbaki. According to one story, he manages to keep the record straight in a truly remarkable manner. The story is that Dieudonné once published, under Bourbaki's name, a note which later was found to contain a mistake. The mistake was corrected in a paper entitled "On an Error of M. Bourbaki" and signed Jean Dieudonné.

The membership of Bourbaki seems to vary between 10 and 20. With one conspicuous exception all the members have always been French. The exception is Samuel Eilenberg (originally from Warsaw, now at Columbia University). Known to the friends of his youth as $S^2P^2$ (for Smart Sammy the Polish Prodigy), Eilenberg is a charming extrovert who learned more about the U.S. within six months of his arrival than most Americans ever find out. (One of the first things he did was to go on an extended hitchhiking tour.) Since he speaks French like a native and knows more about algebraic topology than any Frenchman, the unwritten rule restricting Bourbaki to Frenchmen was waived to admit him.

The French orientation of Bourbaki is not mere chauvinism but a linguistic necessity (since Frenchmen started it). When a collection of prima donnas such as Weil, Dieudonné, Claude Chevalley, and Henri Cartan get together with their colleagues, the rate and volume of the flow of French is impressive. To follow and take part in the conversation under such circumstances you must not only speak French fast and loud, but you must know the latest Parisian student slang. Even if everyone in the room fulfills these conditions, it is difficult to see how any work ever gets done at the famous Bourbaki congresses. But it does get done. The members convene each year, usually in some pleasant French vacation spot, to make major policy decisions. Since their treatise has proved a commercial success (to Bourbaki's considerable surprise), there is ample royalty money to pay travel expenses and to provide the French food and wines that lubricate the proceedings. (The commercial success, by the way, is due mainly to the American market. Four of the five senior members of Bourbaki are now residents of the U.S.)

A lot of work goes into the preparation of a Bourbaki volume. Once a particular project has been decided on, some member agrees to write the first draft. In so doing he lets himself in for a trying experience. When his draft is finished, it is duplicated and copies are sent around to the others. At the next congress the draft is mercilessly criticized, and, quite possibly, completely rejected. The first draft of the Bourbaki book on integration, for instance, was written by Dieudonné and became known as "Dieudonné's monster". Rumor has it that in spirit and content Dieudonné's monster was very similar to a well-known American book on the subject, written by an author whose name we shall simply give here as Blank. Dieudonné's monster was never published; his confreres hooted it down. What settled the matter was Weil's snort: "If we're going to do that sort of thing, let's just translate Blank's book into French and have done with it."

After the first draft has been dealt with, a second draft is begun, possibly by a different member. The process goes on and on: six or seven drafts are not unknown. The result of this painstaking work is not a textbook that it is safe to put into a beginner's hands (even Bourbaki admits that), but it is a reference book, almost an encyclopedia, without which 20th-century mathematics would be, for better or for worse, quite different from what it is.

Bourbaki's youthful exuberance augurs well for the future of their labors, but it is one of the main annoyances to their enemies. The officials of the American Mathematical Society were not amused when they received an application for membership signed N. Bourbaki. They considered the joke sophomoric, and they rejected the application. The secretary of the Society coldly suggested that Bourbaki might apply for an institutional membership. Since the dues for institutional membership are substantially higher than those for individuals, and since Bourbaki did not wish to admit that he does not exist, nothing more was heard of the matter.

Yes, the joke may be sophomoric, but sophomores are young, and mathematics is a young man's profession. Bourbaki's emphasis on youth is laudable. Upon reaching the age of 50 recently, Dieudonné and Weil, though founding fathers of Bourbaki, announced their retirement from the group. They had declared their intention to get out at 50, and they kept their promise.

It is appropriate to conclude by warning the reader to be on the lookout for Bourbaki-inspired rumors about the author of this article, and to be prepared to take such rumors with a generous grain of salt. The corporation does not like to have its secrets told in public, and it has demonstrated its ability to take effective measures against informers. To be sure, the fiction has been exposed in print before this. In 1949 André Delachet, in his little book on mathematical analysis, referred to the "polycephalic mathematician" N. Bourbaki, and went so far as to mention some of the heads by name. A year or two before that, the *Book of the Year* of the *Encyclopaedia Britannica* had a brief paragraph about Bourbaki as a group. The author of the paragraph was Ralph P. Boas, then executive editor of the journal *Mathematical Reviews*, now a colleague of Dieudonné at Northwestern. Soon afterward, the editors of the *Britannica* received an injured letter signed by N. Bourbaki, protesting against Boas's allegation of Bourbaki's nonexistence. The editors' confusion and Boas's embarrassment were not reduced when a member of the University of Chicago mathematics department wrote a truthful but shrewdly worded letter, implying, but not saying, that Bourbaki did indeed exist. The situation was cleared up for the editors by a letter from the secretary of the American Mathematical Society (the same secretary who had refused to approve Bourbaki's membership application).

Bourbaki got its revenge. Calling forth all its polycephalic, international resources, the corporation circulated a rumor that *Boas* did not exist. Boas, said Bourbaki, is the collective pseudonym of a group of young American mathematicians who act jointly as the editors of *Mathematical Reviews*.

Reprinted from

AMERICAN SCIENTIST, *56*, **4**, pp. 375–389, 1968

# MATHEMATICS AS A CREATIVE ART

By P. R. HALMOS*

Do you know any mathematicians—and, if you do, do you know anything about what they do with their time? Most people don't. When I get into conversation with the man next to me in a plane, and he tells me that he is something respectable like a doctor, lawyer, merchant, or dean, I am tempted to say that I am in roofing and siding. If I tell him that I am a mathematician, his most likely reply will be that he himself could never balance his check book, and it must be fun to be a whiz at math. If my neighbor is an astronomer, a biologist, a chemist, or any other kind of natural or social scientist, I am, if anything, worse off—this man *thinks* he knows what a mathematician is, and he is probably wrong. He thinks that I spend my time (or should) converting different orders of magnitude, comparing binomial coefficients and powers of 2, or solving equations involving rates of reactions.

C. P. Snow points to and deplores the existence of two cultures; he worries about the physicist whose idea of modern literature is Dickens, and he chides the poet who cannot state the second law of thermodynamics. Mathematicians, in converse with well-meaning, intelligent, and educated laymen (do you mind if I refer to all non-mathematicians as laymen?) are much worse off than physicists in converse with poets. It saddens me that educated people don't even know that my subject exists. There is something that they call mathematics, but they neither know how the professionals use that word, nor can they conceive why anybody should do it. It is, to be sure, possible that an intelligent and otherwise educated person doesn't know that egyptology exists, or haematology, but all you have to tell him is that it does, and he will immediately understand in a rough general way why it should and he will have some empathy with the scholar of the subject who finds it interesting.

Usually when a mathematician lectures, he is a missionary. Whether he is talking over a cup of coffee with a collaborator, lecturing to a graduate class of specialists, teaching a reluctant group of freshman engineers, or addressing a general audience of laymen—he is still preaching and seeking to make converts. He will state theorems and he will discuss proofs and he will hope that when he is done his audience will know more mathematics than they did before. My aim today is different—I am not

* This article is based on an informal lecture delivered at the University of Illinois on December 12, 1967, as part of the celebration of the Centennial year of the University. The author, an alumnus of Illinois, was a Professor of Mathematics at the University of Michigan and is now at the University of Hawaii.

here to proselyte but to enlighten—I seek not converts but friends. I do not want to teach you what mathematics is, but only *that* it is.

I call my subject mathematics—that's what all my colleagues call it, all over the world—and there, quite possibly, is the beginning of confusion. The word covers two disciplines—many more, in reality, but two, at least two, in the same sense in which Snow speaks of two cultures. In order to have some words with which to refer to the ideas I want to discuss, I offer two temporary and ad hoc neologisms. Mathematics, as the word is customarily used, consists of at least two distinct subjects, and I propose to call them *mathology* and *mathophysics*. Roughly speaking, mathology is what is usually called pure mathematics, and mathophysics is called applied mathematics, but the qualifiers are not emotionally strong enough to disguise that they qualify the same noun. If the concatenation of syllables I chose here reminds you of other words, no great harm will be done; the rhymes alluded to are not completely accidental. I originally planned to entitle this lecture something like "Mathematics is an art," or "Mathematics is not a science," or "Mathematics is useless," but the more I thought about it the more I realized that I mean that "Mathology is an art," "Mathology is not a science," and "Mathology is useless." When I am through, I hope you will recognize that most of you have known about mathophysics before, only you were probably calling it mathematics; I hope that all of you will recognize the distinction between mathology and mathophysics; and I hope that some of you will be ready to embrace, or at least applaud, or at the very least, recognize mathology as a respectable human endeavor.

In the course of the lecture I'll have to use many analogies (literature, chess, painting), each imperfect by itself, but I hope that in their totality they will serve to delineate what I want delineated. Sometimes in the interest of economy of time, and sometimes doubtless unintentionally, I'll exaggerate; when I'm done, I'll be glad to rescind anything that was inaccurate or that gave offense in any other way.

### *What Mathematicians Do*

As a first step toward telling you what mathematicians do, let me tell you some of the things they do not do. To begin with, mathematicians have very little to do with numbers. You can no more expect a mathematician to be able to add a column of figures rapidly and correctly than you can expect a painter to draw a straight line or a surgeon to carve a turkey—popular legend attributes such skills to these professions, but popular legend is wrong. There is, to be sure, a part of mathematics called number theory, but even that doesn't deal with numbers in the legendary sense—a number theorist and an adding machine would find very little to talk about. A machine might enjoy proving that $1^3 + 5^3 + 3^3 = 153$, and it might even go on to discover that there are only five positive integers with the property that the equation indicates (1, 370, 371, 407), but most mathematicians couldn't care less; many mathematicians enjoy and respect the theorem that every positive integer is the sum of not more than four squares, whereas the infinity involved in the word "every" would frighten and paralyze any ordinary office machine, and, in any case, that's probably not the sort of thing that the person who relegates mathematicians to numbers had in mind.

Not even those romantic objects of latter day science fiction, the giant brains, the computing machines that run our lives these days—not even they are of interest to the mathematician as such. Some mathematicians are interested in the logical problems involved in the reduction of difficult questions to the sort of moronic baby talk that machines understand: the logical design of computing machines is definitely mathematics. Their construction is not, that's engineering, and their product, be it a payroll, a batch of sorted mail, or a supersonic plane, is of no mathematical interest or value.

Mathematics is not numbers or machines; it is also not the determination of the heights of mountains by trigonometry, or compound interest by algebra, or moments of inertia by calculus. Not today it isn't. At one point in history each of those things, and others like them, might have been an important and non-trivial research problem, but once the problem is solved, its repetitive application has as much to do with mathematics as the work of a Western Union messenger boy has to do with Marconi's genius.

There are at least two other things that mathematics isn't; one of them is something it never was, and the other is something it once included and by now has sloughed off. The first is physics. Some laymen confuse mathematics and theoretical physics and speak, for instance, of Einstein as a great mathematician. There is no doubt that Einstein was a great man, but he was no more a great mathematician than he was a great violinist. He used mathematics to find out facts about the universe, and that he successfully used certain parts of differential geometry for that purpose adds a certain piquancy to the appeal of differential geometry. Withal, relativity theory and differential geometry are not the same thing. Einstein, Schrödinger, Heisenberg, Fermi, Wigner, Feynman—great men all, but not mathematicians; some of them, in fact, strongly antimathematical, preach against mathematics, and would regard it as an insult to be called a mathematician.

What once was mathematics remains mathematics always, but it can become so thoroughly worked out, so completely understood, and, in the light of millenia of contributions, with hindsight, so trivial, that mathematicians never again need to or want to spend time on it. The celebrated Greek problems (trisect the angle, square the circle, duplicate the cube) are of this kind, and the irrepressible mathematical amateur to the contrary notwithstanding, mathematicians are no longer trying to solve them. Please understand, it isn't that they have given up. Perhaps you have heard that, according to mathematicians, it is impossible to square a circle, or trisect an angle, and perhaps you have heard or read that, therefore, mathematicians are a pusillanimous chicken-hearted lot, who give up easily, and use their ex-cathedra pronouncements to justify their ignorance. The conclusion may be true, and you may believe it if you like, but the proof is inadequate. The point is a small one but a famous one and one of historical interest: let me digress to discuss it for a moment.

*A Short Digression*

The problem of trisecting the angle is this: given an angle, construct another one that is just one third as large. The problem is perfectly easy, and several methods for solving it are known. The catch is that the original Greek formulation of the problem is more stringent: it requires a construction that uses ruler and compasses only. Even that can be done, and I could show you a perfectly simple method in one minute and convince you that it works in two more minutes. The real difficulty is that the precise formulation of the problem is more stringent still. The precise formulation demands a construction that uses a ruler and compasses only and, moreover, severely restricts how they are to be used; it prohibits, for instance, marking two points on the ruler and using the marked points in further constructions. It takes some careful legalism (or some moderately pedantic mathematics) to formulate really precisely just what was and what wasn't allowed by the Greek rules. The modern angle trisector either doesn't know those rules, or he knows them but thinks that the idea is to get a close approximation, or he knows the rules and knows that an exact solution is required but lets wish be father to the deed and simply makes a mistake. Frequently his attitude is that of the visitor from outer space to golf. (If all you want is to get that little white ball in that little green hole, why don't you just go and put it there?)

Allow me to add a short digression to the digression. I'd like to remind you that when a mathematician says that something is impossible, he doesn't mean that it is very very difficult, beyond his powers, and probably beyond the powers of all humanity for the foreseeable future. That's what is often meant when some one says it's impossible to travel at the speed of sound five miles above the surface of the earth, or instantaneously to communicate with someone a thousand miles away, or to tamper with the genetic code so as to produce a race of citizens who are simultaneously intelligent and peace-loving. That's what is belittled by the classic business braggadocio (the impossible takes a little longer). The mathematical impossible is different: it is more modest and more secure. The mathematical impossible is the logical impossible. When the mathematician says that it is impossible to find a positive number whose sum with 10 is less than 10, he merely reminds us that that's what the words mean (positive, sum, 10, less); when he says that it is impossible to trisect every angle by ruler and compasses, he means exactly the same sort of thing, only the number of technical words involved is large enough and the argument that strings them together is long enough that they fill a book, not just a line.

### *The Start of Mathematics*

No one knows when and where mathematics got started, or how, but it seems reasonable to guess that it emerged from the same primitive physical observations (counting, measuring) with which we all begin our own mathematical insight (ontogeny recapitulates phylogeny). It was probably so in the beginning, and it is true still, that many mathematical ideas originate not from pure thought but from material necessity; many, but probably not all. Almost as soon as a human being finds it necessary to count his sheep (or sooner?) he begins to wonder about numbers and shapes and motions and arrangements—curiosity about such things seems to be as necessary to the human spirit as curiosity about earth, water, fire, and air, and curiosity—sheer pure intellectual curiosity—about stars and about life. Numbers and shapes and motions and arrangements, and also thoughts and their order, and concepts such as "property" and "relation"—all such things are the raw material of mathematics. The technical but basic mathematical concept of "group" is the best humanity can do to understand the intuitive concept of "symmetry" and the people who study topological spaces, and ergodic paths, and oriented graphs are making precise our crude and vague feelings about shapes, and motions, and arrangements.

Why do mathematicians study such things, and why should they? What, in other words, motivates the individual mathematician, and why does society encourage his efforts, at least to the extent of providing him with the training and subsequently the livelihood that, in turn, give him the time he needs to think? There are two answers to each of the two questions: because mathematics is practical and because mathematics is an art. The already existing mathematics has more and more new applications each day, and the rapid growth of desired applications suggests more and more new practical mathematics. At the same time, as the quantity of mathematics grows and the number of people who think about it keeps doubling over and over again, more new concepts need explication, more new logical interrelations cry out for study, and understanding, and simplification, and more and more the tree of mathematics bears elaborate and gaudy flowers that are, to many beholders, worth more than the roots from which it all comes and the causes that brought it all into existence.

*Mathematics Today*

Mathematics is very much alive today. There are more than a thousand journals that publish mathematical articles; about 15,000 to 20,000 mathematical articles are printed every year. The mathematical achievements of the last 100 years are greater in quantity and in quality than those of all previous history. Difficult mathematical problems, which stumped Hilbert, Cantor, or Poincaré, are being solved, explained, and generalized by beardless (and bearded) youths in Berkeley and in Odessa.

Mathematicians sometimes classify themselves and each other as either problem-solvers or theory-creators. The problem-solvers answer yes-or-no questions and discuss the vital special cases and concrete examples that are the flesh and blood of mathematics; the theory creators fit the results into a framework, illuminate it all, and point it in a definite direction—they provide the skeleton and the soul of mathematics. One and the same human being can be both a problem-solver and a theory-creator, but, usually, he is mainly one or the other. The problem-solvers make geometric constructions, the theory-creators discuss the foundations of Euclidean geometry; the problem-solvers find out what makes switching diagrams tick, the theory-creators prove representation theorems for Boolean algebras. In both kinds of mathematics and in all fields of mathematics the progress in one generation is breath-taking. No one can call himself a mathematician nowadays who doesn't have at least a vague idea of homological algebra, differential topology, and functional analysis, and every mathematician is probably somewhat of an expert on at least one of these subjects—and yet when I studied mathematics in the 1930's none of those phrases had been invented, and the subjects they describe existed in seminal forms only.

Mathematics is abstract thought, mathematics is pure logic, mathematics is creative art. All these statements are wrong, but they are all a little right, and they are all nearer the mark than "mathematics is numbers" or "mathematics is geometric shapes." For the professional pure mathematician, mathematics is the logical dovetailing of a carefully selected sparse set of assumptions with their surprising conclusions via a conceptually elegant proof. Simplicity, intricacy, and above all, logical analysis are the hallmark of mathematics.

The mathematician is interested in extreme cases—in this respect he is like the industrial experimenter who breaks lightbulbs, tears shirts, and bounces cars on ruts. How widely does a reasoning apply, he wants to know, and what happens when it doesn't? What happens when you weaken one of the assumptions, or under what conditions can you strengthen one of the conclusions? It is the perpetual asking of such questions that makes for broader understanding, better technique, and greater elasticity for future problems.

Mathematics—this may surprise you or shock you some—is never deductive in its creation. The mathematician at work makes vague guesses, visualizes broad generalizations, and jumps to unwarranted conclusions. He arranges and rearranges his ideas, and he becomes convinced of their truth long before he can write down a logical proof. The conviction is not likely to come early—it usually comes after many attempts, many failures, many discouragements, many false starts. It often happens that months of work result in the proof that the method of attack they were based on cannot possibly work, and the process of guessing, visualizing, and conclusion-jumping begins again. A reformulation is needed—and—and this too may surprise you—more experimental work is needed. To be sure, by "experimental work" I do not mean test tubes and cyclotrons. I mean thought-experiments. When a mathematician wants to prove a theorem about an infinite-dimensional Hilbert space, he examines its finite-dimensional analogue, he looks in detail at the 2- and 3-dimensional cases, he often tries out a particular numerical case, and he hopes that he will gain thereby an insight that pure definition-juggling has not yielded. The deductive stage, writing the result down, and writing down its rigorous proof are relatively trivial once the real insight arrives; it is more like the draftsman's work, not the architect's.

### *The Mathematical Fraternity*

The mathematical fraternity is a little like a self-perpetuating priesthood. The mathematicians of today train the mathematicians of tomorrow and, in effect, decide whom to admit to the priesthood. Most people do not find it easy to join—mathematical talent and genius are apparently exactly as rare as talent and genius in painting and music—but anyone can join, everyone is welcome. The rules are nowhere explicitly formulated, but they are intuitively felt by everyone in the profession. Mistakes are forgiven and so is obscure exposition—the indispensable requisite is mathematical insight. Sloppy thinking, verbosity without content, and polemic have no role, and—this to me is one of the most wonderful aspects of mathematics—they are much easier to spot than in the non-mathematical fields of human endeavor (much easier than, for instance, in literature among the arts, in art criticism among the humanities, and in your favorite abomination among the social sciences).

Although most of mathematical creation is done by one man at a desk, at a blackboard, or taking a walk, or, sometimes, by two men in conversation, mathematics is nevertheless a sociable science. The creator needs stimulation while he is creating and he needs an audience after he has created. Mathematics is a sociable science in the sense that I don't think it can be done by one man on a desert island (except for a very short time), but it is not a mob science, it is not a team science. A theorem is not a pyramid; inspiration has never been known to descend on a committee. A great theorem can no more be obtained by a "project" approach than a great painting; I don't think a team of little Gausses could have obtained the theorem about regular polygons under the leadership of a rear admiral anymore than a team of little Shakespeares could have written *Hamlet* under such conditions.

*A Tiny and Trivial Mathematical Problem*

I have been trying to give you a description of what mathematics is and how mathematicians do it, in broad general terms, and I wouldn't blame you if you had been finding it thoroughly unsatisfactory. I feel a little as if I had been describing snow to a Fiji Islander. If I told him snow was white like an egg, wet like mud, and cold like a mountain water fall, would he then understand what it's like to ski in the Alps? To show him a spoonful of scrapings from the just defrosted refrigerator of His Excellency the Governor is not much more satisfactory—but it is a little. Let me, therefore, conclude this particular tack by mentioning a tiny and trivial mathematical problem and describing its solution—possibly you'll then get (if you don't already have) a little feeling for what attracts and amuses mathematicians and what is the nature of the inspiration I have been talking about.

Imagine a society of 1025 tennis players. The mathematically minded ones among you, if you haven't already heard about this famous problem, have immediately been alerted by the number. It is known to any one who ever kept on doubling something, anything, that 1024 is $2^{10}$. All cognoscenti know, therefore, that the presence in the statement of a problem of a number like $1 + 2^{10}$ is bound to be a strong hint to its solution; the chances are, and this can be guessed even before the statement of the problem is complete, that the solution will depend on doubling—or halving—something ten times. The more knowledgeable cognoscenti will also admit the possibility that the number is not a hint but a trap. Imagine then that the tennis players are about to conduct a gigantic tournament, in the following manner. They draw lots to pair off as far as they can, the odd man sits out the first round, and the paired players play their matches. In the second round only the winners of the first round participate, and the whilom odd man. The procedure is the same for the second round as for the first—pair off and play at random, with the new odd man (if any) waiting it out. The rules demand that this procedure be continued, over and over again, until the champion of the society is selected. The champion, in this sense, didn't exactly beat everyone else, but he can say, of each of his fellow players, that he beat some one, who beat some one, ..., who beat some one, who beat that player. The question is: how many matches were played altogether, in all the rounds of the whole tournament?

There are several ways of attacking the problem, and even the most naive one works. According to it, the first round has 512 matches (since

1025 is odd and 512 is a half of 1024), the second round has 256 (since the 512 winners in the first round, together with the odd man of that round, make 513, which is odd again, and 256 is a half of 512), etc. The "etcetera" yields, after 512 and 256, the numbers 128, 64, 32, 16, 8, 4, 2, 1, and 1 (the very last round, consisting of only one match, is the only one where there is no odd man), and all that is necessary is to add them up. That's a simple job that pencil and paper can accomplish in a few seconds; the answer (and hence the solution of the problem) is 1024.

The mathematical wiseacre would proceed a little differently. He would quickly recognize, as advertised, that the problem has to do with repeated halvings, so that the numbers to be added up are the successive powers of 2, from the ninth down to the first,—no, from ninth down to the zeroth!—together with the last 1 caused by the obviously malicious attempt of the problem-setter to confuse the problem-solver by using 1025 instead of 1024. The wiseacre would then proudly exhibit his knowledge of the formula for the sum of a geometric progression, he would therefore know (without addition) that the sum of 512, 256,. . ., 8, 4, 2, and 1 is 1023, and he would then add the odd 1 to get the same total of 1024.

The trouble with the wiseacre's solution is that it's much too special. If the number of tennis players had been 1000 instead of 1025, the wiseacre would be no better off than the naive layman. The wiseacre's solution works, but it is as free of inspiration as the layman's. It is shorter but it is still, in the mathematician's contemptuous word, computational.

The problem has also an inspired solution, that requires no computation, no formulas, no numbers—just pure thought. Reason like this: each match has a winner and a loser. A loser cannot participate in any later rounds; every one in the society, except only the champion, loses exactly one match. There are, therefore, exactly as many matches as there are losers, and, consequently, the number of matches is exactly one less than the membership of the society. If the number of members is 1025, the answer is 1024. If the number had been 1000, the answer would be 999, and, obviously, the present pure thought method gives the answer, with no computation, for every possible number of players.

That's it: that's what I offer as a microcosmic example of a pretty piece of mathematics. The example is bad because, after all my warning that mathematicians are interested in other things than counting, it deals with counting; it's bad because it does not, cannot, exhibit any of the conceptual power and intellectual technique of non-trivial mathematics; and it's bad because it illustrates applied mathematics (that is, mathematics as applied to a "real life" problem) more than it illustrates pure mathematics (that is, the distilled form of a question about the logical interrelations of concepts—concepts, not tennis players, and tournaments, and matches). For an example, for a parable, it does pretty well nevertheless; if your imagination is good enough mentally to reconstruct the ocean from a drop of water, then you can reconstruct mathematics from the problem of the tennis players.

### *Mathology vs. Mathophysics*

I've been describing mathematics, but, the truth to tell, I've had mathology (pure) in mind, more than mathophysics (applied). For some reason the practitioners of mathophysics tend to minimize the differences between the two subjects and the others, the mathologists, tend to emphasize them. You've long ago found me out, I am sure. Every mathematician is in one camp or the other (well, almost every—a few are in both camps), and I am a mathologist by birth and training. But in a report such as this one, I must try not to exaggerate my prejudices, so I'll begin by saying that the similarities between mathology and mathophysics are great indeed. It is a historical fact that ultimately all mathematics comes to us, is suggested to us, by the physical universe: in that sense all mathematics is applied. It is, I believe, a psychological fact that even the purest of the pure among us is just a wee bit thrilled when his thoughts make a new and unexpected contact with the non-mathematical universe. The kind of talent required to be good in mathology is intimately related to the kind that mathophysics demands. The articles that mathophysicists write are frequently indistinguishable from those of their mathological colleagues.

As I see it, the main difference between mathophysics and mathology is the *purpose* of the intellectual curiosity that motivated the work—or, perhaps, it would be more accurate to say that it is the *kind* of intellectual curiosity that is relevant. Let me ask you a peculiar but definitely mathematical question. Can you load a pair of dice so that all possible rolls—better: all possible sums that can show on one roll, all the numbers between 2 and 12 inclusive—are equally likely? The question is a legitimate piece of mathematics; the answer to it is known, and it is not trivial. I mention it here so that you may perform a quick do-it-yourself psychoanalysis on yourself. When I asked the question, did you think of homogeneous and non-homogeneous distributions of mass spread around in curious ways through two cubes, or did you think of sums of products of twelve numbers (the twice six probabilities associated with the twice six faces of the two dice)? If the former, you are a crypto-mathophysicist, if the latter you are a potential mathologist.

How do you choose your research problem, and what about it attracts you? Do you want to know about nature or about logic? Do you prefer concrete facts to abstract relations? If it's nature you want to study, if the concrete has the greater appeal, then you are a mathophysicist. In mathophysics the question always comes from outside, from the "real world," and the satisfaction the scientist gets from the solution comes, to a large extent, from the light it throws on *facts*.

Surely no one can object to mathophysics or think less of it for that; and yet many do. I did not mean to identify "concrete" with "practical" and thereby belittle it, and equally I did not mean to identify "abstract" with "useless." (That $2^{11213} - 1$ is a prime is a concrete fact, but surely a useless one; that $E = mc^2$ is an abstract relation but unfortunately a practical one.) Nevertheless, such identifications—applied-concrete-practical-crude and pure-abstract-pedantic-useless—are quite common in both camps. To the applied mathematician the antonym of "applied" is "worthless," and to the pure mathematician the antonym of "pure" is "dirty."

History doesn't help the confusion. Historically, pure and applied mathematics (mathology and mathophysics) have been much closer together than they are today. By now the very terminology (pure mathematics versus applied mathematics) makes for semantic confusion: it implies identity with small differences, instead of diversity with important connections.

From the difference in purposes follows a difference in tastes and hence of value judgments. The mathophysicist wants to know the facts, and he has, sometimes at any rate, no patience for the hair-splitting pedantry of the mathologist's rigor (which he derides as *rigor mortis*). The mathologist wants to understand the ideas, and he places great value on the aesthetic aspects of the understanding and the way that understanding is arrived at; he uses words such as "elegant" to describe a proof. In motivation, in purpose, frequently in method, and almost always in taste, the mathophysicist and the mathologist differ.

When I tell you that I am a mathologist, I am not trying to defend useless knowledge, or convert you to the view that it's the best kind. I would, however, be less than honest with you if I didn't tell you that I believe that. I like the idea of things being done for their own sake. I like it in music, I like it in the crafts, and I like it even in medicine. I never quite trust a doctor who says that he chose his profession out of a desire to benefit humanity; I am uncomfortable and skeptical when I hear such things. I much prefer the doctor to say that he became one because he liked the idea, because he thought he would be good at it, or even because he got good grades in high school zoology. I like the subject for its own sake, in medicine as much as in music; and I like it in mathematics.

Let me digress for a moment to a brief and perhaps apocryphal story about David Hilbert, probably the greatest mathematician of both the nineteenth and the twentieth centuries. When he was preparing a public address, Hilbert was asked to include a reference to the conflict (even then!) between pure and applied mathematics, in the hope that if anyone could take a step toward resolving it, he could. Obediently, he is said to have begun his address by saying "I was asked to speak about the conflict between pure and applied mathematics. I am glad to do so, because it is, indeed, a lot of nonsense—there should be no conflict, there can be no conflict—there is no conflict—in fact the two have nothing whatsoever to do with one another.!"

It is, I think, undeniable that a great part of mathematics was born, and lives in respect and admiration, for no other reason than that it is interesting—it is interesting in itself. The angle trisection of the Greeks, the celebrated four-color map problem, and Gödel's spectacular contribution to mathematical logic are good because they are beautiful, because they are surprising, because we want to know. Don't all of us feel the irresistible pull of the puzzle? Is there really something wrong with saying that mathematics is a glorious creation of the human spirit and deserves to live even in the absence of any practical application?

### *Mathematics is a Language*

Why does mathematics occupy such an isolated position in the intellectual firmament? Why is it good form, for intellectuals, to shudder and announce that they can't bear it, or, at the very least, to giggle and announce that they never could understand it? One reason, perhaps, is that mathematics is a language. Mathematics is a precise and subtle language designed to express certain kinds of ideas more briefly, more accurately, and more usefully than ordinary language. I do not mean here that mathematicians, like members of all other professional cliques, use jargon. They do, at times, and they don't most often, but that's a personal phenomenon, not the professional one I am describing. What I do mean by saying that mathematics is a language is sketchily and inadequately illustrated by the difference between the following two sentences. (1) If each of two numbers is multiplied by itself, the difference of the two results is the same as the product of the sum of the two given numbers by their difference. (2) $x^2 - y^2 = (x + y)(x - y)$. (Note: the longer formulation is not only awkward, it is also incomplete.)

One thing that sometimes upsets and repels the layman is the terminology that mathematicians employ. Mathematical words are intended merely as labels, sometimes suggestive, possibly facetious, but always precisely defined; their everyday connotations must be steadfastly ignored. Just as nobody nowadays infers from the name Fitzgerald that its bearer is the illegitimate son of Gerald, a number that is called irrational must not be thought unreasonable; just as a dramatic poem called *The Divine Comedy* is not necessarily funny, a number called imaginary has the same kind of mathematical existence as any other. (Rational, for numbers, refers not to the Latin *ratio*, in the sense of reason, but to the English "ratio," in the sense of quotient.)

Mathematics is a language. None of us feels insulted when a sinologist uses Chinese phrases, and we are resigned to living without Chinese, or else spending years learning it. Our attitude to mathematics should be the same. It's a language, and it takes years to learn to speak it well. We all speak it a little, just because some of it is in the air all the time, but we speak it with an accent and frequently inaccurately; most of us speak it, say, about as well as one who can only say *"Oui, monsieur"* and *"S'il vous plaît"* speaks French. The mathematician sees nothing wrong with this as long as he's not upbraided by the rest of the intellectual community for keeping secrets. It took him a long time to learn his language, and he doesn't look down on the friend who, never having studied it, doesn't speak it. It is however sometimes difficult to keep one's temper with the cocktail party acquaintance who demands that he be taught the language between drinks and who regards failure or refusal to do so as sure signs of stupidity or snobbishness.

### *Some Analogies*

A little feeling for the nature of mathematics and mathematical thinking can be got by the comparison with chess. The analogy, like all analogies, is imperfect, but it is illuminating just the same. The rules for chess are as arbitrary as the axioms of mathematics sometimes seem to be. The game of chess is as abstract as mathematics. (That chess is played with solid pieces, made of wood, or plastic, or glass, is not an intrinsic feature of the game. It can just as well be played with pencil and paper, as mathematics is, or blindfold, as mathematics can.) Chess also has its elaborate technical language, and chess is completely deterministic.

There is also some analogy between mathematics and music. The mathologist feels the need to justify pure mathematics exactly as little as the musician feels the need to justify music. Do practical men, the men who meet payrolls, demand only practical music—soothing jazz to make an assembly line worker turn nuts quicker, or stirring marches to make a soldier kill with more enthusiasm? No, surely none of us believes in that kind of justification; music, and mathematics, are of human value because human beings feel they are.

The analogy with music can be stretched a little further. Before a performer's artistic contribution is judged, it is taken for granted that he hits the right notes, but merely hitting the right notes doesn't make him a musician. We don't get the point of painting if we compliment the nude Maya on being a good likeness, and we don't get the point of a historian's work if all we can say is that he didn't tell lies. Mere accuracy in performance, resemblance in appearance, and truth in storytelling doesn't make good music, painting, history: in the same way, mere logical correctness doesn't make good mathematics.

Goodness, high quality, are judged on grounds more important than validity, but less describable. A good piece of mathematics is connected with much other mathematics, it is new without being silly (think of a "new" western movie in which the names and the costumes are changed, but the plot isn't), and it is deep in an ineffable but inescapable sense—the sense in which Johann Sebastian is deep and Carl Philip Emmanuel is not. The criterion for quality is beauty, intricacy, neatness, elegance, satisfaction, appropriateness—all subjective, but all somehow mysteriously shared by all.

Mathematics resembles literature also, differently from the way it resembles music. The writing and reading of literature are related to the writing and reading of newspapers, advertisements, and road signs the way mathematics is related to practical arithmetic. We all need to read and write and figure for daily life: but literature is more than reading and writing, and mathematics is more than figuring. The literature analogy can be used to help understand the role of teachers and the role of the pure-applied dualism.

Many whose interests are in language, in the structure, in the history, and in the aesthetics of it, earn their bread and butter by teaching the rudiments of language to its future practical users. Similarly many, perhaps most, whose interests are in the mathematics of today, earn their bread and butter by teaching arithmetic, trigonometry, or calculus. This is sound economics: society abstractly and impersonally is willing to subsidize pure language and pure mathematics, but not very far. Let the would-be purist pull his weight by teaching the next generation the applied aspects of his craft; then he is permitted to spend a fraction of his time doing what he prefers. From the point of view of what a good teacher must be, this is good. A teacher must know more than the bare minimum he must teach; he must know more in order to avoid more and more mistakes, to avoid the perpetuation of misunderstanding, to avoid catastrophic educational inefficiency. To keep him alive, to keep him from drying up, his interest in syntax, his burrowing in etymology, or his dabbling in poetry play a necessary role.

The pure-applied dualism exists in literature too. The source of literature is human life, but literature is not the life it comes from, and writing with a grim purpose is not literature. Sure there are borderline cases: is Upton Sinclair's "Jungle" literature or propaganda? (For that matter, is Chiquita Banana an advertising jingle or charming light opera?) But the fuzzy boundary doesn't alter the fact that in literature (as in mathematics) the pure and the applied are different in intent, in method, and in criterion of success.

Perhaps the closest analogy is between mathematics and painting. The origin of painting is physical reality, and so is the origin of mathematics—but the painter is not a camera and the mathematician is not an engineer. The painter of "Uncle Sam Wants You" got his reward from patriotism, from increased enlistments, from winning the war—which is probably different from the reward Rembrandt got from a finished work. How close to reality painting (and mathematics) should be is a delicate matter of judgment. Asking a painter to "tell a concrete story" is like asking a mathematician to "solve a real problem." Modern painting and modern mathematics are far out—too far in the judgment of some. Perhaps the ideal is to have a spice of reality always present, but not to crowd it the way descriptive geometry, say, does in mathematics, and medical illustration, say, does in painting.

Talk to a painter (I did) and talk to a mathematician, and you'll be amazed at how similarly they react. Almost every aspect of the life and of the art of a mathematician has its counterpart in painting, and vice versa. Every time a mathematician hears "I could never make my checkbook balance" a painter hears "I could never draw a straight line"—and the comments are equally relevant and equally interesting. The invention of perspective gave the painter a useful technique, as did the invention of 0 to the mathematician. Old art is as good as new; old mathematics is as good as new. Tastes change, to be sure, in both subjects, but a twentieth century painter has sympathy for cave paintings and a twentieth century mathematician for the fraction juggling of the Babylonians. A painting must be painted and then looked at; a theorem must be printed and then read. The painter who thinks good pictures, and the mathematician who dreams beautiful theorems are dilettantes; an unseen work of art is incomplete. In painting and in mathematics there are some objective standards of good—the painter speaks of structure, line, shape, and texture, where the mathematician speaks of truth, validity, novelty, generality—but they are relatively the easiest to satisfy. Both painters and mathematicians debate among themselves whether these objective standards should even be told to the young—the beginner may misunderstand and overemphasize them and at the same time lose sight of the more important subjective standards of goodness. Painting and mathematics have a history, a tradition, a growth. Students, in both subjects, tend to flock to the newest but, except the very best, miss the point; they lack the vitality of what they imitate, because, among other reasons, they lack the experience based on the traditions of the subject.

I've been talking *about* mathematics, but not *in* it, and, consequently, what I've been saying is not capable of proof in the mathematical sense of the word. I hope just the same, that I've shown you that there is a subject called mathematics (mathology?), and that that subject is a creative art. It is a creative art because mathematicians create beautiful new concepts; it is a creative art because mathematicians live, act, and think like artists; and it is a creative art because mathematicians regard it so. I feel strongly about that, and I am grateful for this opportunity to tell you about it. Thank you for listening.

Reprinted from the AMERICAN MATHEMATICAL MONTHLY,
Vol. 80, No. 4, April, 1973
pp. 382–394

# THE LEGEND OF JOHN VON NEUMANN

P. R. HALMOS, Indiana University

John von Neumann was a brilliant mathematician who made important contributions to quantum physics, to logic, to meteorology, to war, to the theory and applications of high-speed computing machines, and, via the mathematical theory of games of strategy, to economics.

**Youth.** He was born December 28, 1903, in Budapest, Hungary. He was the eldest of three sons in a well-to-do Jewish family. His father was a banker who received a minor title of nobility from the Emperor Franz Josef; since the title was hereditary, von Neumann's full Hungarian name was Margittai Neumann János. (Hungarians put the family name first. Literally, but in reverse order, the name means John Neumann of Margitta. The "of", indicated by the final "i", is where the "von" comes from; the place name was dropped in the German translation. In ordinary social intercourse such titles were never used, and by the end of the first world war their use had gone out of fashion altogether. In Hungary von Neumann is and always was known as Neumann János and his works are alphabetized under N. Incidentally, his two brothers, when they settled in the U.S., solved the name problem differently. One of them reserves the title of nobility for ceremonial occasions only, but, in daily life, calls himself Neumann; the other makes it less conspicuous by amalgamating it with the family name and signs himself Vonneuman.)

Even in the city and in the time that produced Szilárd (1898), Wigner (1902), and Teller (1908), von Neumann's brilliance stood out, and the legends about him started accumulating in his childhood. Many of the legends tell about his memory. His love of history began early, and, since he remembered what he learned, he ultimately became an expert on Byzantine history, the details of the trial of Joan of Arc, and minute features of the battles of the American Civil War.

---

Paul Halmos claims that he took up mathematics because he flunked his master's orals in philosophy.

He received his Univ. of Illinois Ph.D. under J.L. Doob. Then he was von Neumann's assistant, followed by positions at Illinois, Syracuse, M. I. T. 's Radiation Lab, Chicago, Michigan, Hawaii, and now is Distinguished Professor at Indiana Univ. He spent leaves at the Univ. of Uruguay, Montevideo, Univ. of Miami, Univ. of California, Berkeley, Tulane, and Univ. of Washington. He held a Guggenheim Fellowship and was awarded the MAA Chauvenet Prize.

Professor Halmos' research is mainly measure theory, probability, ergodic theory, topological groups, Boolean algebra, algebraic logic, and operator theory in Hilbert space. He has served on the Council of the AMS for many years and was Editor of the Proceedings of the AMS and Mathematical Reviews. His eight books, all widely used, include *Finite-Dimensional Vector Spaces* (Van Nostrand, 1958), *Measure Theory* (Van Nostrand, 1950), *Naive Set Theory* (Van Nostrand, 1960), and *Hilbert Space Problem Book* (Van Nostrand, 1967).

The present paper is the original uncut version of a brief article commissioned by the Encyclopaedia Britannica. *Editor.*

He could, it is said, memorize the names, addresses, and telephone numbers in a column of the telephone book on sight. Some of the later legends tell about his wit and his fondness for humor, including puns and off-color limericks. Speaking of the Manhattan telephone book he said once that he knew all the numbers in it — the only other thing he needed, to be able to dispense with the book altogether, was to know the names that the numbers belonged to.

Most of the legends, from childhood on, tell about his phenomenal speed in absorbing ideas and solving problems. At the age of 6 he could divide two eight-digit numbers in his head; by 8 he had mastered the calculus; by 12 he had read and understood Borel's *Théorie des Fonctions.*

These are some of the von Neumann stories in circulation. I'll report others, but I feel sure that I haven't heard them all. Many are undocumented and unverifiable, but I'll not insert a separate caveat for each one: let this do for them all. Even the purely fictional ones say something about him; the stories that men make up about a folk hero are, at the very least, a strong hint to what he was like.)

In his early teens he had the guidance of an intelligent and dedicated high-school teacher, L. Rátz, and, not much later, he became a pupil of the young M. Fekete and the great L. Fejér, "the spiritual father of many Hungarian mathematicians". ("Fekete" means "Black", and "Fejér" is an archaic spelling, analogous to "Whyte".)

According to von Kármán, von Neumann's father asked him, when John von Neumann was 17, to dissuade the boy from becoming a mathematician, for financial reasons. As a compromise between father and son, the solution von Kármán proposed was chemistry. The compromise was adopted, and von Neumann studied chemistry in Berlin (1921–1923) and in Zürich (1923–1925). In 1926 he got both a Zürich diploma in chemical engineering and a Budapest Ph.D. in mathematics.

**Early work.** His definition of ordinal numbers (published when he was 20) is the one that is now universally adopted. His Ph.D. dissertation was about set theory too; his axiomatization has left a permanent mark on the subject. He kept up his interest in set theory and logic most of his life, even though he was shaken by K. Gödel's proof of the impossibility of proving that mathematics is consistent.

He admired Gödel and praised him in strong terms: "Kurt Gödel's achievement in modern logic is singular and monumental — indeed it is more than a monument, it is a landmark which will remain visible far in space and time. ... The subject of logic has certainly completely changed its nature and possibilities with Gödel's achievement." In a talk entitled "The Mathematician", speaking, among other things, of Gödel's work, he said: "This happened in our lifetime, and I know myself how humiliatingly easily my own values regarding the absolute mathematical truth changed during this episode, and how they changed three times in succession!"

He was Privatdozent at Berlin (1926–1929) and at Hamburg (1929–1930). During this time he worked mainly on two subjects, far from set theory but near to one another: quantum physics and operator theory. It is almost not fair to call them two subjects: due in great part to von Neumann's own work, they can be viewed as two aspects of the same subject. He started the process of making precise mathematics out of quantum theory, and (it comes to the same thing really) he was inspired by the new physical concepts to make broader and deeper the purely mathematical study of infinite-dimensional spaces and operators on them. The basic insight was that the geometry of the vectors in a Hilbert space has the same formal properties as the structure of the states of a quantum-mechanical system. Once that is accepted, the difference between a quantum physicist and a mathematical operator-theorist becomes one of language and emphasis only. Von Neumann's book on quantum mechanics appeared (in German) in 1932. It has been translated into French (1947), Spanish (1949), and English (1955), and it is still one of the standard and one of the most inspiring treatments of the subject. Speaking of von Neumann's contributions to quantum mechanics, E. Wigner, a Nobel laureate, said that they alone "would have secured him a distinguished position in present day theoretical physics".

**Princeton.** In 1930 von Neumann went to Princeton University for one term as visiting lecturer, and the following year he became professor there. In 1933, when the Institute for Advanced Study was founded, he was one of the original six professors of its School of Mathematics, and he kept that position for the rest of his life. (It is easy to get confused about the Institute and its formal relation with Princeton University, even though there is none. They are completely distinct institutions. The Institute was founded for scholarship and research only, not teaching. The first six professors in the School of Mathematics were J. W. Alexander, A. Einstein, M. Morse, O. Veblen, J. von Neumann, and H. Weyl. When the Institute began it had no building, and it accepted the hospitality of Princeton University. Its members and visitors have, over the years, maintained close professional and personal relations with their colleagues at the University. These facts kept contributing to the confusion, which was partly clarified in 1940, when the Institute acquired a building of its own, about a mile from the Princeton campus.)

In 1930 von Neumann married Marietta Kövesi; in 1935 their daughter Marina was born. (In 1956 Marina von Neumann graduated from Radcliffe *summa cum laude*, with the highest scholastic record in her class. In 1972 Marina von Neumann Whitman was appointed by President Nixon to the Council of Economic Advisers.) In the 1930's the stature of von Neumann, the mathematician, grew at the rate that his meteoric early rise had promised, and the legends about Johnny, the human being, grew along with it. He enjoyed life in America and lived it in an informal manner, very differently from the style of the conventional German professor. He was not a refugee and he didn't feel like one. He was a cosmopolite in attitude and a U.S. citizen by choice.

The parties at the von Neumanns' house were frequent, and famous, and long. Johnny was not a heavy drinker, but he was far from a teetotaller. In a roadside restaurant he once ordered a brandy with a hamburger chaser. The outing was in honor of his birthday and he was feeling fine that evening. One of his gifts was a toy, a short prepared tape attached to a cardboard box that acted as sounding board; when the tape was pulled briskly past a thumbnail, it would squawk "Happy birthday!" Johnny squawked it often. Another time, at a party at his house, there was one of those thermodynamic birds that dips his beak in a glass of water, straightens up, teeter-totters for a while, and then repeats the cycle. A temporary but firm house rule was quickly passed: everyone had to take a drink each time that the bird did.

He liked to drive, but he didn't do it well. There was a "von Neumann's corner" in Princeton, where, the story goes, his cars repeatedly had trouble. One often quoted explanation that he allegedly offered for one particular crack-up goes like this: "I was proceeding down the road. The trees on the right were passing me in orderly fashion at 60 miles an hour. Suddenly one of them stepped in my path. Boom!"

He once had a dog named "Inverse". He played poker, but only rarely, and he usually lost.

In 1937 the von Neumanns were divorced; in 1938 he married Klára Dán. She learned mathematics from him and became an expert programmer. Many years later, in an interview, she spoke about him. "He has a very weak idea of the geography of the house. ...Once, in Princeton, I sent him to get me a glass of water; he came back after a while wanting to know where the glasses were. We had been in the house only seventeen years. ...He has never touched a hammer or a screwdriver; he does nothing around the house. Except for fixing zippers. He can fix a broken zipper with a touch."

Von Neumann was definitely not the caricatured college professor. He was a round, pudgy man, always neatly, formally dressed. There are, to be sure, one or two stories of his absentmindedness. Klári told one about the time when he left their Princeton house one morning to drive to a New York appointment, and then phoned her when he reached New Brunswick to ask: "Why am I going to New York?" It may not be strictly relevant, but I am reminded of the time I drove him to his house one afternoon. Since there was to be a party there later that night, and since I didn't trust myself to remember exactly how I got there, I asked how I'd be able to know his house when I came again. "That's easy," he said; "it's the one with that pigeon sitting by the curb."

Normally he was alert, good at rapid repartee. He could be blunt, but never stuffy, never pompous. Once the telephone interrupted us when we were working in his office. His end of the conversation was very short; all he said between "Hello" and "Goodbye" was "Fekete pestis!", which means "Black plague!" Remembering, after he hung up, that I understood Hungarian, he turned to me, half apologetic and half exasperated, and explained that he wasn't speaking of one of the horsemen of the Apocalypse, but merely of some unexpected and unwanted dinner guests that his wife just told him about.

On a train once, hungry, he asked the conductor to send the man with the sandwich tray to his seat. The busy and impatient conductor said "I will if I see him". Johnny's reply: "This train is linear, isn't it?"

**Speed.** The speed with which von Neumann could think was awe-inspiring. G. Pólya admitted that "Johnny was the only student I was ever afraid of. If in the course of a lecture I stated an unsolved problem, the chances were he'd come to me as soon as the lecture was over, with the complete solution in a few scribbles on a slip of paper." Abstract proofs or numerical calculations — he was equally quick with both, but he was especially pleased with and proud of his facility with numbers. When his electronic computer was ready for its first preliminary test, someone suggested a relatively simple problem involving powers of 2. (It was something of this kind: what is the smallest power of 2 with the property that its decimal digit fourth from the right is 7? This is a completely trivial problem for a present-day computer: it takes only a fraction of a second of machine time.) The machine and Johnny started at the same time, and Johnny finished first.

One famous story concerns a complicated expression that a young scientist at the Aberdeen Proving Ground needed to evaluate. He spent ten minutes on the first special case; the second computation took an hour of paper and pencil work; for the third he had to resort to a desk calculator, and even so took half a day. When Johnny came to town, the young man showed him the formula and asked him what to do. Johnny was glad to tackle it. "Let's see what happens for the first few cases. If we put $n = 1$, we get..." — and he looked into space and mumbled for a minute. Knowing the answer, the young questioner put in "2.31 ?" Johnny gave him a funny look and said "Now if $n = 2$,...",and once again voiced some of his thoughts as he worked. The young man, prepared, could of course follow what Johnny was doing, and, a few seconds before Johnny finished, he interrupted again, in a hesitant tone of voice: "7.49 ?" This time Johnny frowned, and hurried on: "If $n = 3$, then...". The same thing happened as before — Johnny muttered for several minutes, the young man eavesdropped, and, just before Johnny finished, the young man exclaimed: "11.06!" That was too much for Johnny. It couldn't be! No unknown beginner could outdo him! He was upset and he sulked till the practical joker confessed.

Then there is the famous fly puzzle. Two bicyclists start twenty miles apart and head toward each other, each going at a steady rate of 10 m.p.h. At the same time a fly that travels at a steady 15 m.p.h. starts from the front wheel of the southbound bicycle and flies to the front wheel of the northbound one, then turns around and flies to the front wheel of the southbound one again, and continues in this manner till he is crushed between the two front wheels. Question: what total distance did the fly cover? The slow way to find the answer is to calculate what distance the fly covers on the first, northbound, leg of the trip, then on the second, southbound, leg, then on the third, etc., etc., and, finally, to sum the infinite series so obtained. The quick way is to observe that the bicycles meet exactly one hour after their start, so that the fly had just an hour for his travels; the answer must therefore be 15 miles. When the question was put to von Neumann, he solved it in an instant, and thereby disappointed the questioner: "Oh, you must have heard the trick before!" "What trick?" asked von Neumann; "all I did was sum the infinite series."

I remember one lecture in which von Neumann was talking about rings of operators. At an appropriate point he mentioned that they can be classified two ways: finite versus infinite, and discrete versus continuous. He went on to say: "This leads to a total of four possibilities, and, indeed, all four of them can occur. Or — let's see — can they?" Many of us in the audience had been learning this subject from him for some time, and it was no trouble to stop and mentally check off all four possibilities. No trouble — it took something like two seconds for each, and, allowing for some fumbling and shifting of gears, it took us perhaps 10 seconds in all. But after two seconds von Neumann had already said "Yes, they can," and he was two sentencse into the next paragraph before, dazed, we could scramble aboard again.

**Speech.** Since Hungarian is not exactly a *lingua franca*, all educated Hungarians must acquire one or more languages with a popular appeal greater than that of their mother tongue. At home the von Neumanns spoke Hungarian, but he was perfectly at ease in German, and in French, and, of course, in English. His English was fast and grammatically defensible, but in both pronunciation and sentence construction it was reminiscent of German. His "Sprachgefühl" was not perfect, and his sentences tended to become involved. His choice of words was usually exactly right; the occasional oddities (like "a self-obvious theorem") disappeared in later years. His spelling was sometimes more consistent than commonplace: if "commit", then "ommit". S. Ulam tells about von Neumann's trip to Mexico, where "he tried to make himself understood by using 'neo-Castilian', a creation of his own — English words with an 'el' prefix and appropriate Spanish endings".

He prepared for lectures, but rarely used notes. Once, five minutes before a non-mathematical lecture to a general audience, I saw him as he was preparing. He sat in the lounge of the Institute and scribbled on a small card a few phrases such as these: "Motivation, 5 min.; historical background, 15 min.; connection with economics, 10 min.;..."

As a mathematical lecturer he was dazzling. He spoke rapidly but clearly; he spoke precisely, and he covered the ground completely. If, for instance, a subject has four possible axiomatic approaches, most teachers content themselves with developing one, or at most two, and merely mentioning the others. Von Neumann was fond of presenting the "complete graph" of the situation. He would, that is, describe the shortest path that leads from the first to the second, from the first to the third, and so on through all twelve possibilities.

His one irritating lecturing habit was the way he wielded an eraser. He would write on the board the crucial formula under discussion. When one of the symbols in it had been proved to be replaceable by something else, he made the replacement not by rewriting the whole formula, suitably modified, but by erasing the replaceable symbol and substituting the new one for it. This had the tendency of inducing symptoms of acute discouragement among note-takers, especially since, to maintain the flow of the argument, he would keep talking at the same time.

His style was so persuasive that one didn't have to be an expert to enjoy his lectures; everything seemed easy and natural. Afterward, however, the Chinese-dinner phenomenon was likely to occur. A couple of hours later the average memory could no longer support the delicate balance of mutually interlocking implications, and, puzzled, would feel hungry for more explanation.

**Style.** As a writer of mathematics von Neumann was clear, but not clean; he was powerful but not elegant. He seemed to love fussy detail, needless repetition, and notation so explicit as to be confusing. To maintain a logically valid but perfectly transparent and unimportant distinction, in one paper he introduced an extension of the usual functional notation: along with the standard $\phi(x)$ he dealt also with something denoted by $\phi((x))$. The hair that was split to get there had to be split again a little later, and there was $\phi(((x)))$, and, ultimately, $\phi((((x))))$. Equations such as

$$(\psi((((a)))))^2 = \phi((((a))))$$

have to be peeled before they can be digested; some irreverent students referred to this paper as von Neumann's onion.

Perhaps one reason for von Neumann's attention to detail was that he found it quicker to hack through the underbrush himself than to trace references and see what others had done. The result was that sometimes he appeared ignorant of the standard literature. If he needed facts, well-known facts, from Lebesgue integration theory, he waded in, defined the basic notions, and developed the theory to the point where he could use it. If, in a later paper, he needed integration theory again, he would go back to the beginning and do the same thing again.

He saw nothing wrong with long strings of suffixes, and subscripts on subscripts; his papers abound in avoidable algebraic computations. The reason, probably, is that he saw the large picture; the trees did not conceal the forest from him. He saw and he relished all parts of the mathematics he was thinking about. He never wrote "down" to an audience; he told it as he saw it. The practice caused no harm; the main result was that, quite a few times, it gave lesser men an opportunity to publish "improvements" of von Neumann.

Since he had no formal connections with educational institutions after he was 30, von Neumann does not have a long list of students; he supervised only one Ph.D. thesis. Through his lectures and informal conversations he acquired, however, quite a few disciples who followed in one or another of his footsteps. A few among them are J. W. Calkin, J. Charney, H. H. Goldstine, P. R. Halmos, I. Halperin, O. Morgenstern, F. J. Murray, R. Schatten, I. E. Segal, A. H. Taub, and S. Ulam.

**Work habits.** Von Neumann was not satisfied with seeing things quickly and clearly; he also worked very hard. His wife said "he had always done his writing at home during the night or at dawn. His capacity for work was practically unlimited." In addition to his work at home, he worked hard at his office. He arrived early, he stayed late, and he never wasted any time. He was systematic in both large things and small; he was, for instance, a meticulous proofreader. He would correct a manuscript, record on the first page the page numbers where he found errors, and, by appropriate tallies, record the number of errors that he had marked on each of those pages. Another example: when requested to prepare an abstract of not more than 200 words, he would not be satisfied with a statistical check — there are roughly 20 lines with about 10 words each — but he would count every word.

When I was his assistant we wrote one paper jointly. After the thinking and the talking were finished, it became my job to do the writing. I did it, and I submitted to him a typescript of about 12 pages. He read it, criticized it mercilessly, crossed out half, and rewrote the rest; the result was about 18 pages. I removed some of the Germanisms, changed a few spellings, and compressed it into 16 pages. He was far from satisfied, and made basic changes again; the result was 20 pages. The almost divergent process continued (four innings on each side as I now recall it); the final outcome was about 30 typescript pages (which came to 19 in print).

Another notable and enviable trait of von Neumann's was his mathematical courage. If, in the middle of a search for a counterexample, an infinite series came up, with a lot of exponentials that had quadratic exponents, many mathematicians would start with a clean sheet of paper and look for another counterexample. Not Johnny! When that happened to him, he cheerfully said: "Oh, yes, a theta function...", and plowed ahead with the mountainous computations. He wasn't afraid of anything.

He knew a lot of mathematics, but there were also gaps in his knowledge, most notably number theory and algebraic toplogy. Once when he saw some of us at a blackboard staring at a rectangle that had arrows marked on each of its sides, he wanted to know that what was. "Oh just the torus, you know — the usual identification convention." No, he didn't know. The subject is elementary, but some of it just never crossed his path, and even though most graduate students knew about it, he didn't.

Brains, speed, and hard work produced results. In von Neumann's *Collected Works* there is a list of over 150 papers. About 60 of them are on pure mathematics (set theory, logic, topological groups, measure theory, ergodic theory, operator theory, and continuous geometry), about 20 on physics, about 60 on applied mathematics (including statistics, game theory, and computer theory), and a small handful on some special mathematical subjects and general non-mathematical ones. A special number of the *Bulletin of the American Mathematical Society* was devoted to a discussion of his life and work (in May 1958).

**Pure mathematics.** Von Neumann's reputation as a mathematician was firmly established by the 1930's, based mainly on his work on set theory, quantum theory, and operator theory, but enough more for about three ordinary careers, in pure mathematics alone, was still to come. The first of these was the proof of the ergodic theorem. Various more or less precise statements had been formulated earlier in statistical mechanics and called the ergodic hypothesis. In 1931 B. O. Koopman published a penetrating remark whose main substance was that one of the contexts in which a precise statement of the ergodic hypothesis could be formulated is the theory of operators on Hilbert space — the very subject that von Neumann used earlier to make quantum mechanics precise and on which he had written several epoch-making papers. It is tempting to speculate on von Neumann's reaction to Koopman's paper. It could have been something like this: "By Koopman's remark the ergodic hypothesis becomes a theorem about Hilbert spaces — and if that's what it is I ought to be able to prove it. Let's see now... ." Soon after the appearance of Koopman's paper, von Neumann formulated and proved the statement that is now known as the mean ergodic theorem for unitary operators. There was some temporary confusion, caused by publication dates, about who did what before whom, but by now it is universally recognized that von Neumann's theorem preceded and inspired G. D. Birkhoff's point ergodic theorem. In the course of the next few years von Neumann published several more first-rate papers on ergodic theory, and he made use of the techniques and results of that theory later, in his studies of rings of operators.

In 1900 D. Hilbert presented a famous list of 23 problems that summarized the state of mathematical knowledge at the time and showed where further work was needed. In 1933 A. Haar proved the existence of a suitable measure (which has come to be called Haar measure) in topological groups; his proof appears in the *Annals of Mathematics*. Von Neumann had access to Haar's result before it was published, and he quickly saw that that was exactly what was needed to solve an important special case (compact groups) of one of Hilbert's problems (the 5th). His solution appears in the same issue of the same journal, immediately after Haar's paper.

In the second half of the 1930's the main part of von Neumann's publications was a sequence of papers, partly in collaboration with F. J. Murray, on what he called rings of operators. (They are now called von Neumann algebras.) It is possible that this is the work for which von Neumann will be remembered the longest. It is a technically brilliant development of operator theory that makes contact with von Neumann's earlier work, generalizes many familiar facts about finite-dimensional algebra, and is currently one of the most powerful tools in the study of quantum physics.

A surprising outgrowth of the theory of rings of operators is what von Neumann called continuous geometry. Ordinary geometry deals with spaces of dimension 1, 2, 3, etc. In his work on rings of operators von Neumann saw that what really determines the dimension structure of a space is the group of rotations that it admits. The group of rotations associated with the ring of *all* operators yields the familiar dimensions. Other groups, associated with different rings, assign to spaces dimensions whose values can vary continuously; in that context it makes sense to speak of a space of dimension 3/4, say. Abstracting from the "concrete" case of rings of operators, von Neumann formulated the axioms that make these continuous-dimensional spaces possible. For several years he thought, wrote, and lectured about continuous geometries. In 1937 he was the Colloquium Lecturer of the American Mathematical Society and chose that subject for his topic.

**Applied mathematics.** The year 1940 was just about the half-way point of von Neumann's scientific life, and his publications show a discontinuous break then. Till then he was a topflight pure mathematician who understood physics; after that he was an applied mathematician who remembered his pure work. He became interested in partial differential equations, the principal classical tool of the applications of mathematics to the physical world. Whether the war made him into an applied mathematician or his interest in applied mathematics made him invaluable to the war effort, in either case he was much in demand as a consultant and advisor to the armed forces and to the civilian agencies concerned with the problems of war. His papers from this point on are mainly on statistics, shock waves, flow problems, hydrodynamics, aerodynamics, ballistics, problems of detonation, meteorology, and, last but not least, two non-classical, new aspects of the applicability of mathematics to the real world: games and computers.

Von Neumann's contributions to war were manifold. Most often mentioned is his proposal of the implosion method for bringing nuclear fuel to explosion (during World War II) and his espousal of the development of the hydrogen bomb (after the war). The citation that accompanied his honorary D.Sc. from Princeton in 1947 mentions (in one word) that he was a mathematician, but praises him for being a physicist, an engineer, an armorer, and a patriot.

**Politics.** His political and administrative decisions were rarely on the side that is described nowadays by the catchall term "liberal". He appeared at times to advocate preventive war with Russia. As early as 1946 atomic bomb tests were already receiving adverse criticism, but von Neumann thought that they were necessary and (in, for instance, a letter to the *New York Times*) defended them vigorously. He disagreed with J. R. Oppenheimer on the H-bomb crash program, and urged that the U.S. proceed with it before Russia could. He was, however, a "pro-Oppenheimer" witness at the Oppenheimer security hearings. He said that Oppenheimer opposed the program "in good faith" and was "very constructive" once the decision to go ahead with the super bomb was made. He insisted that Oppenheimer was loyal and was not a security risk.

As a member of the Atomic Energy Commission (appointed by President Eisenhower, he was sworn in on March 15, 1955), having to "think about the unthinkable", he urged a United Nations study of world-wide radiation effects. "We willingly pay 30,000–40,000 fatalities per year (2% of the total death rate)," he wrote, "for the advantages of individual transportation by automobile." He mentioned a fall-out accident in an early Pacific bomb test that resulted in one fatality and danger to 200 people, and he compared it with a Japanese ferry accident that "killed about 1,000 people, including 20 Americans — yet the...fall-out was what attracted almost world-wide attention." He asked: "Is the price in international popularity worth paying?" And he answered: "Yes: we have to accept it as part payment for our more advanced industrial position."

**Game theory.** At about the same time that he began to apply his analytic talents to the problems of war, von Neumann found time and energy to apply his combinatorial insight to what he called the theory of games, whose major application was to economics. The mathematical cornerstone of the theory is one statement, the so-called minimax theorem, that von Neumann proved early (1928) in a short article (25 pages); its elaboration and applications are in the book he wrote jointly with O. Morgenstern in 1944. The minimax theorem says about a large class of two-person games that there is no point in playing them. If either player considers, for each possible strategy of play, the *maximum* loss that he can expect to sustain with that strategy, and then chooses the "optimal" strategy that *minimizes* the maximum loss, then he can be statistically sure of not losing more than that minimax value. Since (and this is the whole point of the theorem) that value is the negative of the one, similarly defined, that his opponent can guarantee for himself, the long-run outcome is completely determined by the rules.

Mathematical economics before von Neumann tried to achieve success by imitating the technique of classical mathematical physics. The mathematical tools used were those of analysis (specifically the calculus of variations), and the procedure relied on a not completely reliable analogy between economics and mechanics. The secret of the success of the von Neumann approach was the abandonment of the mechanical analogy and its replacement by a fresh point of view (games of strategy) and new tools (the ideas of combinatorics and convexity).

The role that game theory will play in the future of mathematics and economics is not easy to predict. As far as mathematics is concerned, it is tenable that the only thing that makes the Morgenstern-von Neumann book 600 pages longer than the original von Neumann paper is the development needed to apply the abstruse deductions of one subject to the concrete details of another. On the other hand, enthusiastic proponents of game theory can be found who go so far as to say that it may be "one of the major scientific contributions of the first half of the 20th century".

**Machines.** The last subject that contributed to von Neumann's fame was the theory of electronic computers and automata. He was interested in them from every point of view: he wanted to understand them, design them, build them, and use them. What are the logical components of the processes that a computer will be asked to perform? What is the best way of obtaining practically reliable answers from a machine with unreliable components? What does a machine need to "remember", and what is the best way to equip it with a "memory"? Can a machine be built that can not only save us the labor of computing but save us also the trouble of building a new machine — is it possible, in other words, to produce a self-reproducing automaton? (Answer: in principle, yes. A sufficiently complicated machine, embedded in a thick chowder of randomly distributed spare parts, its "food", would pick up one part after another till it found a usable one, put it in place, and continue to search and construct till its descendant was complete and operational.) Can a machine successfully imitate "randomness", so that when no formulae are available to solve a concrete physical problem (such as that of finding an optimal bombing pattern), the machine can perform a large number of probability experiments and yield an answer that is statistically accurate? (The last question belongs to the concept that is sometimes described as the Monte Carlo method.) These are some of the problems that von Neumann studied and to whose solutions he made basic contributions.

He had close contact with several computers — among them the MANIAC (Mathematical Analyzer, Numerical Integrator, Automatic Calculator), and the affectionately named JONIAC. He advocated their use for everything from the accumulation of heuristic data for the clarification of our intuition about partial differential equations to the accurate long-range prediction and, ultimately, control of the weather. One of the most striking ideas whose study he suggested was to dye the polar icecaps so as to decrease the amount of energy they would reflect — the result could warm the earth enough to make the climate of Iceland approximate that of Hawaii.

The last academic assignment that von Neumann accepted was to deliver and prepare for publication the Silliman lectures at Yale. He worked on that job in the hospital where he died, but he couldn't finish it. His notes for it were published, and even they make illuminating reading. They contain tantalizing capsule statements of insights, and throughout them there shines an attitude of faith in and dedication to knowledge. While physicists, engineers, meteorologists, statisticians, logicians, and computers all proudly claim von Neumann as one of theirs, the Silliman lectures prove, indirectly by their approach and explicitly in the author's words, that von Neumann was first, foremost, and always a mathematician.

**Death.** Von Neumann was an outstanding man in tune with his times, and it is not surprising that he received many awards and honors. There is no point in listing them all here, but a few may be mentioned. He received several honorary doctorates, including ones from Princeton (1947), Harvard (1950), and Istanbul (1952). He served a term as president of the American Mathematical Society (1951–1953), and he was a member of several national scientific academies (including, of course, that of the U. S.). Somewhat to his embarrassment, he was elected to the East German Academy of Science, but the election didn't seem to take — in later years no mention is made of it in the standard biographical reference works. He received the Enrico Fermi award in 1956, when he already knew that he was incurably ill.

Von Neumann became ill in 1955. There was an operation, and the result was a diagnosis of cancer. He kept on working, and even travelling, as the disease progressed. Later he was confined to a wheelchair, but still thought, wrote, and attended meetings. In April 1956 he entered Walter Reed Hospital, and never left it. Of his last days his good friend Eugene Wigner wrote: "When von Neumann realized he was incurably ill, his logic forced him to realize that he would cease to exist, and hence cease to have thoughts. ...It was heartbreaking to watch the frustration of his mind, when all hope was gone, in its struggle with the fate which appeared to him unavoidable but unacceptable."

Von Neumann was baptized a Roman Catholic (in the U. S.), but, after his divorce, he was not a practicing member of the church. In the hospital he asked to see a priest — "one that will be intellectually compatible". Arrangements were made, he was given special instruction, and, in due course, he again received the sacraments. He died February 8, 1957.

The heroes of humanity are of two kinds: the ones who are just like all of us, but very much more so, and the ones who, apparently, have an extra-human spark. We can all run, and some of us can run the mile in less than 4 minutes; but there is nothing that most of us can do that compares with the creation of the Great G-minor Fugue. Von Neumann's greatness was the human kind. We can all think clearly, more or less, some of the time, but von Neumann's clarity of thought was orders of magnitude greater than that of most of us, all the time. Both Norbert Wiener and John von Neumann were great men, and their names will live after them, but for different reasons. Wiener saw things deeply but intuitively; von Neumann saw things clearly and logically.

What made von Neumann great? Was it the extraordinary rapidity with which he could understand and think and the unusual memory that retained everything he had once thought through? No. These qualities, however impressive they might have been, are ephemeral; they will have no more effect on the mathematics and the mathematicians of the future than the prowess of an athlete of a hundred years ago has on the sport of today.

The "axiomatic method" is sometimes mentioned as the secret of von Neumann's success. In his hands it was not pedantry but perception; he got to the root of the matter by concentrating on the basic properties (axioms) from which all else follows. The method, at the same time, revealed to him the steps to follow to get from the foundations to the applications. He knew his own strengths and he admired, perhaps envied, people who had the complementary qualities, the flashes of irrational intuition that sometimes change the direction of scientific progress. For von Neumann it seemed to be impossible to be unclear in thought or in expression. His insights were illuminating and his statements were precise.

Reprinted from
MATHEMATICS TOMORROW
L. A. Steen (Ed.), pp. 9–20, Springer-Verlag New York Inc., 1981

# Applied Mathematics Is Bad Mathematics

*Paul R. Halmos*

It isn't really (applied mathematics, that is, isn't really bad mathematics), but it's different.

Does that sound as if I had set out to capture your attention, and, having succeeded, decided forthwith to back down and become conciliatory? Nothing of the sort! The "conciliatory" sentence is controversial, believe it or not; lots of people argue, vehemently, that it (meaning applied mathematics) is not different at all, it's all the same as pure mathematics, and anybody who says otherwise is probably a reactionary establishmentarian and certainly wrong.

If you're not a professional mathematician, you may be astonished to learn that (according to some people) there are different kinds of mathematics, and that there is anything in the subject for anyone to get excited about. There are; and there is; and what follows is a fragment of what might be called the pertinent sociology of mathematics: what's the difference between pure and applied, how do mathematicians feel about the rift, and what's likely to happen to it in the centuries to come?

## What is it?

There is never any doubt about what mathematics encompasses and what it does not, but it is not easy to find words that describe precisely what it is. In many discussions, moreover, mathematics is not described as a whole

---

*Paul R. Halmos* is Distinguished Professor of Mathematics at Indiana University, and Editor-Elect of the *American Mathematical Monthly*. He received his Ph.D. from the University of Illinois, and has held positions at Illinois, Syracuse, Chicago, Michigan, Hawaii, and Santa Barbara. He has published numerous books and nearly 100 articles, and has been the editor of many journals and several book series. The Mathematical Association of America has given him the Chauvenet Prize and (twice) the Lester Ford award for mathematical exposition. His main mathematical interests are in measure and ergodic theory, algebraic logic, and operators on Hilbert space.

but is divided into two parts, and not just in one way; there are two kinds of mathematics according to each of several different systems of classification.

Some of the dichotomies are well known, and others less so. Mathematics studies sizes and shapes, or, in other words, numbers (arithmetic) and figures (geometry); it can be discrete or continuous; it is sometimes finite and sometimes infinite; and, most acrimoniously, some of it is pure (useless?) and some applied (practical?). Different as these classification schemes might be, they are not unrelated. They are, however, not of equal strengths; the size-shape division, for instance, is much less clear-cut, and much less divisive, than the pure-applied one.

Nobody is forced to decide between vanilla ice cream and chocolate once and for all, and it is even possible to mix the two, but most people usually ask for the same one. A similar (congenital?) division of taste exists for mathematicians. Nobody has to decide once and for all to like only algebra (discrete) or only topology (continuous), and there are even flourishing subjects called algebraic topology and topological algebra, but most mathematicians do in fact lean strongly toward either the discrete or the continuous.

## Squares and spheres

It would be a shame to go on and on about mathematics and its parts without looking at a few good concrete examples, but genuine examples are much too technical to describe in the present context. Here are a couple of artificial ones (with some shortcomings, which I shall explain presently).

Suppose you want to pave the floor of a room whose shape is a perfect square with tiles that are themselves squares so that no two tiles are exactly the same size. Can it be done? In other words, can one cover a square with a finite number of non-overlapping smaller squares all of which have different side-lengths? This is not an easy question to answer.

Here is another puzzle: if you have a perfect sphere, like a basketball, what's the smallest number of points you can mark on it so that every point on the surface is within an inch of one of the marked ones? In other words, what's the most economical way to distribute television relay stations on the surface of the globe?

Is the square example about sizes (numbers) or shapes (figures)? The answer seems to be that it's about both, and so is the sphere example. In this respect the examples give a fair picture; mixed types are more likely to occur (and are always more interesting) than the ones at either extreme. The examples have different flavors, however. The square one is more nearly arithmetic, discrete, finite, pure, and the one about spheres leans toward being geometric, continuous, infinite, applied.

The square problem is of some mild interest, and it has received attention in the professional literature several times, but it doesn't really have the respect of most mathematicians. The reason is not that it is obviously useless in the practical sense of the word, but that it is much too special (petty?, trivial?), in the sense of being isolated from most of the rest of mathematics and requiring *ad hoc* methods for its solution. It is not really a fair example of pure mathematics.

The sphere example, on the other hand, is of a great deal of practical use, but, nevertheless, it is not a fair example of applied mathematics: it is much easier (and much purer) than most applied problems, and, in particular, it does not involve motion, which plays the central role in the classical conception of what applied mathematics is all about.

Still, for what they are worth, here they are, and it might help to keep them in mind as the discussion proceeds.

## Fiction or action

The pure and applied distinction is visible in the arts and in the humanities almost as clearly as in the sciences: witness Mozart versus military marches, Rubens versus medical illustrations, or Virgil's *Aeneid* versus Cicero's *Philippics*. Pure literature deals with abstractions such as love and war, and it tells about imaginary examples of them in emotionally stirring language. Pure mathematics deals with abstractions such as the multiplication of numbers and the congruence of triangles, and it reasons about Platonically idealized examples of them with intellectually convincing logic.

There is, to be sure, one sense of the word in which all literature is "applied". Shakespeare's sonnets have to do with the everyday world, and so does Tolstoy's *War and Peace*, and so do Caesar's commentaries on the wars he fought; they all start from what human beings see and hear, and all speak of how human beings move and feel. In that same somewhat shallow sense all mathematics is applied. It all starts from sizes and shapes (whose study leads ultimately to algebra and geometry), and it reasons about how sizes and shapes change and interact (and such reasoning leads ultimately to the part of the subject that the professionals call analysis).

There can be no doubt that the fountainhead, the inspiration, of all literature is the physical and social universe we live in, and the same is true about mathematics. There is also no doubt that the physical and social universe daily affects each musician, and painter, and writer, and mathematician, and that therefore a part at least of the raw material of the artist is the world of facts and motions, sights and sounds. Continual contact between the world and art is bound to change the latter, and perhaps even to improve it.

The ultimate goal of "applied literature", and of applied mathematics, is action. A campaign speech is made so as to cause you to pull the third lever on a voting machine rather than the fourth. An aerodynamic equation is solved so as to cause a plane wing to lift its load fast enough to avoid complaints from the home owners near the airport. These examples are crude and obvious; there are subtler ones. If the biography of a candidate, a factually correct and honest biography, does not directly mention the forthcoming election, is it then pure literature? If a discussion of how mathematically idealized air flows around moving figures of various shapes, a logically rigorous and correct discussion, does not mention airplanes or airports, is it then pure mathematics? And what about the in-between cases: the biography that, without telling lies, is heavily prejudiced, and the treatise on aerodynamics that, without being demonstrably incorrect, uses cost-cutting rough approximations—are they pure or applied?

## Continuous spectrum

Where are the dividing lines in the chain from biography to interpretive history to legend to fiction? We might be able to tell which of Toynbee, Thucydides, Homer, and Joyce is pure and which is applied, but if we insert a dozen names between each pair of them, as we pass from interpreted fact to pure fancy, the distinctions become blurred and perhaps impossible to define. The mathematical analogy is close: if we set out to sort a collection of articles that range from naval architecture to fluid dynamics to partial differential equations to topological vector spaces, the pure versus applied decisions that are clear at the two ends of the spectrum become fuzzy in the middle.

*Pure mathematics can be practically useful and applied mathematics can be artistically elegant.*

To confuse the issue still more, pure mathematics can be practically useful and applied mathematics can be artistically elegant. Pure mathematicians, trying to understand involved logical and geometrical interrelations, discovered the theory of convex sets and the algebraic and topological study of various classes of functions. Almost as if by luck, convexity has become the main tool in linear programming (an indispensable part of modern economic and industrial practice), and functional analysis has become the main tool in quantum theory and particle physics. The physicist regards the applicability of von Neumann algebras (a part of functional analysis) to elementary particles as the only justification of the former; the mathematician regards the connection as the only interesting aspect of the latter. *De gustibus non disputandum est*?

Just as pure mathematics can be useful, applied mathematics can be more beautifully useless than is sometimes recognized. Applied mathematics is not engineering; the applied mathematician does not design airplanes or atomic bombs. Applied mathematics is an intellectual discipline, not a part of industrial technology. The ultimate goal of applied mathematics is action, to be sure, but, before that, applied mathematics is a part of theoretical science concerned with the general principles behind what makes planes fly and bombs explode.

The differences between people are sometimes as hard to discern as the differences between subjects, and it can even happen that one and the same person is both a pure and an applied mathematician. Some applied mathematicians (especially the better ones) have a sound training in pure mathematics, and some pure mathematicians (especially the better ones) have a sound training in applicable techniques. When the occasion for a crossover arises (a pure mathematician successfully solves a special case of the travelling salesman problem that arises in operations research, a relativity theorist brilliantly derives a formula in 4-dimensional differential geometry), each one is secretly more than a little proud: "See! I can do that stuff too!"

## Doers and knowers

What I have said so far is that in some sense all mathematics is applied mathematics and that on some level it is not easy to tell the pure from the applied. Now I'll talk about the other side: pure and applied are different indeed, and if you know what to look for, and have the courage of your convictions, you can always tell which is which. My purpose is to describe something much more than to prove something. My hope is neither to convert the pagan nor to convince the agnostic, but just to inform the traveller from another land: there are two sects here, and these are the things that they say about each other.

The difference of opinion is unlike the one in which one sect says "left" and the other says "right"; here one sect says "we are all one sect" and the other says "oh no, we're not, we are two". That kind of difference makes it hard to present the facts in an impartial manner; the mere recognition that a conflict exists amounts already to taking sides. There is no help for that, so I proceed, with my own conclusions admittedly firm, to do the best I can for the stranger in our midst.

Human beings want to know and to do. People want to know what their forefathers did and said, they want to know about animals and vegetables and minerals, and they want to know about concepts and numbers and sights and sounds. People want to grow food and to sew clothes, they want to build houses and to design machines, and they want to cure diseases and to speak languages.

The doers and the knowers frequently differ in motivation, attitude, technique, and satisfaction, and these differences are visible in the special case of applied mathematicians (doers) and pure mathematicians (knowers). The motivation of the applied mathematician is to understand the world and perhaps to change it; the requisite attitude (or, in any event, a customary one) is one of sharp focus (keep your eye on the problem); the techniques are chosen for and judged by their effectiveness (the end is what's important); and the satisfaction comes from the way the answer checks against reality and can be used to make predictions. The motivation of the pure mathematician is frequently just curiosity; the attitude is more that of a wide-angle lens than a telescopic one (is there a more interesting and perhaps deeper question nearby?); the choice of technique is dictated at least in part by its harmony with the context (half the fun is getting there); and the satisfaction comes from the way the answer illuminates unsuspected connections between ideas that had once seemed to be far apart.

*The challenge [for the pure mathematician] is the breathtakingly complicated logical structure of the universe, and victory is permanent.*

The last point deserves emphasis, especially if you belong to the large group of people who proudly dislike mathematics and regard it as inglorious drudgery. For the pure mathematician, his subject is an inexhaustible source of artistic pleasure: not only the excitement of the puzzle and the satisfaction of the victory (if it ever comes!), but mostly the joy of contemplation. The challenge doesn't come from our opponent who can win only if we lose, and victory doesn't disappear as soon as it's achieved (as in tennis, say); the challenge is the breathtakingly complicated logical structure of the universe, and victory is permanent (more like recovering precious metal from a sunken ship).

The basic differences in motivation, attitude, technique, and satisfaction are probably connected with more superficial but more noticeable differences in exposition. Pure and applied mathematicians have different traditions about clarity, elegance, and perhaps even logical rigor, and such differences frequently make for unhappy communication.

The hows and the whys listed above are not offered as a checklist to be used in distinguishing applied science from pure thought; that is usually done by a sort of intuitive absolute pitch. The word "spectrum" gives an analogical hint to the truth. In some sense red and orange are the same—just waves whose lengths differ a bit—and it is impossible to put your finger on the spot in the spectrum where red ends and orange begins—but, after that is granted, red and orange are still different, and the task of telling them apart is almost never a difficult one.

## Beauty and boredom

Many pure mathematicians regard their specialty as an art, and one of their terms of highest praise for another's work is "beautiful". Applied mathematicians seem sometimes to regard their subject as a systematization of methods; a suitable word of praise for a piece of work might be "ingenious" or "powerful".

*Mathematics, . . . despite its many subdivisions and their enormous rates of growth, . . . is an amazingly unified intellectual structure.*

Here is another thing that has frequently struck me: mathematics (pure mathematics), despite its many subdivisions and their enormous rate of growth (started millennia ago and greater today than ever before), is an amazingly unified intellectual structure. The mathematics that is alive and vigorous today has so many parts, and each is so extensive, that no one can possibly know them all. As a result, we, all of us, often attend colloquium lectures on subjects about which we know much less than an average historian, say, knows about linguistics. It doesn't matter, however, whether the talk is about unbounded operators, commutative groups, or parallelizable surfaces; the interplay between widely separated parts of mathematics always shows up. The concepts and methods of each one illuminate all others, and the unity of the structure as a whole is there to be marvelled at.

That unity, that common aesthetic insight, is mostly missing between pure and applied mathematics. When I try to listen to a lecture about fluid mechanics, I soon start wondering and puzzling at the (to me) *ad hoc* seeming approach; then the puzzlement is replaced by bewilderment, boredom, confusion, acute discomfort, and, before the end, complete chaos. Applied mathematicians listening to a lecture on algebraic geometry over fields of non-zero characteristic go through a very similar sequence of emotions, and they describe them by words such as inbred, artificial, baroque folderol, and unnecessary hairsplitting.

It might be argued that from the proper, Olympian, impartial scientific point of view both sides are wrong, but perhaps to a large extent both are right—which would go to prove that we are indeed looking at two subjects, not one. To many pure mathematicians applied mathematics is nothing but a bag of tricks, with no merit except that they work, and to many applied mathematicians much of pure mathematics deserves to be described as meaningless abstraction for its own sake with no merit at all. (I mention in passing that in moments of indulgent self-depreciation the students of one particular branch of pure mathematics, category theory, refer to their branch as "abstract nonsense"; applied mathematicians tend to refer to it the same way, and they seem to mean it.)

## New heresy

Some say that the alleged schism between pure and applied mathematics is a recent heresy at which the founding greats would throw up their hands in horror—the world is going to the dogs! There is a pertinent quotation from Plato's *Philebus*, which doesn't quite refute that statement, but it's enough to make you think about it again.

"Socrates: Are there not two kinds of arithmetic, that of the people and that of the philosophers? . . . And how about the arts of reckoning and measuring as they are used in building and in trade when compared with philosophical geometry and elaborate computations—shall we speak of each of these as one or two?

"Protarchus: . . . I should say that each of them was two."

Is the distinction that Socrates is driving at exactly the pure-applied one? If not that, then what?

The only other curiosity along these lines that I'll mention is that you can usually (but not always) tell an applied mathematician from a pure one just by observing the temperature of his attitude toward the same-different debate. If he feels strongly and maintains that pure and applied are and must be the same, that they are both mathematics and the distinction is meaningless, then he is probably an applied mathematician. About this particular subject most pure mathematicians feel less heat and speak less polemically: they don't really think pure and applied are the same, but they don't care all that much. I think what I have just described is a fact, but I confess I can't help wondering why it's so.

## New life

The deepest assertion about the relation between pure and applied mathematics that needs examination is that it is symbiotic, in the sense that neither can survive without the other. Not only, as is universally admitted, does the applied need the pure, but, in order to keep from becoming inbred, sterile, meaningless, and dead, the pure needs the revitalization and the contact with reality that only the applied can provide.

The first step in the proof of the symbiosis is historical: all of pure mathematics, it is said, comes from the real world, the way geometry, according to legend, comes from measuring the effect of the floods of the Nile. (If that's false, if geometry existed before it was needed, the symbiosis argument begins on a shaky foundation. If it's true, the argument tends to prove only that applied mathematics cannot get along without pure, as an anteater cannot get along without ants, but not necessarily the reverse.)

Insofar as all mathematics comes from the study of sizes (of *things*) and shapes (of *things*), it is true that all mathematics comes from the things of the real world. Whether renewed contact with physics or psychology or biology or economics was needed to give birth to some of the greatest parts of 20th century mathematics (such as Cantor's continuum problem, the Riemann hypothesis, and the Poincaré conjecture) is dubious.

The crux of the matter is, however, not historical but substantive. By way of a parable, consider chess. Mathematicians usually but sometimes grudgingly admit that chess is a part of mathematics. They do so grudgingly because they don't consider chess to be "good" mathematics; from the mathematical point of view it is "trivial". No matter: mathematics it is, and pure mathematics at that.

---

*Applied mathematics cannot get along without pure, as an anteater cannot get along without ants, but not necessarily the reverse.*

---

Chess has not been conceptually revitalized in many hundreds of years, but is vigorously alive nevertheless. There are millions of members of chess clubs, and every now and then the whole civilized world spends days watching Bobby Fischer play Boris Spassky. Chess fires the imagination of a large part of humanity; it shows them aesthetic lights and almost mystic insights.

Not only does chess (like many parts of mathematics) not need external, real-world revitalization, but, in fact, every now and then it spontaneously revitalizes itself. The most recent time that occurred was when retrograde chess analysis began to be studied seriously. (Sample problem: a chess position is given; you are to decide which side of the board White started from and whether Black had ever castled.) And here's a switch: not only was the real world not needed to revitalize chess, but, in fact, the life giving went the other way. Retrograde chess analysis challenged computer scientists with a new kind of problem, and it now constitutes a small but respectable and growing part of applied mathematics.

A revitalist might not be convinced by all this; he might point to the deplorable tendency of mathematics to become ultra-abstract, ultra-complicated, and involutedly ugly, and say that contact with the applications remedies that. The disease exists, that is well known, but, fortunately, so does nature's built-in cure. Several parts of mathematics have become cancerously overgrown in the course of the centuries; certain parts of elementary Euclidean geometry form a probably non-controversial example. When that happens, a wonderful remission always follows. Old mathematics never dies what the Greeks bequeathed us 2500 years ago is still alive and true and interesting—but the outgrowths get simplified, their valuable core becomes integrated into the main body, and the nasty parts get sloughed off.

(Parenthetically: the revitalization argument could, in principle, be applied to painting, but so far as I know no one has applied it. Painting originates in the real world, it has been known to leave that world for realms of abstraction and complication that some find repulsive, but the art as a whole continues alive and well through all that.

A time argument is sometimes mentioned as a good feature of the contact of pure mathematics with applied. Example: if only pure mathematicians had paid closer attention to Maxwell, they would have discovered topological groups much sooner. Perhaps so—but just what did we lose by discovering them later? Would the world be better off if Rembrandt had been born a century earlier? What's the hurry?)

Whether contact with applications can prevent or cure the disease of elaboration and attenuation in mathematics is not really known; what is known is that many of the vigorous and definitely non-cancerous parts have no such contact (and probably, because of their level of abstraction, cannot have any). Current examples: analytic number theory and algebraic geometry.

When I say that mathematics doesn't *have* to be freshened by periodic contact with reality, of course I do not mean that it *must not* be: many of the beautiful concepts of pure mathematics were first noticed in the study of one or another part of nature. Perhaps they would not have been discovered without external stimulation, or perhaps they would—certainly many things were.

As far as the interaction between pure and applied mathematics is concerned, the truth seems to be that it exists, in both directions, but it is much stronger in one direction than in the other. For pure mathematics the applications are a great part of the origin of the subject and continue to be an occasional source of inspiration—they are, however, not indispensable. For applied mathematics, the pure concepts and deductions are a tool, an organizational scheme, and frequently a powerful hint to truths about the world—an indispensable part of the applied organism. It's the ant and the anteater again: arguably, possibly, the anteater is of some ecological value to the ant, but, certainly, indisputably, the ant is necessary for the anteater's continued existence and success.

## What's next?

The most familiar parts of mathematics are algebra and geometry, but for the profession there is a third one, analysis, that plays an equally important role. Analysis starts from the concept of change. It's not enough just to study sizes and shapes; it is necessary also to study how sizes and shapes vary. The natural way to measure change is to examine the difference between the old and the new, and that word, "difference", leads in an etymologically straight line to the technical term "differential equation".

Most of the classical parts of applied mathematics are concerned with change—motion—and their single most usable tool is the theory and technique of differential equations.

Phenomena in the real world are likely to depend on several variables: the success of the stew depends on how long you cook it, how high the temperature is, how much wine you add, etc. To predict the outcome correctly, the variables must be kept apart: how does the outcome change when a part of the data is changed? That's why much of applied mathematics is inextricably intertwined with the theory of *partial* differential equations; for some people, in fact, the latter phrase is almost a synonym of applied mathematics.

Are great breakthroughs still being made and will they continue? Is a Shakespeare of mathematics (such as Archimedes or Gauss) likely to be alive and working now, or to be expected ever again? Algebra, analysis, and geometry—what's the mathematics of the future, and how will the relations between pure and applied develop?

I don't know the answers, nobody does, but the past and the present give some indications; based on them, and on the hope that springs eternal, I'll hazard a couple of quick guesses. The easiest question is about great breakthroughs: yes, they are still being made. Answers to questions raised many decades and sometimes centuries ago are being found almost every year. If Cantor, Riemann, and Poincaré came alive now, they would be excited and avid students, and they would learn much that they wanted to know.

---

*In the foreseeable future discrete mathematics will be an increasingly useful tool in the attempt to understand the world, and . . . analysis will therefore play a proportionally smaller role.*

---

Is there an Archimedes alive now? Probably not. Will there ever be another Gauss? I don't see why not, I hope so, and that's probably why I think so.

I should guess that in the foreseeable future (as in the present) discrete mathematics will be an increasingly useful tool in the attempt to understand the world, and that analysis will therefore play a proportionally smaller role. That is not to say that analysis in general and partial differential equations in particular have had their day and are declining in power; but, I am guessing, not only combinatorics but also relatively sophisticated number theory and geometry will displace some fraction of the many pages that analysis has been occupying in all books on applied mathematics.

Applied mathematics is bound to change, in part because the problems change and in part because the tools for their solution change. As we learn more and more about the world, and learn how to control some of it, we need to ask new questions, and as pure mathematics grows, sloughs off the excess, and becomes both deeper and simpler thereby, it offers applied mathematics new techniques to use. What will all that do to the relation between the ant and the anteater? My guess is, nothing much. Both kinds of curiosity, the pure and the practical, are bound to continue, and the Socrates of 2400 years from now will probably see the difference between them as clearly as did the one 2400 years ago.

So, after all that has been said, what's the conclusion? Perhaps it is in the single word "taste".

A portrait by Picasso is regarded as beautiful by some, and a police photograph of a wanted criminal can be useful, but the chances are that the Picasso is not a good likeness and the police photograph is not very inspiring to look at. Is it completely unfair to say that the portrait is a bad copy of nature and the photograph is bad art?

Much of applied mathematics has great value. If an intellectual technique teaches us something about how blood is pumped, how waves propagate, and how galaxies expand, then it gives us science, knowledge, in the meaning of the word that deserves the greatest respect. It is no insult to the depth and precision and social contribution of great drafters of legislative prose (with their rigidly traditional diction and style) to say that the laws they write are bad literature. In the same way it is no insult to the insight, technique, and scientific contribution of great applied mathematicians to say of their discoveries about blood, and waves, and galaxies that those discoveries are first-rate applied mathematics; but, usually, applied mathematics is bad mathematics just the same.

Reprinted from the
TWO-YEAR COLLEGE MATHEMATICS JOURNAL
Vol. 13, No. 14, pp. 226–242, Sept. 1982

# Paul Halmos: Maverick Mathologist

Donald J. Albers

Paul R. Halmos is Distinguished Professor of Mathematics at Indiana University and Editor of the *American Mathematical Monthly*. He received his Ph.D. from the University of Illinois and has held positions at Illinois, Syracuse, Chicago, Michigan, Hawaii, and Santa Barbara. He has authored ten books and 100 articles. He is a member of the Royal Society of Edinburgh and of the Hungarian Academy of Science.

The writings of Halmos have had a large impact on both research in mathematics and the teaching of mathematics. He has won several awards for his mathematical exposition, including the Chauvenet Prize, and has twice won the Lester R. Ford Award.

In August of 1981, I interviewed Paul Halmos in Pittsburgh at the combined annual summer meetings of the Mathematical Association of America and the American Mathematical Society. During the course of the interview, Halmos confessed to being a *maverick mathologist.* A *mathologist* is a pure mathematician and is to be distinguished from a *mathophysicist*, who is an applied mathematician. (Both terms were coined by Halmos.) A few of his statements from the interview help to underscore his maverick nature:

> "I don't think mathematics needs to be supported."
>
> "If the NSF had never existed, if the government had never funded American mathematics, we would have half as many mathematicians as we now have, and I don't see anything wrong with that."
>
> "The computer is important, but not to mathematics."

In the pages that follow, Halmos in his inimitable style talks about teaching mathematics, writing mathematics, and doing mathematics. After a short time with him, I was convinced that he is a *maverick* and a *mathologist*.

**Albers:** *You have described yourself as a downward-bound philosopher. What does that mean?*

**Halmos:** Most mathematicians think of a hierarchy in which mathematics is above physics, and physics is higher than engineering. If they do that, then they are honor-bound to admit that philosophy is higher than mathematics. I started graduate school with the idea of studying philosophy. I had studied enough mathematics and philosophy for a major in either one. My first choice was philosophy, but I kept a parallel course with mathematics until I flunked my master's exams in philosophy. I couldn't answer all the questions on the history of philosophy that they asked, so I said the hell with it—I'm going into mathematics. I made philosophy my minor, but even that didn't help; I flunked the minor exams too.

**A:** *So mathematics was not your original calling, if you like?*

**H:** As a philosophy student, I played around with symbolic logic and was fascinated with all the symbols in *Principia Mathematica*. Even as a philosopher, I tended toward math.

**A:** *You are the third Hungarian we have interviewed.*

**H:** I reject the appellation.

**A:** *We know that you were born in Hungary and that you lived there until the age of* 13, *but you still reject the appellation.*

**H:** I don't feel Hungarian. I speak Hungarian, but by culture, education, world view, and everything else I can think of—I feel American. When I go to Hungary, I feel like an American tourist, a stranger. I speak English with an accent, but I speak it infinitely better than I speak Hungarian. I can control it, and I cannot do that in Hungarian. In every respect, except accent, I am an American.

**A:** *You may not claim Hungary, but I wouldn't be surprised if Hungary claims you. In fact, you are a member of the Hungarian Academy of Science.*

**H:** I was elected a member of the Hungarian Academy of Science only a couple of years ago, in recognition of my work, I hope, but I am sure that my having been born in Hungary helped. In theory, it needn't help, as there are a certain number of foreign members elected each year. But if they are in some sense ex-Hungarians or have Hungarian roots, that doesn't hurt. I am not ashamed of my Hungarian connection, but just as a matter of fact I try to straighten out my friends and tell them that they shouldn't attribute to my country of origin whatever properties they ascribe to me.

**A:** *How did you come to leave Hungary?*

**H:** I give full credit to my father. In 1924, when he was in his early forties, he left Hungary, where he had been a practicing physician with a flourishing practice. The country was at peace and in good shape, but he thought it was a sinking ship. He arranged for his practice to be taken over by another physician, who was also foster father to his three boys, of whom I am the youngest. (My mother died when I was six months old, and I never knew her.) He came to this country with the feeble English that he had learned. After working as an intern at an Omaha hospital for a year in order to prepare for and pass the state and national boards, he started a practice in Chicago. Five years later he became a citizen and imported his sons. Coming to America wasn't a decision on *my* part; it was a decision on *his* part. It turned out to be a very smart move.

**A:** *Did you have any glimmerings of strong mathematical interests as a child? We know the Hungarians do a remarkably good job of producing superior mathematics students.*

**H:** Yes and no. I cannot give credit to the Hungarian system, which I admire and about which I am somewhat puzzled (as are most Americans), as to how they produce Erdös's, Pólya's, and Szegö's, and dozens more that most of us can rattle off. I know the rumor that they look for them in high school and encourage them and conduct special examinations to find them. Nothing like that had a chance to happen to me. By the age of thirteen I was exposed to a lot more mathematics than American students are exposed to nowadays, but not more than American students were exposed to in those days. I was exposed to parentheses and quadratic equations, two linear equations in two unknowns, a few applied ideas, and the basic things in physics. I remember that I enjoyed drawing the design of a water pump and other things like that. I was good at it, the way good students in calculus are good at calculus in our classrooms, but not a genius. I just enjoyed it and fooled around with it. In mathematics classes, I usually was above average. I was bored when class was going on, and I did things like take logarithms of very large numbers for fun.

The American system in those days was eight years of elementary school and four years of high school. In Europe it was the other way around—four years of elementary school followed by eight years of secondary school, adding up to the same thing. I left Hungary when I was in the third year of secondary school, which would have been the equivalent of the seventh grade in this country.

**A:** *So the thirteen-year-old Halmos came to the U.S. and entered a high school in the Chicago area. You spoke Hungarian and German and knew a little Latin, and yet instruction was in English. That must have posed a few problems.*

**H:** For the first six months it was a hell of a problem. On my first day, somebody showed me to a classroom in which, I still remember, a very nice man was talking about physics. I listened dutifully for the first hour and didn't understand a single word of what was being said. At the end of the hour, everyone got up and went to some other room, but I didn't know where to go, so I just sat there. The instructor, Mr. Payne, came over to my seat and asked me something and I shrugged my shoulders helplessly. We tried various languages. I didn't know much English, and he didn't know German. We both knew a few Latin words and a few French words, and he finally succeeded in telling me that I had to go to Room 252. I went to Room 252, and that was my first day in an American high school. Six months later I spoke rapid, incorrect, ungrammatical, colloquial English.

**A:** *Were there any special events in high school that stand out in your memory?*

**H:** Well, there was a little chicanery surrounding my admission. I explained this business of eight years followed by four in this country, and four years followed by eight in Europe. There was some confusion about that. I *hinted* to the school authorities that I had completed three years of *secondary* school, and I was *believed*. There was, to be sure, a perfunctory examination of my record, and, after being translated by an official in the Hungarian consulate, it said three years of secondary school. That means in effect that I skipped four grades at once, and I went from what was the equivalent of the seventh grade to the eleventh grade; and a year and a half later, at the age of fifteen, I graduated from high school.

**A:** *So you were a very young high-school graduate.*

**H:** Yes. I entered the University of Illinois at the age of fifteen.

**A:** *That's very young to be entering college. Did that produce any difficulty?*

**H:** There were no problems. I was tall for my age and cocky. I pretended to be older and got along fine.

**A:** *When did you become interested in mathematics and philosophy?*

**H:** I started out in chemical engineering, and at the end of one year decided that it was for the birds; I got my hands too dirty. That's how mathematics and philosophy came into the act.

**A:** *Can you remember what attracted you to mathematics and philosophy? Can you separate them?*

**H:** It is difficult. I remember calculus was not easy for me. I was a routine calculus student—I think I got B's. I didn't understand about limits. I doubt that they taught it. At that time, they probably wouldn't have dared. But I was good at integrating and differentiating things in a mechanical sense. Somehow I liked it. I kept fooling around with it. In philosophy, it was symbolic logic that interested me. What attracted me is hard to say, just as it is hard for any of us to say what attracts us to a subject. There was something about abstraction. I liked the cleanness, the security of the ideas. When I learned something about history, I was at the least very suspicious; and strange as it may sound, when I learned something about physics and chemistry, I was most suspicious: I was practically doubtful, and I thought it might not even be true. In mathematics and in that kind of philosophy (logic), I knew exactly what was going on.

**A:** *Was there some point when you decided that you were going to be a mathematician?*

**H:** There was *no* point when I decided that I was going to be an *academic*. That somehow was just taken for granted, not by anybody else, but by me. I just wanted to take courses and see what happened. I was studying for a master's and flunking the master's exam in philosophy, but nothing would stop me. I continued taking courses. I finished my bachelor's quickly, in three years instead of four. As a first-year graduate student, I took a course from Pierce Ketchum in complex function theory. I had absolutely no idea of what was going on. I didn't know what epsilons were, and when he said take the unit circle, and some other guy in class said "open or closed," I thought that silly guy was hair-splitting, and what was he fussing about. What difference did it make? I really didn't understand it.

Then one afternoon something happened. I remember standing at the blackboard in Room 213 of the mathematics building talking with Warren Ambrose and suddenly I understood epsilons. I understood what limits were, and all of the stuff that people had been drilling into me became clear. I sat down that afternoon with the calculus textbook by Granville, Smith, and Longley. All of that stuff that previously had not made any sense became obvious; I could prove the theorems. That afternoon I became a mathematician.

**A:** *So there* **was** *a critical point. You even remember the room number.*

**H:** I *think* I remember the room number.

**A:** *After earning your Ph.D., you became a fellow at the Institute for Advanced Study, where you served as an assistant to Johnny von Neumann. How did you come to be his assistant? What was it like being an assistant to someone with that kind of power?*

**H:** Let me back up a little. I got my Ph.D. in 1938, and preparatory to graduation, I applied for jobs. Xerox was not known in those days, and secretarial service was not available to starving graduate students. I typed 120 letters of application, mailed them out, and got two answers, both *no*. I got no job. The University of Illinois took pity on me and kept me on for one year as an instructor. So in '38–'39 I had a job, but I kept applying. I did get a job around February or March at a state university. I accepted it without an interview. It was accomplished with correspondence and some letters of recommendation. Two months later my very good friend Warren Ambrose, who was one year behind me, got his degree. He had been an alternate for a fellowship at the Institute for Advanced Study; and when the first choice declined, he got the scholarship, and that made me mad. I wanted to go, too! I resigned my job, making the department head, whom I had never met, very unhappy, of course. In April, I resigned my job, and went to my father and asked to borrow a thousand dollars, which in those days was a lot of money. The average annual salary of a young Ph.D. was then $1,800. I wrote Veblen and asked if I could be a member of the Institute for Advanced Study even though I had no fellowship. It took him three months to answer. He answered during summer vacation, and said, "Dear Halmos, I just found your letter, and I guess you mean for me to answer. Yes, of course, you are welcome." That's all it took; I moved to Princeton.

But, of course, Veblen wasn't giving me anything except a seat in the library. Six months after I got there, the Institute took pity on me and gave me a fellowship. During the first year, I attended Johnny von Neumann's lectures, and in my second year I became his assistant. I followed his lectures and took careful notes. The system of the Institute was that each professor had an assistant assigned to him. The duties of the assistant depended upon the professor. Einstein's assistant's duties were to walk him home every day and talk German to him. Morse's assistant's duties were to do research with him—eight hours a day sitting with Morse and listening to him talk and talk. Von Neumann's assistant had very little to do—just go to the lectures and take notes; and sometimes those notes were typed up and duplicated. Von Neumann's assistant that year was Hugh Dowker, who is a mathematician *par excellence*, but not in the least interested in matrices and operator theory and all those things that von Neumann lectured on. On the other hand, I was fascinated by them; that was my subject. So, I took careful notes and Dowker used them and took them to Johnny. There was no duplicity about it. He told Johnny what he was doing. When his job was up, I became Johnny's assistant.

How was it? Scary. The most spectacular thing about Johnny was not his power as a mathematician, which was great, or his insight and his clarity, but his rapidity; he was very, very fast. And like the modern computer, which no longer bothers to retrieve the logarithm of 11 from its memory (but, instead, computes the logarithm of 11 each time it is needed), Johnny didn't bother to remember things. He computed them. You asked him a question, and if he didn't know the answer, he thought for three seconds and would produce the answer.

**A:** *You have described an inspirational day with Warren Ambrose when you decided to become a mathematician. Are there other individuals who have been inspirations for you?*

**H:** I'm not prepared for this question. Therefore, my answer is bound to be more honest than for any other question. The first two names that occur to me are two obvious ones. The first is my supervisor, Joe Doob, who is only six years older than I. I was 22 when I finished my Ph.D., and he was 28, both young boys from my present point of view. He arrived at the University of Illinois when he himself was about 25. I was already at the stage where I was signed up to do a Ph.D. thesis with another professor. I remember having lunch with Joe one day at a drugstore and hearing him talk about mathematics. My eyes were opened. I was inspired. He showed me a kind of mathematics, a way to talk mathematics, a way to think about mathematics that wasn't visible to me before. With great trepidation, I approached my Ph.D. supervisor and asked to switch to Joe Doob, and I was off and running.

The other was Johnny von Neumann. The first day that I met him he asked if it would be more comfortable for me to speak Hungarian, which was his best language, and I said it would not. So we spoke English all the time. And as I said before, his speed, plus depth, plus insight, plus inspiration turned me on. They—Doob and von Neumann—were my two greatest inspirations.

**A:** *In* 1942, *you produced a monograph called "Finite-Dimensional Vector Spaces." Was it a result in part of notes that you had taken?*

**H:** Yes. Von Neumann planned a sequence of courses that was going to take him four years. He began at the beginning with the theory of linear algebra—finite-dimensional vector spaces from the advanced point of view. And just as van der Waerden's book was based on Artin's lectures, my book was based on von Neumann's lectures and inspired completely by him. That's what got me started writing books.

**A:** *Most people who read that book remark that it is written in an unusual way; the Halmos style is quite distinctive. I studied from your book, and I still remember that it gave me fits because your problems were not of the classical type. You didn't set* ***prove*** *or* ***show*** *exercises; more often than not you gave statements that the student was to prove if true or disprove if false. I am sure that it was deliberate, and it seems to underscore a philosophy of teaching that you have spoken about in a recent article in the MONTHLY, "The Heart of Mathematics." In that article, you said that it is better to do substantial problems on a lesser number of topics than to do oodles of lesser exercises on a larger number of topics. Had you thought a great deal about that before writing problems for "Finite-Dimensional Vector Spaces"?*

**H:** No. That wasn't a result of thought; it was just instinctive somehow. I felt it was the right way to go, and thirty years later I summarized in expository articles what I have been doing all along. You said it very well. I strongly believe that the way to learn things is to do things—the easiest way to learn to swim is to swim—you can't learn it from lectures about swimming. I also strongly believe that the secret of mathematical exposition, be it just a single lecture, be it a whole course, be it a book, or be it a paper, is not the beautifully written sentence, or even the well-thought-out paragraph, but the architecture of the whole thing. You must have in mind what the lecture or the whole course is going to be. You should get across *one* thing. Determine that thing and then design the whole approach to get at it. Instinctively in that book, and I must repeat it was inspired by von Neumann, I was driving at one thing—that matrix theory is operator theory in the most important and the most translucent special case. Every single step, and in particular every exercise (they were not different from any other step), was designed to shed light on that end.

**A:** *In the article, "The Heart of Mathematics," you discussed courses that went down as low as calculus. What do you think about that approach for precalculus or for high-school algebra? Would you also advocate that approach for such courses?*

**H:** Yes and no. I think, and I repeat, the only way to learn anything is do that thing. The only way to learn to bicycle is ride a bicycle. The only way to teach bicycling is to put challenges in front of the prospective bicyclist and make him conquer them. So, yes, I believe in it. I have tried it, not only in calculus, but in as low-level college courses as precalculus and high school trigonometry with a great deal of joy and enthusiasm many times. To the extent possible, I have tried to follow that kind of system. But let's be honest. The so-called Moore method, which is a way to describe the Socratic question-asking, problem-challenging approach to teaching, doesn't work well when you have forty people in the class, let alone when you have one hundred and forty. It is beautiful if you have two people sitting at two ends of a log, or ten or eleven sitting in a classroom facing you. Obviously, there are practical problems that you have to solve, but they can be solved. Moore, for instance, did teach first-year calculus that way. So a one-word answer to your question is, *yes*, I do advocate it in all teaching; but *no*, one has to be careful. One has to be wise. One has to face realities and adapt to economic circumstances.

**A:** *A few have said that you have been a strong exponent of what is called the New Mathematics.*

**H:** Absolutely not! I was a reactionary all the time. The old mathematics was just fine. I think high-school students should be taught high-school geometry *à la* Euclid. You should teach them step one, reason; step two, reason; and all that stuff. I thought that was wonderful. I got my training that way. Morris Kline and I hardly know each other, but we seem to disagree on everything; and he is (a) strongly against the New Math, and (b) strongly against many things I advocate. It's quite possible that people who agree with him identify me as a champion of the New Math because we disagree on most things.

**A:** *You say you think that you and Kline disagree on just about everything. He stood in strong opposition to the New Math, and you just said that you were absolutely not an exponent of the New Math. Now there is some agreement there.*

**H:** I hate to admit it. He may be against the right thing, but he certainly is against it for the wrong reason.

**A:** *It seems to me that you and Kline have another strong point of agreement, which surprised me a bit. In your article, "Mathematics as a Creative Art," that appeared in American Scientist back in the late sixties, you surprised me by saying that virtually all of mathematics is rooted in the physical world. Kline, as you know, wrote a book entitled "Mathematics and the Physical World."*

**H:** I think we understand different things by it. I get the feeling that Kline either thinks, or would love to think, that all mathematics is not only rooted in the physical world but must aim toward it, must be applicable to it, and must touch that base periodically.

**A:** *So he is what you would call a* ***mathophysicist****?*

**H:** And how! But it's another thing to say, almost a shallow, meaningless thing to say, that we are human beings with eyes, and we can see things that we think are outside of us. Our mathematics—our instinctive, unformulated, undefined terms—come from our sense impressions; and in that sense at least, a trivial sense, mathematics has its basis in the physical world. But that is an uninformative, unhelpful, shallow statement.

**A:** *This prompts the next question for which I can't expect you to give a complete answer in such a short time, but I will ask it anyhow. What is mathematics to you?*

**H:** It is security. Certainty. Truth. Beauty. Insight. Structure. Architecture. I see mathematics, the part of human knowledge that I call mathematics, as one thing—one great, glorious thing. Whether it is differential topology, or functional analysis, or homological algebra, it is all one thing. They all have to do with each other, and even though a differential topologist may not know any functional analysis, every little bit he hears, every rumor that comes to him about that other subject, sounds like something else that he does know. They are intimately interconnected, and they are all facets of the same thing. That interconnection, that architecture, is secure truth and is beauty. That's what mathematics is to me.

**A:** *In "Mathematics as a Creative Art," you were addressing lay readers when you said: "I don't want to teach you* ***what mathematics is, but that it is.****" This reflects a concern that you had at the time about mathematics in the mind of the layman. (You said, "A layman is anyone who is not a mathematician.") Is your concern still there—that a great body of intelligent, well-educated people don't know perhaps that your subject* ***is****? Is that concern stronger or weaker than it was in '68, when you wrote the article?*

**H:** The same, I would say. Let me first of all explain that I am a maverick among mathematicians. I don't think it is vital and important to explain to members of Congress and administrators in the National Science Foundation what mathematics is and how important it is and how much money it must be given. I think we have been given too much money. I don't think mathematics needs to be supported. I think the phrase is almost offensive. Mathematics gets along fine, thank you, without money, and I look back with nostalgia to the good old days, three or four hundred years ago, when only those did mathematics who were willing to do it on their own time.

In the fifties and sixties, a lot of people went into mathematics for the wrong reasons, namely that it was glamorous, socially respected, and well-paying. The Russians fired off Sputnik, the country became hysterical, and then NSF came along with professional, national policies. Anything and everything was tried; nothing was too much. We had to bribe people to come to mathematics classes to make it appear respectable, glamorous, and well-paying. So we did. One way we did it, for instance, was to use a completely dishonest pretense—the mission attitude towards mathematics. The way it worked was that I would propose a certain piece of research, and then if it was judged to be a good piece of research to do, I would get some money. That's so dishonest it sickens me. None of it was true! We got paid for doing research because the country wanted to spend money training mathematicians to help fight the Russians.

Many young people of that period were brought up with this Golden Goose attitude and now regard an NSF grant as their perfect right. Consequently, more and more there tends to be control by the government of mathematical research. There isn't strong control yet, and perhaps I'm just building a straw man to knock down. But time and effort reporting is a big, bad symptom, and other symptoms are coming I am sure. Thus, I say that it was on balance a bad thing. If the NSF had never existed, if the government had never funded American mathematics, we would have half as many mathematicians as we now have, and I don't see anything wrong with that. Mathematics departments would not have as many as eighty-five and one hundred people in some places. They might have fifteen or twenty people in them, and I don't see anything wrong with that. Mathematics got along fine for many thousands of years without special funding.

**A:** *But we certainly have seen a great increase in demand for mathematical skills which means a need for people who are able to teach mathematics. You certainly need someone to deliver the mathematics.*

**H:** That's a different subject. The demand for teaching mathematics that seems to be growing is again because of a perceived threat by the Russians and Chinese. In other words, we want people in computer science; we want people in statistics; and we want people in various industrial and other applications of mathematics. We have to teach them trigonometry and other subjects so that they can do those things. That's not mathematics; that's a trade. It isn't doing mathematicians any good, and it's doing mathematicians good only insofar as it enables them to buy an extra color TV set or more diapers for the baby.

**A:** *Can we return to your concern about letting laymen know that mathematics* ***is****?*

**H:** My interest was more on the intellectual level. I have absolutely no idea of what paleontology is; and if somebody would spend an hour with me, or an hour a day for a week, or an hour a day for a year, teaching it to me, my soul would be richer. In that sense, I was doing the same thing for my colleague, the paleontologist: I was telling him what mathematics is. That, I think, is important. All educable human beings should know what mathematics *is* because their souls would grow by that. They would enjoy life more, they would understand life more, they would have greater insight. They should, in that sense, understand all human activity such as paleontology and mathematics.

**A:** *So you were performing a service to paleontologists perhaps by explaining to them* ***that*** *your subject* ***is*** *rather than* ***what it is****. How do you explain the motivation for your other writing activities? Writing is hard work. In fact, when I reread your "How to Write Mathematics" last night, I was more convinced than ever that you must work very hard when you write. Why do you do it? Now you aren't talking to paleontologists; you are talking to mathematicians.*

**H:** It is the same thing. Why do I do it? It is a many-faceted question with many answers. Yes, writing is very hard work, and so is playing the piano for Rubinstein and Horowitz, but I am sure they love it. So is playing the piano for a first-year student at the age of ten, but many of them love it. Writing is very hard work for me, but I love it. And why do I do it? For the same reason that I explain mathematics to the paleontologist. The answer is the same—it is all communication. That's important to me. I want to make things clear. I enjoy making things clear. I find it very difficult to make things clear, but I enjoy trying, and I enjoy it even more on the rare occasions when I succeed. Whether it is making clear to a medical doctor or to a paleontologist how to solve a problem in the summation of geometric series, or explaining to a graduate student who has had a course in measure theory why $L_2$ is an example of a Hilbert space, I regard them as identical problems. They are problems in communication, explanation, organization, architecture, and structure.

**A:** *So you enjoy doing it. It makes you feel good, and it makes you feel perhaps even better if you can sense that the receiver understands. That sounds like a teacher—the classical reasons that people give for teaching—the joy of seeing the look of understanding.*

**H:** That's very good. Yes, I accept the word. I am proud to be a teacher and get paid for being a teacher, as do many of us who make a living out of mathematics. But it is also something else. Mathematics is, as I once maintained, a creative art and so is the exposition of mathematics. Teaching is an ephemeral subject. It is like playing the violin. The piece is over, and it's gone. The student is taught, and the teaching is gone. The student remains for a while, but after a while he too is gone. But writing is permanent. The book, the paper, the symbols on the sheets of papyrus are always there, and that creation is also the creation of the rounded whole.

**A:** *You've also written about talking mathematics. Based on what you have just said, my strong suspicion is that you get a greater joy out of writing than talking, although you also seem to have a lot of joy when you talk about mathematics.*

**H:** They're nearly the same thing, but writing is more precise. By more precise, I mean the creator has more control over it. I myself feel that I am a pretty good writer, A − or B +; and a good, but less good speaker, B or B −; and therefore, I enjoy writing more. But they are similar and are part of the art of communication.

**A:** *A short time ago, someone talked with me about your book "Naive Set Theory". She said that it has a smooth, conversational style, and is in some ways like a bedtime story. What motivated you to write it?*

**H:** *Naive Set Theory* was the fastest book I ever wrote. *Measure Theory* took eighteen months of practically full-time work and *Naive Set Theory* took six months. Bedtime story is an apt description, for most of it was written while perched on the edge of a bed in a rented house in Seattle, Washington. It was being written because I had just recently learned about axiomatic set theory, which was a tremendous inspiration to me. It was a novelty to me. I didn't realize that it existed and what it meant, and at once I wanted to go out and tackle it. So I wrote it down. It wrote itself. It seemed 100% clear to me that you have to start here and you have to take the next step there, and the third step suggested itself after the first two. I had almost no choice, and, as I keep emphasizing, that's the biggest problem of writing, of communication—the organization of the whole thing. Individual words that you choose you can change around; you can change the sentence around. The structure of the whole thing you cannot change around—that's what was created while I perched on the edge of the bed the first day, and from then on the book wrote itself.

**A:** *Was that a unique writing experience for you?*

**H:** In that respect, yes, because it had a much better defined subject than usual. When I wrote on Hilbert space theory, I had, subjectively speaking, an infinite area from which to carve out a small chunk. Here was an absolute, definite thing. There is much more to axiomatic set theory than I exposed in *Naive Set Theory*, but it was a clearly defined part of it that I wanted to expose, and I did.

**A:** *You have recently written an article with another intriguing title, "Applied Mathematics is Bad Mathematics".*

**H:** I have been sitting here for the last 55 minutes dreading when this question was going to be asked.

**A:** *What do you mean when you say applied is bad?*

**H:** First, it isn't. Second, it is. I chose the title to be provocative. Many mathematicians, whom everybody else respects and whom I respect, agree with the following attitude: There is something called mathematics—put the adjective "pure" in front of it if you prefer. It all hangs together. Be it topology, or algebra, or functional analysis, or combinatorics, it is the same subject with the same facets of the same diamond; it's beautiful and it's a work of art. In all parts of the subject the language is the same; the attitude is the same; the way the researcher feels when he sits down at his desk is the same; the way he feels when he starts a problem is the same. The subject is closely related to two others. One of them is usually called applied mathematics, and its adherents frequently deny that it exists. They say there is no such thing as applied mathematics and that there isn't any difference between applied mathematics and pure mathematics. But, nevertheless, there is a difference in language and attitude. I am about to say a bad word about applied mathematicians, but, believe me, I mean it in a genuinely humble way. They are sloppy. They are sloppy in perhaps the same way that you and I are sloppy, as ordinary mathematicians are sloppy compared with the requirements of a formal logician. And a formal logician would probably be called sloppy by a computing machine.

There are at least three different kinds of language, which can roughly be arranged in a hierarchy: *formal logic* (that's equated nowadays with computer science), *mathematics*, and *applied mathematics*. They have different objectives; they are different facets of beauty; they have different reasons for existence, and have different manners of expression and communication. Since communication is so important to me, that is the first thing that jumps to my eye. A logician just cannot talk the way a topologist talks. And an algebraist couldn't make like an applied mathematician to save his life. Some geniuses like Abraham Robinson can be both. But they are different people being those different things. So, in that sense, what I wanted to say in that article is that there are at least two subjects. Now I am saying there are three or more, and I wanted to call attention to what I think the differences are.

There is a sense in which applied mathematics is like topology, or algebra, or analysis, but (and shoot if you must this old grey head) there is also a sense in which applied mathematics is just bad mathematics. It's a good contribution. It serves humanity. It solves problems about waterways, sloping beaches, airplane flights, atomic bombs, and refrigerators. But just the same, much too often it is bad, ugly, badly arranged, sloppy, untrue, undigested, unorganized, and unarchitected mathematics.

**A:** *Computers are still relatively new objects within our lifetimes and intimately linked to what many call applied mathematics. What do you think of them? Are they important to you?*

**H:** Who am I? A citizen or mathematician? As a mathematician, no, not in the least.

**A:** *Let's take something specific, the work of Appel and Haken and the computer.*

**H:** On the basis of what I read and pick up as hearsay, I am much less likely now, after their work, to go looking for a counterexample to the four-color conjecture than I was before. To that extent, what has happened convinced me that the four-color theorem is true. I have a religious belief that some day soon, maybe six months from now, maybe sixty years from now, somebody will write a proof of the four-color theorem that will take up sixty pages in the *Pacific Journal of Mathematics*. Soon after that, perhaps six months or sixty years later, somebody will write a four-page proof, based on concepts that in the meantime we will have developed and studied and understood. The result will belong to the grand, glorious, architectural structure of mathematics (assuming, that is, that Haken and Appel and the computer haven't made a mistake).

I admit that for a number of my friends, mostly number theorists and topologists, who fool around with small numbers and low-dimensional spaces, the computer is a tremendous scratch pad. But those same friends, perhaps in other bodies, got along just fine twenty-five years ago, before the computer became a scratch pad, using a different scratch pad. Maybe they weren't as efficient, but mathematics isn't in a hurry. Efficiency is meaningless. Understanding is what counts. So, is the computer important to mathematics? My answer is *no*. It is important, but not to mathematics.

**A:** *Do you sense the same attitude about computers among most of your colleagues?*

**H:** I think the ones who share my attitude are perhaps in the minority.

**A:** *There are now mathematicians who seem to have a hybrid nature. Let's take someone like Don Knuth, who earned his Ph.D. in mathematics, and along the way discovered the art of computing.*

**H:** It's not fair arguing by citing examples of great men. How could I possibly disagree? Don Knuth is a great man. Computer science is a great science. What else is there to say? In many respects, that science touches mathematics and uses mathematical ideas. The extent to which the big architecture of mathematics uses the ideas rather than the scratch pad aspect of that science is, however, vanishingly small.

Nevertheless, the connection between computer science and the big body of pure mathematics is sufficiently close that it cannot be ignored, and I advise all of my students to learn computer science for two reasons. First, even though efficiency is not important to mathematics, it may be important to them; if they can't get jobs as pure mathematicians, they need to have something else to do. Second, to the layman, the difference between this part and that part and the third part of something, all of which looks like mathematics to him, looks like hair-splitting. My students and all of us should represent this science in the outside world.

**A:** *It is rumored that you're one of the world's great walkers.*

**H:** That's certainly false. I enjoy walking very much. It is the only exercise I take. I do it very hard. I walk four miles every day at a minimum. I just came from a ten-day holiday, most days of which I walked fast for 10, 12, 15 miles, and I got hot and sweaty. I love it because I am alone; I can think and daydream; and because I feel my body is working up to a healthy state. To call me one of the world's great walkers is an exaggeration, I'm sure.

**A:** *Have you been a walker, in a strong sense, for many years?*

**H:** Twenty-five years. When I was forty, I had every disease in the book, *I thought* hypochondriacally. I went to the doctor with a brain tumor, with heart disease, with cancer, and everything else, *I thought*. He examined me and said, "Halmos, there isn't anything wrong with you. Go take a long walk." So I took a five-minute walk. And then the next week, I increased it to six minutes, and seven, and eight, and nine, until I got to sixty, and I stopped. On weekends, I walk greater distances. When I was young, I drank like a fish, smoked heavily, and had every other vice that you can imagine. Then when I started worrying about such things, I really started worrying about such things. How old are you?

**A:** *I just turned* 40.

**H:** Then start worrying!

**A:** *I fully expect you soon to write another article that would pick up on two previous articles. You have done "How to Write Mathematics" and "How to Talk Mathematics." May I soon expect to see "How to Dream Mathematics"?*

**H:** I'm ahead of you, but I haven't written the article. I have half-planned an entire book. If I live long enough and really have the guts to stand up in public to do it, I might write a book on how to be a mathematician. I have outlined it on paper.

It will include all aspects of the profession, except how to do research. I won't pretend to tell anyone how to do that. What I think I can do is describe the mechanical steps that people go through (and apparently we all have to go through) to do research, to be a referee, to be an author, to write papers, to teach classes, to deal with students—in short, to be a member of the profession—the most glorious profession of all.